KEY ELEMENTS IN POLYMERS FOR ENGINEERS AND CHEMISTS

From Data to Applications

KEY ELEMENTS IN POLYMERS FOR ENGINEERS AND CHEMISTS

From Data to Applications

Edited by

**Alexandr A. Berlin, DSc, Viktor F. Kablov, DSc,
Andrey A. Pimerzin, DSc, and Simon S. Zlotsky, PhD**

Gennady E. Zaikov, DSc, and A. K. Haghi, PhD
Reviewers and Advisory Board Members

Apple Academic Press

TORONTO   NEW JERSEY

Apple Academic Press Inc. | Apple Academic Press Inc.
3333 Mistwell Crescent | 9 Spinnaker Way
Oakville, ON L6L 0A2 | Waretown, NJ 08758
Canada | USA

©2014 by Apple Academic Press, Inc.

First issued in paperback 2021

Exclusive worldwide distribution by CRC Press, a member of Taylor & Francis Group
No claim to original U.S. Government works

ISBN 13: 978-1-77463-308-3 (pbk)
ISBN 13: 978-1-926895-80-2 (hbk)

Library of Congress Control Number: 2014937165

Library and Archives Canada Cataloguing in Publication

Key elements in polymers for engineers and chemists: from data to applications/edited by Alexandr A. Berlin, DSc, Viktor F. Kablov, DSc, Andrey A. Pimerzin, DSc, and Simon S. Zlotsky, PhD; Gennady E. Zaikov, DSc, and A. K. Haghi, PhD, Reviewers and Advisory Board Members.

Includes bibliographical references and index.
ISBN 978-1-926895-80-2 (bound)
1. Polymers. 2. Polymerization. I. Kablov, Viktor F. (Viktor Fedorovich), editor II. Berlin, Alexandr A., editor III. Pimerzin, Andrey A., editor IV. Zlotsky, Simon S., editor

QD381.K49 2014   547'.7   C2014-902294-8

Apple Academic Press also publishes its books in a variety of electronic formats. Some content that appears in print may not be available in electronic format. For information about Apple Academic Press products, visit our website at **www.appleacademicpress.com** and the CRC Press website at **www.crcpress.com**

ABOUT THE EDITORS

Alexandr A. Berlin, DSc

Professor Alexandr A. Berlin, DSc, is Director of the N. N. Semenov Institute of Chemical Physics at the Russian Academy of Sciences, Moscow, Russia. He is a member of the Russian Academy of Sciences and many national and international associations. Dr. Berlin is world-renowned scientist in the field of chemical kinetics (combustion and flame), chemical physics (thermodynamics), chemistry and physics of oligomers, polymers, and composites and nanocomposites. He is the contributor of 100 books and volumes and 1000 original papers and reviews.

Viktor F. Kablov, DSc

Viktor F. Kablov, DSc, was appointed Director of VolzhskyPolytechnical Institute (branch) of VSTU–VPI (branch) of VSTU in 2000, and in 2002, he was elected Head of the Department of Polymer Chemical Technology and Industrial Ecology at the Institute. Professor Kablov has organized new training courses in modeling, computer-assisted methods and information systems in polymer engineering, engineering creativity methods, polymer chemistry and physics, and biotechnology. In 2010, he opened a Master's degree program in elastomer processing technology. He has also authored over 150 inventions, more than 300 research publications, 6 textbooks and 3 monographs.

Prof. Kablov's area of research relates to developing a scientific basis for obtaining elastomer materials operating in extreme conditions (heat-, fire-, and corrosion-resistant materials, antifriction composite materials); developing nano- and microheterogenic processes; modifying and operational additives to improve adhesive; thermal oxidation; and processing properties of polymer materials, as well as increasing the fire resistance of materials; developing materials for medical applications and fundamentally new hybrid polymers based on biopolymers and synthetic polymers; obtaining high selectivity sorbents based on template synthesis; developing waste bio recycling methods, as well as research in other areas of technology.

Andrey Pimerzin, DSc

AndreyPimerzin, DSc, was appointed Vice Rector of the Samara State Technical University, Samara, Russia, in 2007. He was also elected as Head of the Department of Chemical Technology Oil Gas Refining at the same university in 2001. In 2009, Professor Pinerzin was invited as a researcher at the Thermodynamics

Research Center NIST in the USA, and in 2011, he was one of the organizers of the international conference RCCT-2011. Under Andrey Pimerzin's supervision, eight PhD and one Doctor of Science theses have been defended. He is the author of more than 250 science publications, including 11 textbooks and 9 patents. His research interests include chemical and statistical thermodynamics of organic compounds, thermochemistry, kinetics, heterogeneous catalysis, and refining technology.

Simon S. Zlotsky, PhD

Professor Simon Solomonovich Zlotsky received his BS in chemistry from Chemistry and Technological Department of Ufa Petroleum Institute in 1968. From 1970 to 1973 he studied and gained his PhD from the Moscow Institute of the Petrochemical and Gas Industry. Dr. Zlotsky worked as a research fellow at the Ufa State Petroleum Technological University (1973–77) and researched his doctoral thesis on chemistry of cyclic acetals. From 1982 to 2010 he was a Professor in the Department of General Chemistry in the Ufa State Petroleum Technological University, and he has headed the department since 2010.

REVIEWERS AND ADVISORY BOARD MEMBERS

Gennady E. Zaikov, DSc

Gennady E. Zaikov, DSc, is Head of the Polymer Division at the N. M. Emanuel Institute of Biochemical Physics, Russian Academy of Sciences, Moscow, Russia, and Professor at Moscow State Academy of Fine Chemical Technology, Russia, as well as Professor at Kazan National Research Technological University, Kazan, Russia. He is also a prolific author, researcher, and lecturer. He has received several awards for his work, including the the Russian Federation Scholarship for Outstanding Scientists. He has been a member of many professional organizations and on the editorial boards of many international science journals.

A. K. Haghi, PhD

A. K. Haghi, PhD, holds a BSc in urban and environmental engineering from University of North Carolina (USA); a MSc in mechanical engineering from North Carolina A&T State University (USA); a DEA in applied mechanics, acoustics and materials from Université de Technologie de Compiègne (France); and a PhD in engineering sciences from Université de Franche-Comté (France). He is the author and editor of 65 books as well as 1000 published papers in various journals and conference proceedings. Dr. Haghi has received several grants, consulted for a number of major corporations, and is a frequent speaker to national and international audiences. Since 1983, he served as a professor at several universities. He is currently Editor-in-Chief of the *International Journal of Chemoinformatics and Chemical Engineering* and *Polymers Research Journal* and on the editorial boards of many international journals. He is also a member of the Canadian Research and Development Center of Sciences and Cultures (CRDCSC), Montreal, Quebec, Canada.

CONTENTS

LIST OF CONTRIBUTORS

M. R. Ahmad
Faculty of Applied Sciences, Universiti Teknologi MARA, 40450 Shah Alam, Malaysia

J. Aneli
R. Dvali Institute of Machine Mechanis, email: JimAneli@yahoo.com

V. A. Babkin
Volgograd State Architect-build University, Sebrykov Departament, Russia

C. H. Chan
Faculty of Applied Sciences, Universiti Teknologi MARA, 40450 Shah Alam, Malaysia

Rustam Ya. Deberdeev
Kazan National Research Technological University, 68 Karl Marx Street, 420015 Kazan, Republic of Tatarstan, Russian Federation, Fax: +7 (843) 231-41-56; E-mail: n.v.ulitin@mail.ru

K. S Dibirova
Dagestan State Pedagogical University, Makhachkala 367003, Yaragskii st., 57, Russian Federation

O. Emamgholipour
University of Guilan, Iran

Akbar Khodaparast Haghi
Department of Textile Engineering, University of Guilan, Rasht, Iran, E-mail: Haghi@Guilan.ac.ir

Sim Lai Har
Centre of Foundation Studies, Universiti Teknologi MARA, 42300 Puncak Alam, Selangor.Malysia

Mahdi Hasanzadeh
Department of Textile Engineering, University of Guilan, Rasht, Iran, Email: m_hasanzadeh@aut. ac.ir, Tel.: +98-21-33516875; fax: +98-182-3228375

Amirah Hashifudin
Faculty of Applied Sciences, Universiti Teknologi MARA, 40450 Shah Alam, Selangor. Malysia

E. Markarashvili
Javakhishvili Tbilisi State University Faculty of Exact and Natural Sciences, Department of Macro-molecular Chemistry. I. Chavchavadze Ave., 3, Tbilisi 0128, Republic of Georgia

O. Mukbaniani
Javakhishvili Tbilisi State University Faculty of Exact and Natural Sciences, Department of Macro-molecular Chemistry. I. Chavchavadze Ave., 3, Tbilisi 0128, Republic of Georgia, email: OmariMui@ yahoo.com

Hans-Werner Kammer
University of Halle, Mansfelder Str. 28, D-01309 Dresden, Germany

E. Klodzinska
Institute for Engineering of Polymer Materials and Dyes, 55 M. Sklodowskiej-Curie str., 87-100 Torun, Poland, E-mail: S.Kubica@impib.pl

G. A. Korablev
[1]Izhevsk State Agricultural Academy, Basic Research and Educational Center of Chemical Physics and Mesoscopy, Udmurt Scientific Center, Ural Division, Russian Academy of Science, Russia, Izhevsk, 426000, e-mail: korablev@udm.net, biakaa@mail.ru

G. V. Kozlov
Dagestan State Pedagogical University, Makhachkala 367003, Yaragskii st., 57, Russian Federation

G. M. Magomedov
Dagestan State Pedagogical University, Makhachkala 367003, Yaragskii st., 57, Russian Federation

Bentolhoda Hadavi Moghadam
Department of Textile Engineering, University of Guilan, Rasht, Iran, E-mail address: motaghitalab@guilan.ac.ir

Siti Nor Hafiza Mohd Yusoff
Faculty of Applied Sciences, Universiti Teknologi MARA, 40450 Shah Alam, Selangor. Malysia

Vahid Mottaghitalab
Department of Textile Engineering, University of Guilan, Rasht, Iran, E-mail address: motaghitalab@guilan.ac.ir

V Nikolai
Kazan National Research Technological University, 68 Karl Marx Street, 420015 Kazan, Republic of Tatarstan, Russian Federation, Fax: +7 (843) 231-41-56, E-mail: n.v.ulitin@mail.ru

Aleksey V. Oparkin
Kazan National Research Technological University, 68 Karl Marx Street, 420015 Kazan, Republic of Tatarstan, Russian FederationFax: +7 (843) 231-41-56; E-mail: n.v.ulitin@mail.ru

N. H. A. Rosli
Faculty of Applied Sciences, Universiti Teknologi MARA, 40450 Shah Alam, Malaysia

Yoga Sugama Salim
Department of Chemistry, University of Malaya, 50603 Kuala Lumpur, Malaysia

Evgenii B. Shirokih
Kazan National Research Technological University, 68 Karl Marx Street, 420015 Kazan, Republic of Tatarstan, Russian Federation, Fax: +7 (843) 231-41-56; E-mail: n.v.ulitin@mail.ru

R. H. Y. Subban
Faculty of Applied Sciences, Universiti Teknologi MARA, 40450 Shah Alam, Malaysia

Ulitin, Timur
Kazan National Research Technological University, 68 Karl Marx Street, 420015 Kazan, Republic of Tatarstan, Russian Federation, Fax: +7 (843) 231-41-56; E-mail: n.v.ulitin@mail.ru

Nikolai V. Ulitin
Kazan National Research Technological University, 68 Karl Marx Street, 420015 Kazan, Republic of Tatarstan, Russian Federation Fax: +7 (843) 231-41-56; E-mail: n.v.ulitin@mail.ru

Hiroshi Uyama
Department of Applied Chemistry, Graduate School of Engineering, Osaka University, Suita, Osaka 565-0871, Japan

Kai Weng Chan
Synthomer (M) Sdn. Bhd., 86000 Kluang, Malaysia

Tan Winie
Faculty of Applied Sciences, Universiti Teknologi MARA, 40450 Shah Alam, Malaysia*. Tel.: +60-3-55435738; fax: +60-3-55444562, E-mail address: tanwinie@salam.uitm.edu.my

Yuanrong Xin
Department of Applied Chemistry, Graduate School of Engineering, Osaka University, Suita, Osaka 565-0871, Japan

Siti Nor Hafiza Mohd Yusoff
Faculty of Applied Sciences, Universiti Teknologi MARA, 40450 Shah Alam, Selangor, Malaysia

Gennady E. Zaikov
N. M. Emanuel Institute of Biochemical Physics, Russian Academy of Sciences, 4 Kosygina st., Moscow, Russian Federation, 119334, Fax: +7(499)137-41-01; E-mail: chembio@sky.chph.ras.ru

Nur Aziemah Zainudin
Faculty of Applied Sciences, Universiti Teknologi MARA, 40500 Shah Alam, Malaysia Chin Han Chan Faculty of Applied Sciences, Universiti Teknologi MARA, 40500 Shah Alam, Malaysia

LIST OF ABBREVIATIONS

AFM	atomic force microscopy
AIBN	azobisisobutyronitrile
ATR	attenuated total reflection
CPE	composite polymer electrolyte
CTMP	chemithermomechanical pulp
DBTC	dibenzyltritiocarbonate
DCAA	dichloroacetic acid
DPNR	deproteinized natural rubber
DSC	differential scanning calorimetry
ENR	epoxidized natural rubber
eV	electron-volts
IS	impedance spectroscopy
ITS	indirect tensile strength
LB	Lattice–Boltzmann
LLDPE	linear low density polyethylene
M&S	modeling and simulation
MDD	maximum dry density
MEMS	microelectromechanical systems
MMA	methyl methacrylate
MMD	molecular-mass distribution
MMD	molecular-mass distribution
MMT	montmorillonite
MNR	modified natural rubber
MOFs	metal-organic frameworks
MWNT	multi-walled nanotubes
NIPS	non-solvent induced phase separation
NR	natural rubber
MST	mechanical stability time
NRL	natural rubber latex
ODE's	ordinary differential equations
OSA	objective-based simulated Annealing
OWC	optimum water content
Pac	polyacrylate
PAN	polyacrylonitrile
PANi	polyaniline

PC	polycarbonate
PD	polydispersity
PEI	polyethylenimine
PEO	poly ethylene oxide
PPy	polypyrrole
PSM	post-synthetic modification
PSW	plastic solid waste
RAFT	reversible addition-fragmentation chain transfer
RDP	radial density profile
RH	rice husk ash
RVP	radial velocity profile
SBUs	secondary building units
SC	clayey sand
SEM	scanning electron microscopy
SEP	spatial-energy parameter
SF	silica fume
SPE	solid polymer electrolyte
SRNF	resistant nanofiltration
SWNT	single-wall nanotubes
TEM	transmission electron microscopy
TEOS	tetraethoxysilane
TEPA	tetraethylenepentamine
THF	tetrahydrofuran
TMTD	thiuramdisulphide
TSC	total solid content

PREFACE

Polymers have played a significant part in the existence of humans. They have a role in every aspect of modern life, including health care, food, information technology, transportation, energy industries, etc. The speed of developments within the polymer sector is phenomenal and, at the same, time, crucial to meet demands of today's and future life. Specific applications for polymers range from using them in adhesives, coatings, painting, foams, and packaging to structural materials, composites, textiles, electronic and optical devices, biomaterials and many other uses in industries and daily life. Polymers are the basis of natural and synthetic materials. They are macromolecules, and in nature are the raw material for proteins and nucleic acids, which are essential for human bodies.

Cellulose, wool, natural rubber and synthetic rubber, plastics are well-known examples of natural and synthetic types. Natural and synthetic polymers play a massive role in everyday life, and a life without polymers really does not exist. A correct understanding of polymers did not exist until 1920s. In 1922, Staudinger published his idea that polymers were long chain molecules with normal chemical bonds holding them together. But for nearly 10 years this idea did not attract much attention. Around this period other researchers like Carothers who tended towards Staudinger's idea, discovered a type of synthetic material, which could be produced by its constituent monomers. Later on it was shown that as well as addition reaction, polymers could be prepared through condensation mechanism.

Previously it was believed that polymers could only be prepared through addition polymerization. The mechanism of the addition reaction was also unknown and hence there was no sound basis of proposing a structure for the polymers. This lack of information was the main controversy existed between Staudinger and his critics. The studies by Carothers and other researchers resulted in theorizing the condensation polymerization. It became clear that difunctional molecules like dihydric alcohols and dicarboxylic acids could react repeatedly with the release of water to form polyesters of high molecular mass. This mechanism became well understood and the structure of the resultant polyester could be specified with greater confidence.

In 1941/42 the world witnessed the infancy of polyethylene terephthalate or better known as the polyester. A decade later for the first time polyester/cotton blends introduced. In those days Terylene and Dacron (commercial names for polyester fibers) were miracle fibers but still overshadowed by nylon. Not many would have predicted those decades later, polyester would have become the world's inexpensive, general purpose fibers as well as becoming a premium fiber for special functions in engineering textiles, fashion and many other technical end-uses. From the time nylon and polyester were first used there have been an amazing technological advances which have made them so cheap to manufacture and widely available.

These developments have made the polymers such as polyesters to contribute enormously in today's modern life. One of the most important applications is the furnishing sector (home, office, cars, aviation industry, etc.), which benefits hugely from the advances in technology. There are a number of requirements for a fabric to function in its chosen end use, for example, resistance to pilling and abrasion, as well as, dimensional stability. Polyester is now an important part of the upholstery fabrics. The shortcomings attributed to the fiber in its early days have mostly been overcome. Now it plays a significant part in improving the life span of a fabric, as well as its dimensional stability, which is due to its heat-setting properties.

About half century has passed since synthetic leather a composite material completely different from conventional ones came to the market. Synthetic leather was originally developed for end-uses such as, the upper of shoes. Gradually other uses like clothing steadily increased the production of synthetic leather and suede. Synthetic leathers and suede have a continuous ultrafine porous structure comprising a three-dimensional entangled nonwoven fabric and an elastic material principally made of polyurethane. Polymeric materials consisting of the synthetic leathers are polyamide and polyethylene terephthalate for the fiber and polyurethanes with various soft segments, such as aliphatic polyesters, polyethers and polycarbonates for the matrix.

The introduction of plastics is associated with the twentieth century but the first plastic material, celluloid, were made in 1865. During the 1970s, clothes of polyester became fashionable but by the 1980s synthetics lost the popularity in favor of natural materials. Although people were less enthusiastic about synthetic fabrics for everyday wear, Gore–Tex and other synthetics became popular for outdoors and workout clothing. At the same time as the use of synthetic materials in clothing declined, alternative uses were found. One great example is the use of polyester for making beverage bottles where it replaced glass with its shatterproof properties as a significant property.

In general it can be said that plastics enhance and even preserve life. Kevlar, for instance, when it is used in making canoes for recreation or when used to make a bulletproof vest. Polyester enhances life, when this highly nonreactive material is used to make replacement human blood vessels or even replacement skin for burn victims. With all the benefits attributed to plastics, they have their negative side. A genuine environmental problem exists due to the fact that the synthetic polymers do not break down easily compared with the natural polymers. Hence the need not only to develop biodegradable plastics, but also to work on more effective means of recycling. A lot of research needed to study the methods of degradation and stabilization of polymers in order to design polymers according to the end-use.

Among the most important and versatile of the hundreds of commercial plastics is polyethylene. Polyethylene is used in a wide variety of applications because it can be produced in many different forms. The first type to be commercially exploited was called low-density polyethylene (LDPE). This polymer is characterized by a large degree of branching, forcing the molecules to pack together rather than loosely forming a low-density material. LDPE is soft and pliable and has applications ranging from plastic bags, containers, textiles, and electrical insulation, to coatings for packaging materials.

Another form of polyethylene differing from LDPE in structure is high-density polyethylene (HDPE). HDPE demonstrates little or no branching, resulting in the molecules to be tightly packed. HDPE is much more rigid than LDPE and is used in applications where rigidity is important. Major uses of HDPE are plastic tubing, bottles, and bottle caps. Other variations of polyethylene include high and ultra-high molecular mass ones. These types are used in applications where extremely tough and resilient materials are needed.

Natural polymers unlike the synthetic ones do possess very complex structure. Natural polymers such as cellulose, wool, and natural rubber are used in many products in large proportions. Cellulose derivatives are one of the most versatile groups of regenerated materials with various fields of application. Cellulose is found in nature in all forms of plant life, particularly in wood and cotton. The purest form of cellulose is obtained from the seed hairs of the cotton plant, which contain up to 95% cellulose. The first cellulose derivatives came to stage around 1845 when the nitration of starch and paper led to discovery of cellulose nitrate. In 1865 for the first time a moldable thermoplastic made of cellulose nitrate and castor oil.

In 1865 the first acetylation of cellulose was carried out but the first acetylation process for use in industry was announced in 1894. In 1905 an acetylation process was introduced which yielded a cellulose acetate soluble in the cheap solvent, acetone. It was during the First World War when cellulose acetate dope found importance for weather proofing and stiffening the fabric of aircraft wings. There was a large surplus production capacity after the war, which led to civilian end uses such as the production of cellulose acetate fibers by 1920's. Cellulose acetate became the main thermoplastic molding material when the first modern injection molding machines were designed. Among the cellulose derivatives, cellulose acetates are produced in the largest volume. Cellulose acetate can be made into fibers, transparent films and the less substituted derivatives are true thermoplastics. Cellulose acetates are moldable and can be fabricated by the conventional processes. They have toughness, good appearance, capable of many color variations including white transparency.

New applications are being developed for polymers at a very fast rate all over the world at various research centers. Examples of these include electro active polymers, nanoproducts, robotics, etc. Electro active polymers are special types of materials, which can be used for example as artificial muscles and facial parts of robots or even in nanorobots. These polymers change their shape when activated by electricity or even by chemicals. They are lightweight but can bear a large force, which is very useful when being utilized for artificial muscles. Electro active polymers together with nanotubes can produce very strong actuators. Currently research works are carried out to combine various types of electro active polymers with carbon nanotubes to make the optimal actuator. Carbon nanotubes are very strong, elastic, and conduct electricity. When they are used as an actuator, in combination with an electro active polymer the contractions of the artificial muscle can be controlled by electricity. Already works are under way to use electro active polymers in space. Various space agencies are investigating the possibility of using these polymers in space. This technology has a lot to offer for the future, and with the ever-increasing work on nanotechnology, electro active materials will play very important part in modern life.

**— Alexandr A. Berlin, DSc, Viktor F. Kablov, DSc,
Andrey A. Pimerzin, DSc, and Simon S. Zlotsky, PhD**

CHAPTER 1

RAFT-POLYMERIZATION OF STYRENE—KINETICS AND MECHANISM

NIKOLAI V. ULITIN, ALEKSEY V. OPARKIN,
RUSTAM YA. DEBERDEEV, EVGENII B. SHIROKIH,
and GENNADY E. ZAIKOV

CONTENTS

1.1   INTRODUCTION

The kinetic modeling of styrene controlled radical polymerization, initiated by 2,2'-asobis(isobutirnitrile) and proceeding by a reversible chain transfer mechanism was carried out and accompanied by "addition-fragmentation" in the presence dibenzyltritiocarbonate. An inverse problem of determination of the unknown temperature dependences of single elementary reaction rate constants of kinetic scheme was solved. The adequacy of the model was revealed by comparison of theoretical and experimental values of polystyrene molecular-mass properties. The influence of process controlling factors on polystyrene molecular-mass properties was studied using the model

The controlled radical polymerization is one of the most developing synthesis methods of narrowly dispersed polymers nowadays [1–3]. Most considerations were given to researches on controlled radical polymerization, proceeding by a reversible chain transfer mechanism and accompanied by "addition-fragmentation" (RAFT – reversible addition-fragmentation chain transfer) [3]. It should be noted that for classical RAFT-polymerization (proceeding in the presence of sulphur-containing compounds, which formula is Z–C(=S)–S–R', where Z – stabilizing group, R' – outgoing group), valuable progress was obtained in the field of synthesis of new controlling agents (RAFT-agents), as well as in the field of research of kinetics and mathematical modeling; and for RAFT-polymerization in symmetrical RAFT-agents' presence, particularly, tritiocarbonates of formula R'–S–C(=S)–S–R', it came to naught in practice: kinetics was studied in extremely general form [4] and mathematical modeling of process hasn't been carried out at all. Thus, the aim of this research is the kinetic modeling of polystyrene controlled radical polymerization initiated by 2,2'-asobis(isobutirnitrile) (AIBN), proceeding by reversible chain transfer mechanism and accompanied by "addition-fragmentation" in the presence of dibenzyltritiocarbonate (DBTC), and also the research of influence of the controlling factors (temperature, initial concentrations of monomer, AIBN and DBTC) on molecular-mass properties of polymer.

1.2   EXPERIMENTAL PART

Prior using of styrene (Aldrich, 99%), it was purified of aldehydes and inhibitors at triple cleaning in a separatory funnel by 10%-th (mass) solution of NaOH (styrene to solution ratio is 1:1), then it was scoured by distilled water to neutral reaction and after that it was dehumidified over $CaCl_2$ and rectified in vacuo.

AIBN (Aldrich, 99%) was purified of methanol by re-crystallization.

DBTC was obtained by the method presented in research [4]. Masses of initial substances are the same as in Ref. [4]. Emission of DBTC was 81%. NMR ^{13}C (CCl_3D) δ, ppm: 41.37, 127.60, 128.52, 129.08, 134.75, and 222.35.

Examples of polymerization were obtained by dissolution of estimated quantity of AIBN and DBTC in monomer. Solutions were filled in tubes, 100 mm long, and having internal diameter of 3 mm, and after degassing in the mode of "freezing-defrosting" to residual pressure 0.01-mmHg column, the tubes were unsoldered. Polymerization was carried out at 60°C.

Research of polymerization's kinetics was made with application of the calorimetric method on Calvet type differential automatic microcalorimeter DAK-1–1 in the mode of immediate record of heat emission rate in isothermal conditions at 60°C. Kinetic parameters of polymerization were calculated basing on the calorimetric data as in the work [5]. The value of polymerization enthalpy $\Delta H = -73.8$ kJ × mol^{-1} [5] was applied in processing of the data in the calculations.

Molecular-mass properties of polymeric samples were determined by gel-penetrating chromatography in tetrahydrofuran at 35°C on chromatograph GPCV 2000 "Waters". Dissection was performed on two successive banisters PLgel MIXED–C 300×7.5 mm, filled by stir gel with 5 μm vesicles. Elution rate – 0.1 mL × min^{-1}. Chromatograms were processed in programme "Empower Pro" with use of calibration by polystyrene standards.

1.2.1 MATHEMATICAL MODELING OF POLYMERIZATION PROCESS

Kinetic scheme, introduced for description of styrene controlled radical polymerization process in the presence of trithiocarbonates, includes the following phases.

1. Real initiation

$$I \xrightarrow{k_d} 2R(0)\cdot$$

2. Thermal initiation [6]. It should be noted that polymer participation in thermal initiation reactions must reduce the influence thereof on molecular-mass distribution (MMD). However, since final mechanism of these reactions has not been ascertained in recording of balance differential equations for polymeric products so far, we will ignore this fact.

$$3M \xrightarrow{k_{i1}} 2R(1),$$

$$2M+P \xrightarrow{k_{i2}} R(1)+R(i),$$

$$2P \xrightarrow{k_{i3}} 2R(i)\cdot$$

In these three reactions summary concentration of polymer is recorded as P.

3. Chain growth

$$R(0)+M \xrightarrow{k_p} R(1),$$

$$R'+M \xrightarrow{k_p} R(1),$$

$$R(i)+M \xrightarrow{k_p} R(i+1).$$

4. Chain transfer to monomer

$$R(i)+M \xrightarrow{k_{tr}} P(i, 0, 0, 0) + R(1).$$

5. Reversible chain transfer [4]. As a broadly used assumption lately, we shall take that intermediates fragmentation rate constant doesn't depend on leaving radical's length [7].

$$R(i)+RAFT(0, 0) \underset{k_f}{\overset{k_{a1}}{\rightleftharpoons}} Int(i, 0, 0) \underset{k_{a2}}{\overset{k_f}{\rightleftharpoons}} RAFT(i, 0)+R' \quad (I)$$

$$R(j)+RAFT(i, 0) \underset{k_f}{\overset{k_{a2}}{\rightleftharpoons}} Int(i, j, 0) \underset{k_{a2}}{\overset{k_f}{\rightleftharpoons}} RAFT(i, j)+R' \quad (II)$$

$$R(k)+RAFT(i, j) \underset{k_f}{\overset{k_{a2}}{\rightleftharpoons}} Int(i, j, k) \quad (III)$$

6. Chain termination [4]. For styrene's RAFT-polymerization in the trithio-carbonates presence, besides reactions of radicals quadratic termination

$$R(0)+R(0) \xrightarrow{k_{tl}} R(0)\text{-}R(0),$$

$$R(0)+R' \xrightarrow{k_{tl}} R(0)\text{-}R',$$

$$R'+R' \xrightarrow{k_{tl}} R'\text{-}R',$$

$$R(0)+R(i) \xrightarrow{k_{tl}} P(i, 0, 0, 0),$$

$$R'+R(i) \xrightarrow{k_{tl}} P(i, 0, 0, 0),$$

$$R(j)+R(i\text{-}j) \xrightarrow{k_{tl}} P(i, 0, 0, 0)$$

are character reactions of radicals and intermediates cross termination.

$$R(0)+Int(i, 0, 0)\xrightarrow{k_{t2}}P(i, 0, 0, 0),$$

$$R(0)+Int(i, j, 0)\xrightarrow{k_{t2}}P(i, j, 0, 0),$$

$$R(0)+Int(i, j, k)\xrightarrow{k_{t2}}P(i, j, k, 0),$$

$$R'+Int(i, 0, 0)\xrightarrow{k_{t2}}P(i, 0, 0, 0),$$

$$R'+Int(i, j, 0)\xrightarrow{k_{t2}}P(i, j, 0, 0),$$

$$R'+Int(i, j, k)\xrightarrow{k_{t2}}P(i, j, k, 0),$$

$$R(j)+Int(i, 0, 0)\xrightarrow{k_{t2}}P(i, j, 0, 0),$$

$$R(k)+Int(i, j, 0)\xrightarrow{k_{t2}}P(i, j, k, 0),$$

$$R(m)+Int(i, j, k)\xrightarrow{k_{t2}}P(i, j, k, m).$$

In the introduced kinetic scheme: I, R(0), R(i), R', M, RAFT(i, j), Int(i, j, k), P(i, j, k, m) – reaction system's components (refer to Table 1); i, j, k, m – a number of monomer links in the chain; kd – a real rate constant of the initiation reaction; ki1, ki2, ki3, – thermal rate constants of the initiation reaction's; kp, ktr, ka1, ka2, kf, kt1, kt2 are the values of chain growth, chain transfer to monomer, radicals addition to low-molecular RAFT-agent, radicals addition to macromolecular RAFT-agent, intermediates fragmentation, radicals quadratic termination and radicals and intermediates cross termination reaction rate constants, respectively.

TABLE 1   Signs of components in a kinetic scheme.

RAFT(i, 0)

P(i, j, k, m)

RAFT(i, j)

The differential equations system describing this kinetic scheme, is as follows:

$$d[I] / dt = -k_d[I];$$

$$d[R(0)]/dt = 2f\,k_d[I] - [R(0)](k_p[M] + k_{t1}(2[R(0)] + [R'] + [R]) + k_{t2}(\sum_{i=1}^{\infty}[Int(i, 0, 0)] +$$

$$+ \sum_{i=1}^{\infty}\sum_{j=1}^{\infty}[Int(i, j, 0)] + \sum_{i=1}^{\infty}\sum_{j=1}^{\infty}\sum_{k=1}^{\infty}[Int(i, j, k)]));$$

$$d[M] / dt = -(k_p([R(0)] + [R'] + [R]) + k_{tr}[R])[M] - 3k_{i1}[M]^3 - 2k_{i2}[M]^2([M]_0 - [M]);$$

$$d[R']/dt = -k_p[R'][M] + 2k_f\sum_{i=1}^{\infty}[Int(i, 0, 0)] - k_{a2}[R']\sum_{i=1}^{\infty}[RAFT(i, 0)] +$$

$$+ k_f\sum_{i=1}^{\infty}\sum_{j=1}^{\infty}[Int(i, j, 0)] - k_{a2}[R']\sum_{i=1}^{\infty}\sum_{j=1}^{\infty}[RAFT(i, j)] - [R'](k_{t1}([R(0)] + 2[R'] + [R]) +$$

$$+ k_{t2}(\sum_{i=1}^{\infty}[Int(i, 0, 0)] + \sum_{i=1}^{\infty}\sum_{j=1}^{\infty}[Int(i, j, 0)] + \sum_{i=1}^{\infty}\sum_{j=1}^{\infty}\sum_{k=1}^{\infty}[Int(i, j, k)]));$$

$$d[RAFT(0,0)] / dt = -k_{a1}[RAFT(0,0)][R] + k_f\sum_{i=1}^{\infty}[Int(i, 0, 0)];$$

$$d[R(1)]/dt = 2k_{i1}[M]^3 + 2k_{i2}[M]^2([M]_0 - [M]) + 2k_{i3}([M]_0 - [M])^3 + k_p[M]([R(0)] + [R'] -$$

$$- [R(1)]) + k_{tr}[R(i)][M] - k_{a1}[R(1)][RAFT(0,0)] + k_f[Int(1,0,0)] -$$

$$- k_{a2}[R(1)]\sum_{i=1}^{\infty}[RAFT(i,0)] + 2k_f[Int(1,1,0)] - k_{a2}[R(1)]\sum_{i=1}^{\infty}\sum_{j=1}^{\infty}[RAFT(i,j)] +$$

$$+ 3k_f[Int(1,1,1)] - [R(1)](k_{t1}([R(0)] + [R'] + [R]) + k_{t2}(\sum_{i=1}^{\infty}[Int(i,0,0)] +$$

$$+ \sum_{i=1}^{\infty}\sum_{j=1}^{\infty}[Int(i,j,0)] + \sum_{i=1}^{\infty}\sum_{j=1}^{\infty}\sum_{k=1}^{\infty}[Int(i,j,k)])), \quad i = 2,...;$$

$$d[R(i)]/dt = k_p[M]([R(i-1)] - [R(i)]) - k_{tr}[R(i)][M] - k_{a1}[R(i)][RAFT(0,0)] + k_f[Int(i,0,0)] -$$

$$- k_{a2}[R(i)]\sum_{i=1}^{\infty}[RAFT(i,0)] + 2k_f[Int(i,j,0)] - k_{a2}[R(i)]\sum_{i=1}^{\infty}\sum_{j=1}^{\infty}[RAFT(i,j)] + 3k_f[Int(i,j,k)] -$$

$$- [R(i)](k_{t1}([R(0)] + [R'] + [R]) + k_{t2}(\sum_{i=1}^{\infty}[Int(i,0,0)] + \sum_{i=1}^{\infty}\sum_{j=1}^{\infty}[Int(i,j,0)] +$$

$$+ \sum_{i=1}^{\infty}\sum_{j=1}^{\infty}\sum_{k=1}^{\infty}[Int(i,j,k)])), \quad i = 2,...;$$

$$d[Int(i,0,0)]/dt = k_{a1}[RAFT(0,0)][R(i)] - 3k_f[Int(i,0,0)] + k_{a2}[R'][RAFT(i,0)] -$$

$$- k_{t2}[Int(i,0,0)]([R(0)] + [R'] + [R]);$$

$$d[Int(i,j,0)]/dt = k_{a2}[RAFT(i,0)][R(j)] - 3k_f[Int(i,j,0)] + k_{a2}[R'][RAFT(i,j)] -$$

$$- k_{t2}[Int(i,j,0)]([R(0)] + [R'] + [R]);$$

$$d[Int(i,j,k)]/dt = k_{a2}[RAFT(i,j)][R(k)] - 3k_f[Int(i,j,k)] - k_{t2}[Int(i,j,k)]([R(0)] + [R'] + [R]);$$

$d[RAFT(i, 0)]/dt=2k_f[Int(i, 0, 0)]-k_{a2}[R'][RAFT(i, 0)]-k_{a2}[RAFT(i, 0)][R]+2k_f[Int(i, j, 0)];$

$d[RAFT(i, j)]/dt=k_f[Int(i, j, 0)]-k_{a2}[R'][RAFT(i, j)]-k_{a2}[RAFT(i, j)][R] + 3k_f[Int(i, j, k)];$

$$d[P(i, 0, 0, 0)] / dt=[R(i)](k_{t1}([R(0)]+[R'])+k_{tr}[M])+\frac{k_{t1}}{2}\sum_{j=1}^{i-1}[R(j)][R(i-j)] +$$

$$+k_{t2}[Int(i, 0, 0)]([R(0)]+[R']);$$

$$d[P(i, j, 0, 0)]/dt = k_{t2}([Int(i, j, 0)]([R(0)]+[R'])+\sum_{i+j=2}[R(j)][Int(i, 0, 0)]);$$

$$d[P(i, j, k, 0)]/dt = k_{t2}([Int(i, j, k)]([R(0)]+[R'])+\sum_{i+j+k=3}[R(k)][Int(i, j, 0)]);$$

$$d[P(i, j, k, m)]/dt = k_{t2}\sum_{i+j+k+m=4}^{\infty}[R(m)][Int(i, j, k)].$$

where f – initiator's efficiency; $[R]=\sum_{i=1}^{\infty}[R(i)]$ – summary concentration of macro-radicals; t – time.

A method of generating functions was used for transition from this equation system to the equation system related to the unknown MMD moments [8].

Number-average molecular mass (Mn), polydispersity index (PD) and weight-average molecular mass (Mw) are linked to MMD moments by the following expressions:

$$M_n = (\Sigma\mu_1 / \Sigma\mu_0)M_{ST}, \ PD = \Sigma\mu_2\Sigma\mu_0 / (\Sigma\mu_1)^2, \ M_w = PD \cdot M_n,$$

where $\Sigma\mu_0$, $\Sigma\mu_1$, $\Sigma\mu_2$ – sums of all zero, first and second MMD moments; $M_{ST} = 104$ g/mol – styrene's molecular mass.

1.2.2 RATE CONSTANTS

1.2.2.1 REAL AND THERMAL INITIATION

The efficiency of initiation and temperature dependence of polymerization real initiation reaction rate constant by AIBN initiator are determined basing on the data in this research, which have established a good reputation for mathematical modeling of leaving in mass styrene radical polymerization [6]:

$$f = 0.5, \ k_d=1.58\cdot10^{15}e^{-15501/T}, \ s^{-1},$$

where T – temperature, K.

As it was established in the research, thermal initiation reactions' rates constants depend on the chain growth reactions rate constants, the radicals' quadratic termination and the monomer initial concentration:

$$k_{i1}=1.95 \cdot 10^{13} \frac{k_{t1}}{k_p^2 M_0^3} e^{-20293/T}, \ L^2 \cdot mol^{-2} \cdot s^{-1};$$

$$k_{i2}=4.30 \cdot 10^{17} \frac{k_{t1}}{k_p^2 M_0^3} e^{-23878/T}, \ L^2 \cdot mol^{-2} \cdot s^{-1};$$

$$k_{i3}=1.02 \cdot 10^{8} \frac{k_{t1}}{k_p^2 M_0^2} e^{-14807/T}, \ L \cdot mol^{-1} \cdot s^{-1}. \tag{6}$$

1.2.2.2 CHAIN TRANSFER TO MONOMER REACTION'S RATE CONSTANT

On the basis of the data in research [6]:

$$k_{tr}=2.31 \cdot 10^{6} e^{-6376/T}, \ L \cdot mol^{-1} \cdot s^{-1}.$$

1.2.2.3 RATE CONSTANTS FOR THE ADDITION OF RADICALS TO LOW–MOLECULAR AND MACROMOLECULAR RAFT–AGENTS

In research [9], it was shown by the example of dithiobenzoates at first that chain transfer to low- and macromolecular RAFT-agents of rate constants are functions of respective elementary constants. Let us demonstrate this for our process. For this record, the change of concentrations [Int(i, 0, 0)], [Int(i, j, 0)], [RAFT(0,0)] and [RAFT(i, 0)] in quasistationary approximation for the initial phase of polymerization is as follows:

$$d[Int(i, 0, 0)]/dt=k_{a1}[RAFT(0,0)][R]-3k_f[Int(i, 0, 0)] \approx 0, \tag{1}$$

$$d[Int(i, j, 0)]/dt=k_{a2}[RAFT(i, 0)][R]-3k_f[Int(i, j, 0)] \approx 0, \tag{2}$$

$$d[RAFT(0,0)] / dt = -k_{a1}[RAFT(0,0)][R] + k_f[Int(i, 0, 0)], \quad (3)$$

$$d[RAFT(i, 0)]/dt = 2k_f[Int(i, 0, 0)] - k_{a2}[RAFT(i, 0)][R] + 2k_f[Int(i, j, 0)]. \quad (4)$$

The Eq. (1) expresses the following concentration $[Int(i, 0, 0)]$:

$$[Int(i, 0, 0)] = \frac{k_{a1}}{3k_f}[RAFT(0,0)][R].$$

Substituting the expansion gives the following $[Int(i, 0, 0)]$ expression to Eq. (3):

$$d[RAFT(0,0)] / dt = -k_{a1}[RAFT(0,0)][R] + k_f \frac{k_{a1}}{3k_f}[RAFT(0,0)][R].$$

After transformation of the last equation, we have:

$$\frac{d[RAFT(0,0)]}{[RAFT(0,0)]} = -\frac{2}{3}k_{a1}[R]dt.$$

Solving this equation (initial conditions: $t = 0$, $[R] = [R]_0 = 0$, $[RAFT(0,0)] = [RAFT(0,0)]_0$), we obtain:

$$\ln \frac{[RAFT(0,0)]}{[RAFT(0,0)]_0} = -\frac{2}{3}k_{a1}[R]t. \quad (5)$$

To transfer from time t, being a part of Eq. (5), to conversion of monomer c_M, we put down a balance differential equation for monomer concentration, assuming that at the initial phase of polymerization, thermal initiation and chain transfer to monomer are not of importance:

$$d[M] / dt = -k_p[R][M]. \quad (6)$$

Transforming the Eq. (6) with its consequent solution at initial conditions $t = 0$, $[R] = [R]_0 = 0$, $[M] = [M]_0$:

$$d[M] / [M] = -k_p[R]dt,$$

$$\ln \frac{[M]}{[M]_0} = -k_p [R]t .$$

(7)

Link rate $[M]/[M]_0$ with monomer conversion (C_M) in an obvious form like this:

$$C_M = \frac{[M]_0 - [M]}{[M]_0} = 1 - \frac{[M]}{[M]_0},$$

$$\frac{[M]}{[M]_0} = 1 - C_M .$$

We substitute the last ratio to Eq. (7) and express time t:

$$t = \frac{-\ln(1 - C_M)}{k_p [R]}.$$

(8)

After substitution of the expression (8) by the Eq. (5), we obtain the next equation:

$$\ln \frac{[RAFT(0,0)]}{[RAFT(0,0)]_0} = \frac{2}{3} \frac{k_{al}}{k_p} \ln(1 - C_M).$$

(9)

By analogy with introduced $[M]/[M]_0$ to monomer conversion, reduce ratio $[RAFT(0,0)] / [RAFT(0,0)]_0$ to conversion of low-molecular RAFT-agent – $C_{RAFT(0,0)}$. As a result, we obtain:

$$\frac{[RAFT(0,0)]}{[RAFT(0,0)]_0} = 1 - C_{RAFT(0,0)}.$$

(10)

Substitute the derived expression for $[RAFT(0,0)] / [RAFT(0,0)]_0$ from Eq. (10) to Eq. (9):

$$\ln(1 - C_{RAFT(0,0)}) = \frac{2}{3}\frac{k_{a1}}{k_p}\ln(1 - C_M). \tag{11}$$

In the research [9], the next dependence of chain transfer to low-molecular RAFT-agent constant C_{tr1} is obtained on the monomer and low-molecular RAFT-agent conversions:

$$C_{tr1} = \frac{\ln(1 - C_{RAFT(0,0)})}{\ln(1 - C_M)}. \tag{12}$$

Comparing Eqs. (12) and (11), we obtain dependence of chain transfer to low-molecular RAFT-agent constant C_{tr1} on the constant of radicals' addition to macromolecular RAFT-agent and chain growth reaction rate constant:

$$C_{tr1} = \frac{2}{3}\frac{k_{a1}}{k_p}. \tag{13}$$

From Eq. (13), we derive an expression for constant k_{a1}, which will be based on the following calculation:

$$k_{a1} = 1.5 C_{tr1} k_p, \; L\cdot mol^{-1}\cdot s^{-1}$$

As a numerical value for C_{tr1}, we assume value 53, derived in research [4] on the base of Eq. (12), at immediate experimental measurement of monomer and low-molecular RAFT-agent conversions. Since chain transfer reaction in RAFT-polymerization is usually characterized by low value of activation energy, compared to activation energy of chain growth, it is supposed that constant C_{tr1} doesn't depend or slightly depends on temperature. We will propose as an assumption that C_{tr1} doesn't depend on temperature [10].

By analogy with k_{a1}, we deduce equation for constant k_{a2}. From Eq. (2) we express such concentration [Int(i, j, 0)]:

$$[Int(i, j, 0)] = \frac{k_{a2}}{3k_f}[RAFT(i, 0)][R].$$

Substitute expressions, derived for [Int(i, 0, 0)] and [Int(i, j, 0)] in Eq. (4):

$$d[RAFT(i, 0)]/dt = \frac{2}{3}k_{a1}[RAFT(0,0)][R] - \frac{1}{3}k_{a2}[RAFT(i, 0)][R]. \tag{14}$$

Since in the end it was found that constant of chain transfer to low-molecular RAFT-agent C_{tr1} is equal to divided to constant k_p coefficient before expression [RAFT(0,0)][R] in the balance differential equation for [RAFT(0,0)], from Eq. (14) for constant of chain transfer to macromolecular RAFT-agent, we obtain the next expression:

$$C_{tr2} = \frac{1}{3}\frac{k_{a2}}{k_p}.$$

From the last equation we obtain an expression for constant k_{a2}, which based on the following calculation:

$$k_{a2} = 3C_{tr2}k_p, \quad L\cdot mol^{-1}\cdot s^{-1}. \tag{15}$$

In research [4] on the base of styrene and DBTK, macromolecular RAFT-agent was synthesized, thereafter with a view to experimentally determine constant C_{tr2}, polymerization of styrene was performed with the use of the latter. In the course of experiment, it may be supposed that constant C_{tr2} depends on monomer and macromolecular RAFT-agent conversions by analogy with Eq. (12). As a result directly from the experimentally measured monomer and macromolecular RAFT-agent conversions, value C_{tr2} was derived, equal to 1,000. On the ground of the same considerations as for that of C_{tr1}, we assume independence of constant C_{tr2} on temperature.

1.2.2.4 RATE CONSTANTS OF INTERMEDIATES FRAGMENTATION, TERMINATION BETWEEN RADICALS AND TERMINATION BETWEEN RADICALS AND INTERMEDIATES

In research [4] it was shown, that RAFT-polymerization rate is determined by this equation:

$$(W_0 / W)^2 = 1 + \frac{k_{t2}}{k_{t1}} K[RAFT(0,0)]_0 + \frac{k_{t3}}{k_{t1}} K^2[RAFT(0,0)]_0^2,$$

where W_0 and W – polymerization rate in the absence and presence of RAFT-agent, respectively, s^{-1}; K – constant of equilibrium (III), $L\cdot mol^{-1}$; k_{t3} – constant of termination between two intermediates reaction rate, $L\cdot mol^{-1}\cdot s^{-1}$ [11].

For initiated AIBN styrene polymerization in DBTC's presence at 80°C, it was shown that intermediates quadratic termination wouldn't be implemented and RAFT-polymerization rate was determined by equation [4]:

$$\left(W_0 \ / \ W\right)^2 = 1 + 8[RAFT(0,0)]_0 .$$

Since $\dfrac{k_{t2}}{k_{t1}} \approx 1$, then at 80°C $K = 8 \ L \cdot mol^{-1}$ [4]. In order to find dependence of con-stant K on temperature, we made research of polymerization kinetics at 60°C. It was found, (Fig. 1), that the results of kinetic measurements well rectify in coordinates $\left(W_0 \ / \ W\right)^2 = f([RAFT(0,0)]_0)$. At 60°C, $K = 345 \ L \cdot mol^{-1}$ was obtained. Finally dependence of equilibrium constant on temperature has been determined in the form of Vant–Goff's equation:

$$K = 4.85 \cdot 10^{-27} \ e^{22123/T} , \ L \cdot mol^{-1}. \tag{16}$$

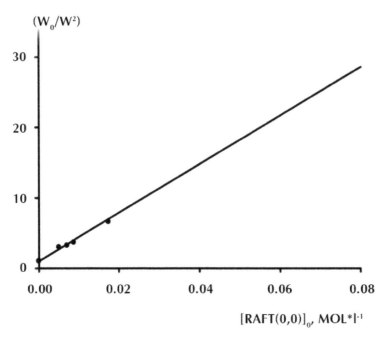

FIGURE 1   Dependence $(W_0 \ / \ W)^2$ on DBTC concentration at 60°C.

In compliance with the equilibrium (III), the constant is equal to

$$K = \frac{k_{a2}}{3k_f}, \; L \cdot mol^{-1}.$$

Hence, reactions of intermediates fragmentation rate constant will be as such:

$$k_f = \frac{k_{a2}}{3K}, \; s^{-1}. \tag{17}$$

The reactions of intermediates fragmentation rate constant was built into the model in the form of dependence (17) considering Eqs. (15) and (16).

As it has been noted above, ratio k_{t2} / k_{t1} equals approximately to one, therefore it will taken, that $k_{t2} \approx k_{t1}$ [4]. For description of gel-effect, dependence as a function of monomer conversion C_M and temperature T (K) [12] was applied:

$$k_{t2} \approx k_{t1} \approx 1.255 \cdot 10^9 e^{-844/T} e^{-2(A_1 C_M + A_2 C_M^2 + A_3 C_M^3)}, \; L \cdot mol^{-1} \cdot s^{-1},$$

where $A_1 = 2.57 - 5.05 \cdot 10^{-3}T$; $A_2 = 9.56 - 1.76 \cdot 10^{-2}T$; $A_3 = -3.03 + 7.85 \cdot 10^{-3}T \cdot$

1.2.2.5   RATE CONSTANT FOR CHAIN GROWTH

The method of polymerization, being initiated by pulse laser radiation [13] is used for determination of rate constant for chain growth k_p lately. It is anticipated that such an estimation method is more correct, than the traditionally used revolving sector method [12]. We made our choice on temperature dependence of the rate constant for chain growth that was derived on the ground of method of polymerization, being initiated by pulse laser radiation:

$$k_p = 4.27 \cdot 10^7 e^{-3910/T}, \; L \cdot mol^{-1} \cdot s^{-1}, \tag{18}$$

since this dependence is more adequately describes the change of polymerization reduced rate with monomer conversion in the network of the developed mathematical model (Fig. 2), than temperature dependence, which is derived by revolving sector method [12]:

$$k_p = 1.057 \cdot 10^7 e^{-3667/T}, \; L \cdot mol^{-1} \cdot s^{-1}. \tag{19}$$

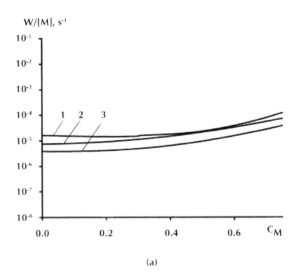

(a)

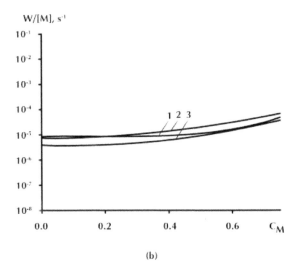

(b)

FIGURE 2 Dependence of initiated AIBN ($[I]_0 = 0.01$ mol·L^{-1}) styrene polymerization reduced rate on monomer conversion at 60°C (1 – experiment; 2 – estimation by introduced in this research mathematical model with temperature dependence of k_p (18); 3 – estimation by introduced in this research mathematical model with temperature dependence of k_p (19): $[RAFT(0,0)]_0 = 0$ mol·L^{-1} (a), 0.007 (b).

1.2.3 MODEL'S ADEQUACY

The results of polystyrene molecular-mass properties calculations by the introduced mathematical model are presented in Figs. 3 and 4. Mathematical model of styrene RAFT-polymerization in the presence of trithiocarbonates, taking into account the radicals and intermediates cross termination, adequately describes the experimental data that prove the process mechanism, built in the model. The essential proof of the mechanism correctness is that in case of conceding the absence of radicals and intermediates cross termination – the experimental data wouldn't substantiate theoretical calculation by the mathematical model, introduced in this assumption (Fig. 5).

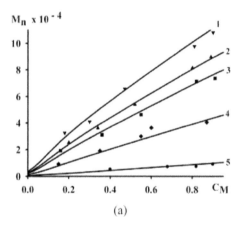

(a)

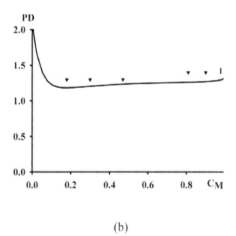

(b)

FIGURE 3 *(Continued)*

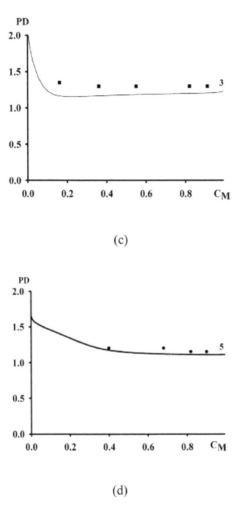

(c)

(d)

FIGURE 3 Dependence of number-average molecular mass (a) and polydispersity index (b)–(d) on monomer conversion for being initiated by AIBN ($[I]_0$=0.01 mol·L^{-1}) styrene bulk RAFT-polymerization at 60°C in the presence of DBTC (lines – estimation by model; points – experiment): $[RAFT(0,0)]_0$ = 0.005 mol·L^{-1} (1), 0.007 (2), 0.0087 (3), 0.0174 (4), 0.087 (5).

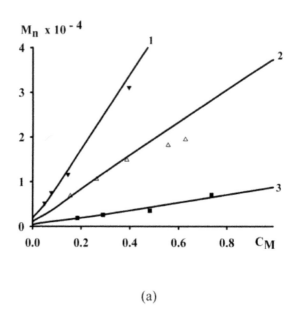

(a)

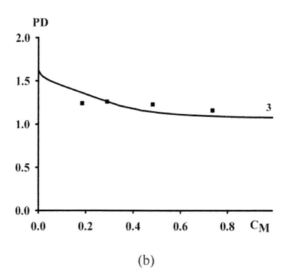

(b)

FIGURE 4   Dependence of number-average molecular mass (a) and polydispersity index (b) on monomer conversion for being initiated by AIBN ($[I]_0$=0.01 mol·L^{-1}) styrene bulk RAFT-polymerization at 80°C in DBTC presence (lines – estimation by model; points – experiment): $[RAFT(0,0)]_0$ = 0.01 mol·L^{-1} (1), 0.02 (2), 0.1 (3) [4].

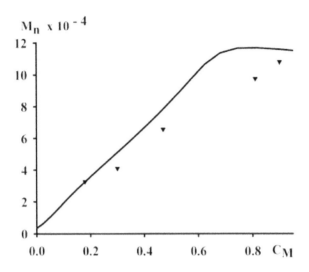

FIGURE 5 Dependence of number-average molecular mass on monomer conversion for initiated AIBN ($[I]_0 = 0.01$ mol·L^{-1}) styrene bulk RAFT-polymerization at 60 °C in DBTC presence $[RAFT(0,0)]0 = 0.005$ mol·L^{-1} (lines – estimation by model assuming that radicals and intermediates cross termination are absent; points – experiment).

Due to adequacy of the model realization at numerical experiment it became possible to determine the influence of process controlling factors on polystyrene molecular-mass properties.

1.2.4 NUMERICAL APPROACH

Research of influence of the process controlling factors on molecular-mass properties of polystyrene, synthesized by RAFT-polymerization method in the presence of AIBN and DBTC, was made in the range of initial concentrations of: initiator – 0–0.1 mol·L^{-1}, monomer – 4.35–8.7 mol·L^{-1}, DBTC – 0.001–0.1 mol·L^{-1}; and at temperatures – 60–120°C.

1.2.4.1 THE INFLUENCE OF AIBN INITIAL CONCENTRATION BY NUMERICAL APPROACH

It was set forth that generally in the same other conditions, with increase of AIBN initial concentration number-average, the molecular mass of polystyrene decreases (Fig. 6). At all used RAFT-agent initial concentrations, there is a linear or close to linear growth of number- average molecular mass of polystyrene with monomer conversion. This means that even the lowest RAFT-agent initial concentrations affect the process of radical polymerization. It should be noted that at high RAFT-agent initial concentrations (Fig. 7) the change of AIBN initial concentration practically doesn't have any influence on number-average molecular mass of polystyrene. But at increased temperatures (Fig. 8), in case of high AIBN initial concentration, it is comparable to high RAFT-agent initial concentration; polystyrene molecular mass would be slightly decreased due to thermal initiation.

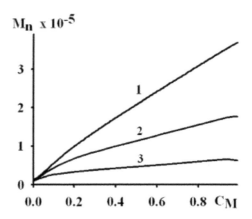

FIGURE 6   Dependence of number-average molecular mass M_n on monomer conversion C_M (60°C) $[M]_0 = 6.1$ mol·L^{-1}, $[RAFT(0, 0)]_0 = 0.001$ mol·L^{-1}, $[I]_0 = 0.001$ mol·L^{-1} (1), 0.01 (2), 0.1 (3).

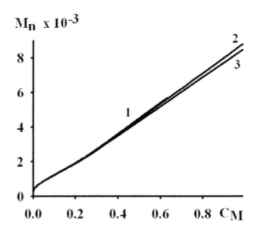

FIGURE 7   Dependence of number-average molecular mass M_n on monomer conversion C_M (60°C) $[M]_0 = 8.7$ mol·L^{-1}, $[RAFT(0, 0)]_0 = 0.1$ mol·L^{-1}, $[I]_0 = 0.001$ mol·L^{-1} (1), 0.01 (2), 0.1 (3).

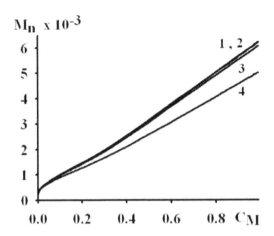

FIGURE 8   Dependence of number-average molecular mass M_n on monomer conversion C_M (120°C) $[M]_0 = 6.1\ mol\cdot L^{-1}$, $[RAFT(0, 0)]_0 = 0.1\ mol\cdot L^{-1}$, $[I]_0 = mol\cdot L^{-1}$ (1), 0.001 (2), 0.01 (3), 0.1 (4).

Since the main product of styrene RAFT-polymerization process, proceeding in the presence of trithiocarbonates, is a narrow-dispersed high-molecular RAFT-agent (marked in kinetic scheme as RAFT(i, j)), which is formed as a result of reversible chain transfer, and widely-dispersed (minimal polydispersity – 1.5) polymer, forming by the radicals quadratic termination, so common polydispersity index of synthesizing product is their ratio. In a broad sense, with increase of AIBN initial concentration, the part of widely-dispersed polymer, which is formed as a result of the radicals quadratic termination, increase in mixture, thereafter general polydispersity index of synthesizing product increases.

However, at high temperatures this regularity can be discontinued – at low RAFT-agent initial concentrations the increase of AIBN initial concentration leads to a decrease of polydispersity index (Fig. 9, curves 3 and 4). This can be related only thereto that at high temperatures thermal initiation and elementary reactions rate constants play an important role, depending on temperature, chain growth and radicals quadratic termination reaction rate constants, monomer initial concentration in a complicated way [6]. Such complicated dependence makes it difficult to analyze the influence of thermal initiation role in process kinetics, therefore the expected width of MMD of polymer, which is expected to be synthesized at high temperatures, can be estimated in every specific case in the frame of the developed theoretical regularities.

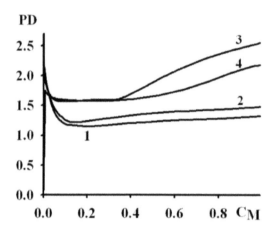

FIGURE 9 Dependence of polydispersity index PD on monomer conversion C_M (120°C) $[M]_0 = 8.7$ mol·L^{-1}, $[RAFT(0, 0)]_0 = 0.001$ mol·L^{-1}, $[I]_0 = 0$ mol·L^{-1} (1), 0.001 (2), 0.01 (3), 0.1 (4).

Special attention shall be drawn to the fact that for practical objectives, realization of RAFT-polymerization process without an initiator is of great concern. In all cases at high temperatures as the result of styrene RAFT-polymerization implementation in the presence of RAFT-agent without AIBN, more high-molecular (Fig. 10) and more narrow-dispersed polymer (Fig. 9, curve 1) is built-up than in the presence of AIBN (Fig. 9, curves 2–4).

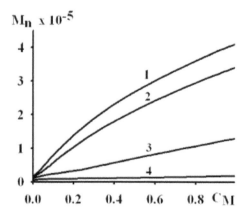

FIGURE 10 Dependence of number-average molecular mass M_n on monomer conversion C_M (120°C) $[M]_0 = 8.7$ mol·L^{-1}, $[RAFT(0, 0)]_0 = 0.001$ mol·L^{-1}, $[I]_0 = 0$ mol·L^{-1} (1), 0.001 (2), 0.01 (3), 0.1 (4).

1.2.4.2 THE INFLUENCE OF MONOMER INITIAL CONCENTRATION BY NUMERICAL EXPERIMENT

In other identical conditions, the decrease of monomer initial concentration re-duces the number-average molecular mass of polymer. Polydispersity index doesn't practically depend on monomer initial concentration.

1.2.4.3 INFLUENCE OF RAFT-AGENT INITIAL CONCENTRATION BY NUMERICAL EXPERIMENT

In other identical conditions, increase of RAFT-agent initial concentration reduc-es the number-average molecular mass and polydispersity index of polymer (Fig. 11).

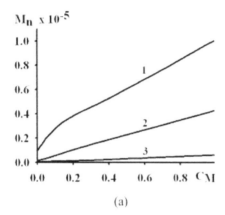

(a)

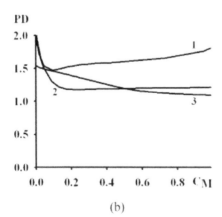

(b)

FIGURE 11 Dependence of number-average molecular mass M_n (a) and polydispersity index PD (b) on monomer conversion C_M (90°C) $[I]_0 = 0.01$ mol·L^{-1}, $[M]_0 = 6.1$ mol·L^{-1}, $[RAFT(0, 0)]_0 = 0.001$ mol·L^{-1} (1), 0.01 (2), 0.1 (3).

1.2.4.4   THE INFLUENCE OF TEMPERATURE BY NUMERICAL EXPERIMENT

Generally, in other identical conditions, the increase of temperature leads to a decrease of number-average molecular mass of polystyrene (Fig. 12 (a)). Thus, polydispersity index increases (Fig. 12 (b)). If RAFT-agent initial concentration greatly exceeds AIBN initial concentration, then the temperature practically doesn't influence the molecular-mass properties of polystyrene.

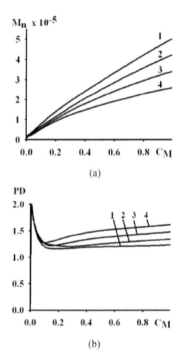

FIGURE 12   Dependence of number-average molecular mass M_n (a) and polydispersity index PD (b) on monomer conversion C_M $[I]_0 = 0.001$ mol·L^{-1}, $[M]_0 = 8.7$ mol·L^{-1}, $[RAFT(0, 0)]_0 = 0.001$ mol·L^{-1}, $T = 60°C$ (1), 90 (2), 120 (3), 150 (4).

1.3   CONCLUSION

The kinetic model developed in this research allows an adequate description of molecular-mass properties of polystyrene, obtained by controlled radical polymerization, which proceeds by reversible chain transfer mechanism and accompanied by "addition-fragmentation." This means, that the model can be used for development of technological applications of styrene RAFT-polymerization in the presence of trithiocarbonates.

Researches were supported by Russian Foundation for Basic Research (project. no. 12–03–97050-r_povolzh'e_a).

KEYWORDS

- **controlled radical polymerization**
- **dibenzyltritiocarbonate**
- **mathematical modeling**
- **polystyrene**
- **reversible addition-fragmentation chain transfer**

REFERENCES

1. Matyjaszewski K.: Controlled/Living Radical Polymerization: Progress in ATRP, D.C.: American Chemical Society, Washington (2009).
2. Matyjaszewski K.: Controlled/Living Radical Polymerization: Progress in RAFT, DT, NMP and OMRP, D.C.: American Chemical Society, Washington (2009).
3. Barner-Kowollik C.: Handbook of RAFT Polymerization, Wiley–VCH Verlag GmbH, Weinheim (2008).
4. Chernikova E.V., Terpugova P.S., Garina E.S., Golubev V.B.: Controlled radical polymerization of styrene and n-butyl acrylate mediated by tritiocarbonates. Polymer Science, Vol. 49(A), №2, 108 (2007).
5. Stephen Z.D. Cheng: Handbook of Thermal Analysis and Calorimetry. Volume 3 – Applications to Polymers and Plastics: New York, Elsevier, (2002).
6. LI I., Nordon I., Irzhak V.I., Polymer Science, 47, 1063 (2005).
7. Zetterlund P.B., Perrier S., Macromolecules, 44, 1340 (2011).
8. Biesenberger J.A., Sebastian D.H.: Principles of polymerization engineering, John Wiley & Sons Inc., New York (1983).
9. Chong Y.K., Krstina J., Le T.P.T., Moad G., Postma A., Rizzardo E., Thang S.H., Macromolecules, 36, 2256 (2003).
10. Goto A., Sato K., Tsujii Y., Fukuda T., Moad G., Rizzardo E., Thang S.H., Macromolecules, 34, 402 (2001).
11. Kwak Y., Goto A., Fukuda T., Macromolecules, 37, 1219 (2004).
12. Hui A.W., Hamielec A.E., J. Appl. Polym. Sci, 16, 749 (1972).
13. Li D., Hutchinson R.A., Macromolecular Rapid Communications, 28, 1213 (2007).

CHAPTER 2

A DETAILED REVIEW ON PORE STRUCTURE ANALYSIS OF ELECTROSPUN POROUS MEMBRANES

BENTOLHODA HADAVI MOGHADAM,
VAHID MOTTAGHITALAB, MAHDI HASANZADEH,
and AKBAR KHODAPARAST HAGHI

CONTENTS

ABSTRACT

Nanoporous membranes are an important class of nanomaterials that can be used in many applications, especially in micro and nanofiltration. Electrospun nanofibrous membranes have gained increasing attention due to the high porosity, large surface area per mass ratio along with small pore sizes, flexibility, and fine fiber diameter, and their production and application in development of filter media. Image analysis is a direct and accurate technique that can be used for characterization of porous media. This technique, due to its convenience in detecting individual pores in a porous media, has some advantages for pore measurement. The three-dimensional reconstruction of porous media, from the information obtained from a two-dimensional analysis of photomicrographs, is a relatively new research area. In the present paper, we have reviewed the recent progress in pore structure analysis of porous membranes with emphasis in image analysis technique. Pore characterization techniques, properties, and characteristics of nanoporous structures are also discussed in this paper.

2.1 INTRODUCTION

Nanofibrous membranes have received increasing attention in recent years as an important class of nanoporous materials. Although there are various techniques to produce polymer nanofiber mats, electrospinning is considered as one of the most efficient ways to obtain nonwoven nanofiber mats with pore sizes ranging from tens of nanometers to tens of micrometers.[1-3]

In recent years, significant progress has been done in the understanding and modeling of pore-scale processes and phenomena. By using increased computational power, realistic pore-scale modeling in recognition of tomographic and structure of porous membranes can be obtained.[4-6] Information about the pore structure of membranes is often obtained by several methods including mercury intrusion porosimetry,[7-9] liquid extrusion porosimetry,[9-12] flow porosimetry[9-15] and image analysis of thin section images.[16-23] Image analysis is a useful technique that is gaining attention due to its convenience in detecting individual pores in the membrane image.

The three-dimensional reconstruction of porous media, such as nanofiber mats, from the information obtained from a two-dimensional micrograph has attracted considerable interest for many applications. A successful reconstruction procedure leads to significant improvement in predicting the macroscopic properties of porous media.[23-44]

This short review intends to introduce recent progress in pore structure analysis of porous membranes, with emphasis on electrospun polymer nanofiber mats. The paper is organized as follows. Section 2.2 present polymer membranes and their types. Section 2.3 deals with the nanofibrous membranes as one of the most important porous media. Section 2.4 presents the porosity of membranes and the techniques used to evaluate the pore characteristics of porous membranes. Finally, Section 2.5 surveys the most characteristic and important recent examples, which image analysis technique was used for characterization of porous media, especially three-dimensional reconstruction of porous structure. The report ends with a conclusion.

2.2 POLYMER MEMBRANES

Membrane technologies are already serving as a useful tool for industrial processes, such as health sector, food industry, sustainable water treatment and energy conversion and storage. Membrane is a selective barrier between two phases and defined as a very thin layer or cluster of layers that allows one or more selective components to permeate through readily when mixtures of different kinds of components are driven to its surface, thereby producing a purified product. The ability to control the diffusion rate of a chemical species through the membrane is key property of membranes.[45–54] Several membrane processes have been proposed based on the barrier structure, including microfiltration,[54–55] ultrafiltration,[56–57] nanofiltration,[54,58] and reverse osmosis.[58] The membrane processes can be categorized based on the barrier structure by different driving forces as shown in Fig. 1.

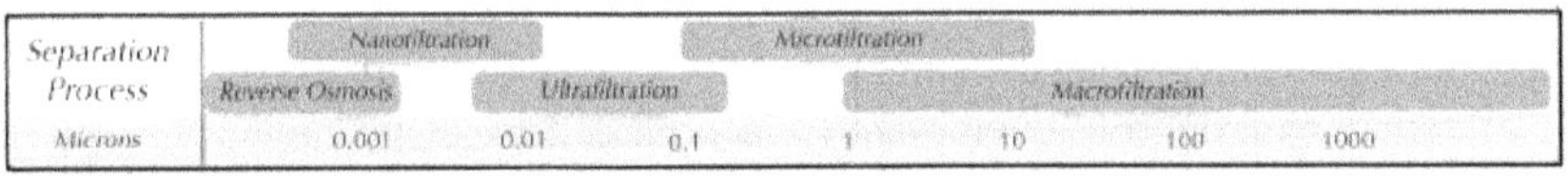

Separation Process	Nanofiltration		Microfiltration			
	Reverse Osmosis	Ultrafiltration		Macrofiltration		
Microns	0.001	0.01	0.1	1	10	100	1000

FIGURE 1   Size range of particles in various membrane separation processes.

Pore size distribution, specific surface area, outer surface, and cross section morphology are some of the most important characteristics of membranes.

2.2.1 TYPES OF MEMBRANES

According to the morphology, membranes can be classified into two main types: 'isotropic' and 'anisotropic'. Isotropic membranes include microporous, nonporous, and dense membranes and are made of single layer with uniform structure through the depth of the membrane. On the other hand, anisotropic membranes consist of more than one layer supported by a porous substrate.

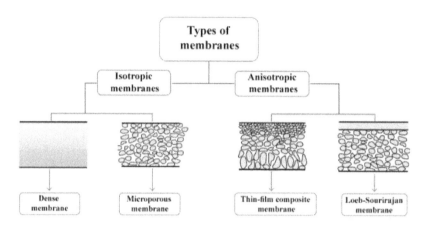

FIGURE 2   Classification of membranes.

2.2.1.1   ISOTROPIC MEMBRANES

Most of the available membranes are porous or consist of a dense top layer on a porous structure. Isotropic dense membranes can be prepared (i) by melt extrusion of polymer or (ii) by solution casting (solvent evaporation).[50–52]

Isotropic microporous membranes have a rigid, highly voided structure, and interconnected pores. These membranes have higher fluxes than isotropic dense membranes. Microporous membranes are prepared by some methods, such as track-etching, expanded-film, and template leaching.[50,59]

2.2.1.2   ANISOTROPIC MEMBRANES

Anisotropic membranes are consists of a very thin (0.1 to 1 pm) selective skin layer on a highly permeable microporous substrate and highly membrane fluxes in which the porosity, pore size, or even membrane composition change from the top to the bottom surface of the membrane. An asymmetrical structure is now produced from a wide variety of polymers, which are currently applied in pressure driven membrane processes, such as reverse osmosis, ultrafiltration, or gas separation.

Anisotropic membranes may be prepared by various techniques, including phase inversion,[50,59] interfacial polymerization,[50] solution coating,[50] plasma deposition,[50] and electrospinning in the laboratory or on a small industrial scale.[60] Nanofibrous membranes, as simple and interesting anisotropic membranes, are described in the following section.

2.3   NANOFIBROUS MEMBRANE

Nanofibrous membranes have high specific surface area, high porosity, small pore size, and flexibility to conform to a wide variety of sizes and shapes. Therefore, they have been suggested as excellent candidate for many applications, especially in micro and nanofiltration. Nanofibrous membrane can be processed by a number of techniques such as drawing,[61] template synthesis,[62] phase separation,[63] self-assembly,[64] and electrospinning.[65] Among them electrospinning has an advantage with its comparative low cost and relatively high production rate.

With regard to the low mechanical properties of nanofibrous membrane, many attempts have been made to improve mechanical properties of nanofibers. In this regard, electrospinning process is the best method to fabricate the mixed membrane of microfiber and nanofibers simultaneously. The good physical and mechanical properties of microfibers (as a substrate) are favorable for enhancing the mechanical performance of nanofibrous membrane.

2.3.1 ELECTROSPINNING PROCESS

Within the past several years, electrospinning process has garnered increasing attention, due to its capability and feasibility to generate large quantities of nanofibers with well-defined structures. Figure 3 shows a schematic illustration of electrospinning setup. In this process, a strong electric field is applied between polymer solution contained in a syringe with a capillary tip and grounded collector. When the electric field overcomes the surface tension force, the charged polymer solution forms a liquid jet and travels towards collection plate. As the jet travels through the air, the solvent evaporates and dry fibers deposits on the surface of a collector.

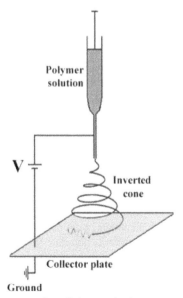

FIGURE 3   Schematic representation of electrospinning process.

The morphology and the structure of the electrospun nanofibrous membrane are dependent upon many parameters which are mainly divided into three categories: solution properties (the concentration, liquid viscosity, surface tension, and dielectric properties of the polymer solution), processing parameters (applied voltage, volume flow rate, tip to collector distance, and the strength of the applied electric field), and ambient conditions (temperature, atmospheric pressure and humidity).[66-80]

2.3.2   POROUS NANOFIBERS

There are various techniques for the fabrication of highly porous nanofiber, including fiber bonding, solvent casting, particle leaching, phase separation, emulsion freeze-drying, gas foaming, and 3–D printing. These porous membranes can also be produced by a combination of electrospinning and phase inversion techniques. Nanofibrous membrane produced by this approach can generate additional space and surface area within as-spun fibrous scaffolds. Due to the fine fiber size and large expected surface area, electrospun nanofibrous membranes have a desirable property for filter media, catalyst immobilization substrates, absorbent media and encapsulated active ingredients, such as activated carbon and various biocides.

Studied showed that the formation of pores on nanofibers during electrospinning process affected by many parameters such as humidity, type of polymer, solvent vapor pressure, electrospinning conditions, etc. Although no generally agreed set of definitions exists, porous materials can be classified in terms of their pore sizes into various categories including capillaries (>200 nm), macropores (50–200 nm), mesopores (2–50 nm) and micropores (0.5–2 nm). According to the literature, the mechanism that forms porous surface on polymer casting film is applicable to the phenomenon on electrospun nanofibers.[81-88] The rapid solvent evaporation and subsequent condensation of moisture into water particles result in the formation of nano or micropores on the fiber surface. When the environment humidity increases, the pore size becomes larger. However, this result was observed only when the solution used a highly volatile organic solvent, such as chloroform, tetrahydrofuran and acetone.

2.3.3   POTENTIAL APPLICATIONS

The research and development of electrospun nanofibrous membrane has evinced more interest and attention in recent years due to the heightened awareness of its potential applications in various fields. The electrospun nanofibrous membrane, due to their high specific surface area, high porosity, flexibility, and small pore size, have been suggested as excellent candidate for many applications including filtration, multifunctional membranes, reinforcements in light weight composites, biomedical agents, tissue engineering scaffolds, wound dressings, full cell and protective clothing.[72,91-92]

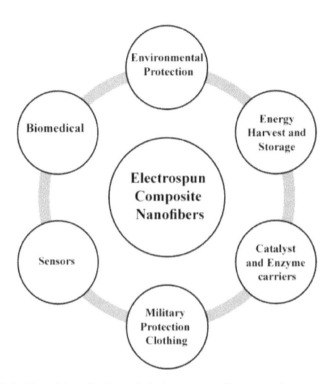

FIGURE 4   Potential applications of electrospun nanofibrous membranes.

Membranes have been widely used for a great variety of applications such as bioseparations, sterile filtration, bioreactors, and so on. Each application depends on the membrane material and structure. For microfiltration and ultrafiltration process, the efficiency was determined by the selectivity and permeability of the membrane.

2.4   MEMBRANE POROSITY

As it has been mentioned earlier, the most important characteristics of membranes are: thickness, pore diameter, solvent permeability and porosity. Moreover, the filtration performance of membranes is also strongly related to their pore structure parameters, that is, percent open area, and pore size distribution. Hence, the porosity and pore structure characteristics play a more important role in membrane process and applications.[93–97]

The porosity, ε_V, is defined as the percentage of the volume of the voids, V_v, to the total volume (voids plus constituent material), V_t, and is given by

$$\varepsilon_V = \frac{V_v}{V_t} \times 100 \tag{1}$$

Similarly, the percent open area, ε_A, that is defined as the percentage of the open area, A_0, to the total area, A_t, is given by

$$\varepsilon_A = \frac{A_0}{A_t} \times 100 \tag{2}$$

Usually, porosity is determined for membranes with a three-dimensional structure (e.g., relatively thick nonwoven fabrics). Nevertheless, for two-dimensional structures such as woven fabrics and relatively thin nonwovens, it is often assumed that porosity and percent open area are equal.[98]

2.4.1 PORE CHARACTERIZATION TECHNIQUES

Development and application of effective procedures for membrane characterization are one of indispensable components of membrane research. Pore structure, as the main characteristics of porous membrane, has significant influence on the performance of membranes. In general, there are three types of pores in membrane: (1) the closed pore that are not accessible; (2) the blind pore that terminate within the material; and (3) the through pore that permit fluid flow through the material and determine the barrier characteristics and permeability of the membrane (*see* Fig. 5).

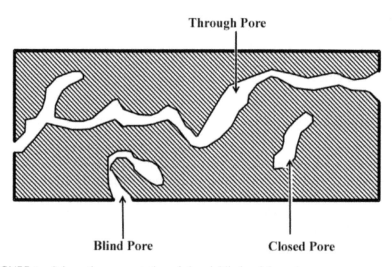

FIGURE 5   Schematic representation of closed, blind and through pores.

Pore structure characteristics, as one of the main tools for evaluating the performance of any porous membrane, can be performed using microscopic and macroscopic approaches. Microscopic techniques usually consist of high-resolution light microscopy, electron microscopy, and X-ray diffraction. The major disadvantage of this technique is that it cannot determine flow properties and also is time consuming and expensive. Various techniques may be used to evaluate the pore characteristics of porous membranes through macroscopic approach, including mercury intrusion porosimetry, liquid extrusion porosimetry and liquid extrusion flow porometry. These techniques are also used for pore structure characterization of nanofibrous membranes, but the low stiffness and high-pressure sensitivity of nanofiber mats limit application of these techniques.[99–108]

2.4.1.1 MERCURY INTRUSION POROSIMETRY

Mercury intrusion porosimetry is a well-known method, which is often used to evaluate pore characteristics of porous membranes. This technique provides statistical information about various aspects of porous media such as pore size distribution, or the volume distribution.

Due to the fact that mercury, as a non-wetting liquid, does not intrude into pore spaces (except under application of sufficient pressure), a relationship between the size of pores and the pressure applied can be found. In this technique, a porous membrane is completely surrounded by mercury and pressure is applied to force the mercury into pores. As mercury pressure increases the large pores are filled with mercury first and the pore sizes are calculated as the mercury pressure increases. At higher pressures, mercury intrudes into the fine pores and, when the pressure reaches a maximum, total pore volume and porosity are calculated.[98]

According to Jena and Gupta,[9–11] the relationship of pressure and pore size is determined by the Laplace equation:

$$D = -\frac{4\gamma\cos\theta}{p} \tag{1}$$

where D is pore diameter, γ is surface tension of mercury, θ is contact angle of mercury and p is pressure on mercury for intrusion into the pore.

In addition to the pore size and its distribution, the total pore volume and the total pore area can also be determined by mercury intrusion porosimetry method. On the other hand, this method gives no information about the number of pores and is generally applicable to porous membrane with pore sizes ranging from 0.0018 µm to 400 µm. Moreover, mercury intrusion porosimetry does not account for closed pores because mercury does not intrude into them. Due to the application of high pressures, sample collapse and compression is possible, hence it is not suitable for fragile compressible materials such as nanofiber sheets. Other concerns include the fact that it is assumed that the pores are cylindrical, which is not the case in reality.[98]

2.4.1.2 Liquid Extrusion Porosimetry

In liquid extrusion porosimetry, a wetting liquid spontaneously intrudes into the pores and then is extrude from pores by a non-reacting gas. The differential pressure p is related to pore diameter D by

$$D = -\frac{4\gamma\cos\theta}{p} \tag{2}$$

where γ is surface tension of wetting liquid, and θ is contact angle of wetting liquid. The volume of extruded liquid and the differential pressure is measured by this technique.

In this method, a membrane is placed under the sample such that the largest pore of the membrane is smaller than the smallest pore of interest in the sample. First, the pores of the sample and the membrane are filled with a wetting liquid and then the pressure on gas is increased to displace the liquid from pores of the sample. Because the gas pressure is inadequate to empty the pores of the membrane, the liquid filled pores of the membrane allow the extruded liquid from the pores of the sample to flow out while preventing the gas to escape.[9–10]

Through pore volume and diameter were determined by measuring the volume of the liquid flowing out of the membrane and differential pressure, respectively. It should be noted that liquid extrusion porosimetry measures only the volume and diameters of through pores (blind pore are not measured), whereas mercury intrusion porosimetry measures all pore diameter.[11–12]

2.4.1.3 FLOW POROSIMETRY (BUBBLE POINT METHOD)

Flow porosimetry or bubble point method is based on the principle that a porous membrane will allow a fluid to pass only when the pressure applied exceeds the capillary attraction of the fluid in the largest pore. In this test method, the pore of the membrane is filled with a liquid and continues airflow is used to remove liquid from the pores. At a critical pressure, the first bubble will come through the largest pore in the wetted specimen. As the pressure increases, the smaller pores are emptied of liquid and gas flow increases. Once the flow rate and the applied pressure are known, particle size distribution, the number of pores, and porosity can be derived. In flow porosimetry, the membrane with pore sizes in the range of 0.013–500 μm can be measured.[98]

It is important to note that flow porosimetry measures only the throat diameter of each through pore and cannot measure the blind pore. This technique is based on the assumption that the pores are cylindrical, which is not the case in reality.[15]

2.4.1.4 *IMAGE ANALYSIS*

Image analysis technique, due to its convenience in detecting individual pores in a nonwoven image, has some advantages for pore measurement. Image analysis technique has been used to measure the pore characteristics of electrospun nanofiberwebs.[98] To measure the pore characteristics of electrospun nanofibrous membranes using image analysis, images (or micrograph) of the nanofiber webs, which are usually obtained by scanning electron microscopy (SEM), transmission electron microscopy (TEM) or atomic force microscopy (AFM), are required. This is highly relevant in that a picture to be used for image analysis must be of high quality and taken under appropriate magnifications.[98] The major advantage of porosity characterization by image analysis technique is that the cross sections provide detailed information about the spatial and size distribution of pores as well as their shape.

In this technique, initial segmentation of the micrographs is required to produce binary images. The typical way of producing a binary image from a greyscale image is by 'global thresholding' in which a single constant threshold is applied to segment the image. All pixels up to and equal to the threshold belong to the object and the remaining belong to the background. Global thresholding is very sensitive to inhomogeneities in the grey-level distributions of object and background pixels. In order to eliminate the effect of inhomogeneities in global thresholding, local thresholding scheme could be used. Firstly, the image is divided into sub-images where the inhomogeneities are negligible. Then the optimal thresholds are found for each sub-image. It can be found that this process is equivalent to segmenting the image with locally varying thresholds.[98] Fig. 6 shows global thresholding and local thresholding of electrospun nanofibrous mat. It is obvious that global thresholding resulted in some broken fiber segments. However, this problem was solved by using local thresholding. It should be mentioned that this process is extremely sensitive to noise contained in the image. So, a procedure to clean the noise and enhance the contrast of the image is necessary before the segmentation.[19,98]

FIGURE 6   (a) SEM image of a real web, (b) global thresholding, (c) local thresholding.

2.4.2 APPROPRIATE TECHNIQUE FOR NANOFIBROUS MEMBRANES

As mentioned above, through pore volume of nanofibrous membranes can be measured by mercury intrusion porosimetry and liquid extrusion porosimetry. Due to the high pressure that is applied to the nanofibrous membranes in mercury intrusion porosimetry method, the pores can get enlarged, which leads to over-estimation of porosity values. Blind pores in the nanofiber mat are negligible. Therefore, porosity of the nanofibrous membranes can be obtained from the measured pore volume and bulk density of the material. Liquid extrusion technique can give liquid permeability and surface area of through pores, which could not be measured by mercury intrusion porosimetry. It is important to note that for many applications such as filtration, pore throat diameters of nanofiber mats are required in addition to pore volume. While mercury intrusion and liquid extrusion porosimetry cannot measure pore throat diameter, flow porosimetry can measures pore throat diameters without distorting pore structure. Therefore, flow porosimetry is more suited for pore characterization of nanofibrous membranes.[18,109–110]

2.5 SUMMARY OF LITERATURE REVIEW

Establishing the quantitative relationships between the microstructure of porous media and their properties are an important goal, with a broad relevance to many scientific sectors and engineering applications. Since variations in pore shape and pore space connectivity are intrinsic features of many porous media, a pore structure model must involve both geometric and topological descriptions of their complex microstructure. Nowadays modeling and simulation of nanoporous membrane is of special interest to many researchers.

According to the literature, there are two types of pore-scale modeling; (1) Lattice–Boltzmann (LB) model and (2) pore network model. LB models capable of simulating flow and transport in the actual pore space. Pore network model has been considered as an effective tools used to investigate macroscopic properties from fundamental pore-scale behavior of processes and phenomena based on geometric volume averaging. This model has been used in chemical engineering, petroleum engineering and hydrology fields to study a wide range of single and multiphase flow processes. Pore network model utilizes an idealization of the complex pore space geometry of the porous media. For this purpose the pore space is represented by pore elements having simple geometric shapes such as pore-bodies and pore-throats that have been represented by spheres and cylinders, respectively.[31,111–113]

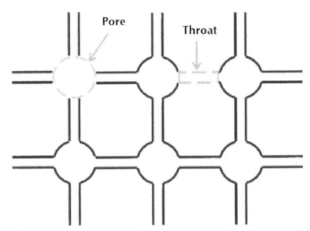

FIGURE 7   Schematic of a pore network illustrating location of pores and throats.

2.5.1   2D IMAGE ANALYSIS OF POROUS MEDIA

Definition of pore network structure, such as pore-body locations, pore-body size distributions, pore-throat size distributions, connectivity, and the spatial correlation between pore-bodies, is an important step towards analysis of porous media. Many valuable attempts have been made for characterization of porous media based on image analysis techniques. For example, Masselin et al.[20] employed image analysis to determine the parameters, such as porosity, pore density, mean pore radius, pore size distribution, and thickness of five asymmetrical ultrafiltration membranes. The results obtained from image analysis for the pore size were found to be in good agreement with rejection data.

Ekneligoda et al.[114] used image analysis technique to extract the area and perimeter of each pore from SEM images of two sandstones, Berea and Fontainebleau. The compressibility of each pore was calculated using boundary elements, and estimated from a perimeter-area scaling law. After the macroscopic bulk modulus of the rock was estimated by the area-weighted mean pore compressibility and the differential effective medium theory, the predicted results were compared with experimental values of the bulk modulus. The resulting predictions are close to the experimental values of the bulk modulus.

A variety of image analysis techniques was used by Lange et al.[19] to characterize the pore structure of cement-based materials, including plain cement paste, pastes with silica fume, and mortars. These techniques include sizing, two-point correlation, and fractal analyses. Backscattered electron images of polished sections were used to observe the pore structure of cement pastes and mortars. They measured pore size distribution of specimen by using image analysis techniques and compared with mercury intrusion porosimetry derived pore size distribution curves. They found that the image-based pore size distribution was able to better describe the large porosity than the mercury intrusion porosimetry.

In the study on pore structure of electrospun nanofibrous membranes by Ziabari et al.[98] a novel image analysis-based method was developed for measuring pore characteristics of electrospun nanofiber webs. Their model was direct, very fast, and presents valuable and comprehensive information regarding pore structure parameters of the webs. In this method, SEM images of nanofiber webs were converted to binary images and used as an input. First, voids connected to the image border are identified and cleared by using morphological reconstruction where the mask image is the input image and marker image is zero everywhere except along the border. Total area, which is the number of pixels in the image, is measured. Then the pores are labelled and each considered as an object. Here the number of pores may be obtained. In the next step, the number of pixels of each object as the area of that object is measured. Having the area of pores, the porosity may be calculated.

They also investigated the effects of web density, fiber diameter and its variation on pore characteristics of the webs by using some simulated images and found that web density and fiber diameter significantly influence the pore characteristics, whereas the effect of fiber diameter variations was insignificant. Furthermore, it seemed that the changes in number of pores were independent of variation of fiber diameter and that this could be attributed to the arrangement of the fibers.

In another study, Ghasemi–Mobarakeh et al.[115] demonstrated the possibility of porosity measurement of various surface layers of nanofibers mat using image analysis. They found that porosity of various surface layers is related to the number of layers of nanofibers mat. This method is not dependent on the magnification and histogram of images. Other methods such as mercury intrusion porosimetry, indirect method, and also calculation of porosity by density measurement cannot be used for porosity measurement of various surface layers and measure the total porosity of nanofibers mat. These methods show high porosity values (higher than 80%) for the nanofibers mat, while the porosity measurements based on thickness and apparent density of nanofibers mat demonstrated the porosity of between 60 and 70%.[116] This value was calculated using the following equation:

$$\varepsilon_V = 1 - \frac{\rho_a}{\rho_b} \times 100 \qquad (1)$$

$$\rho_a = \frac{m}{T \times A} \qquad (2)$$

where ρ_a and ρ_b are apparent density and bulk density of nanofiber mat, m is nanofiber mat mass, A is nanofiber mat area, and T is thickness of nanofiber mat.[116]

2.5.2 *3D IMAGE ANALYSIS OF POROUS MEDIA*

Several instrumental characterization techniques have been suggested to obtain 3D volume images of pore space, such as X-ray computed micro tomography and magnetic resonance computed micro tomography. However, these techniques may be limited by their resolution. So, the 3D stochastic reconstruction of porous media from statistical information (produced by analysis of 2D photomicrographs) has been suggested. Although pore network models can be two- or three-dimensional, 2D image analysis, due to their restricted information about the whole microstructure, was unable to predict morphological characteristics of porous membrane. Therefore, 3D reconstruction of porous structure will lead to significant improvement in predicting the pore characteristics. Recently research work has focused on the 3D image analysis of porous membranes.

Wiederkehr et al.[117] in their study of three-dimensional reconstruction of pore network, utilized an image morphing technique to construct a three-dimensional multiphase model of the coating from a number of such cross section images. They show that the technique can be successfully applied to light microscopy images to reconstruct 3D pore networks. The reconstructed volume was converted into a tetrahedron-based mesh representation suited for the use in finite element applications using a marching cubes approach. Comparison of the results for three-dimensional data and two-dimensional cross-section data suggested that the 3D-simulation should be more realistic due to the more exacter representation of the real microstructure.

Delerue et al.[118] utilized skeletization method to obtain a reconstructed image of the spatialized pore sizes distribution, i.e. a map of pore sizes, in soil or any porous media. The Voronoi diagram, as an important step towards the calculation of pore size distribution both in2D and 3D media, was employed to determine the pore space skeleton. Each voxel has been assigned a local pore size and a reconstructed image of a spatialized local pore size distribution was created. The reconstructed image not only provides a means for calculating the global volume versus size pore distribution, but also performs fluid invasion simulation, which take into account the connectivity of and constrictions in the pore network. In this case, mercury intrusion in a 3D soil image was simulated.

Al–Raoush et al.[31] employed a series of algorithms, based on the three-dimensional skeletonization of the pore space in the form of nodes connected to paths, to extract pore network structure from high-resolution three-dimensional synchrotron microtomography images of unconsolidated porous media systems. They used dilation algorithms to generate inscribed spheres on the nodes and paths of the medial axis to represent pore-bodies and pore-throats of the network, respectively. The authors have also determined the pore network structure, i.e. three-dimensional spatial distribution (x-, y-, and z-coordinates) of pore-bodies and pore-throats, pore-body and pore-throat sizes, and the connectivity, as well as the porosity, specific surface area, and representative elementary volume analysis on the porosity. They show that X-ray microtomography is an effective tool to non-destructively extract the structure of porous media. They concluded that spatial correlation between pore-bodies in the network is important and controls many processes and phenomena in single and multiphase flow and transport problems. Furthermore, the impact of resolution on the properties of the network structure was also investigated and the results showed that it has a significant impact and can be controlled by two factors: the grain size/resolution ratio and the uniformity of the system.

In another study, Liang et al.[5] proposed a truncated Gaussian method based on Fourier transform to generate 3D pore structure from 2D images of the sample. The major advantage of this method is that the Gaussian field is directly generated from its autocorrelation function and also the use of a linear filter transform is avoided Moreover, it is not required to solve a set of nonlinear equations associated with this transform. They show that the porosity and autocorrelation function of the reconstructed porous media, which are measured from a 2D binarized image of a thin section of the sample, agree with measured values. By truncating the Gaussian distribution, 3D porous media can be generated. The results for a Berea sandstone sample showed that the mean pore size distribution, taken as the result of averaging between several serial cross-sections of the reconstructed 3D representation, is in good agreement with the original thresholded 2D image. It is believed that by 3D reconstruction of porous media, the macroscopic properties of porous structure such as permeability, capillary pressure, and relative permeability curves can be determined.

Diógenes et al.[119] reported the reconstruction of porous bodies from 2D photomicrographic images by using simulated annealing techniques. They proposed the following methods to reconstruct a well-connected pore space: (i) Pixel-based Simulated Annealing (PSA), and (ii) Objective-based Simulated Annealing (OSA). The difference between the present methods and other research studies, which tried to reconstruct porous media using pixel-movement based simulated techniques, is that this method is based in moving the microstructure grains (spheres) instead of the pixels. They applied both methods to reconstruct reservoir rocks microstructures, and compared the 2D and 3D results with microstructures reconstructed by truncated Gaussian methods. They found that PSA method is not able to reconstruct connected porous media in 3D, while the OSA reconstructed microstructures with good pore space connectivity. The OSA method also tended to have better permeability determination results than the other methods. These results indicated that the OSA method can reconstruct better microstructures than the present methods.

In another study, a 3D theoretical model of random fibrous materials was employed by Faessel et al.[120] They used X-ray tomography to find 3D information on real networks. Statistical distributions of fibers morphology properties (observed at microscopic scale) and topological characteristics of networks (derived from mesoscopic observation), is built using mathematical morphology tools. The 3D model of network is assembled to simulate fibrous networks. They used a number of parameter describing a fiber, such as length, thickness, parameters of position, orientation and curvature, which derived from the morphological properties of the real network.

2.6 CONCLUSION

In recent years, great efforts have been devoted to nanoporous membranes. As a conclusion, much progress has been made in the preparation and characterization of porous media. Among several porous membranes, electrospun nanofibrous membranes, due to the high porosity, large surface area-to-volume ratios, small pores, and fine fiber diameter, have gained increasing attention. Useful techniques for evaluation of the pore characteristics of porous membranes are reviewed. Image analysis techniques have been suggested as a useful method for characterization of porous media due to its convenience in detecting individual pores. It is believed that the three-dimensional reconstruction of porous media, from the information obtained from a two-dimensional analysis of photomicrographs, will bring a promising future tonanoporous membranes.

KEYWORDS

- **image analysis**
- **nanofibrous membrane**
- **porous media**
- **three-dimensional analysis**

REFERENCE

1. Rutledge G.C., Li Y., Fridrikh S., *National Textile Center Research Briefs – Materials Competency*, 2004.
2. Yao C., Li X., Song T., *J. Appl. Polym. Sci.*, **103**, 380 (2007).
3. Kim G., Kim W., J. Biomed. Mater. Res B: *Appl. Bio. Mater.*, **81B**, 104 (2007).
4. Silina D., Patzekb T., *Phys. A: Stat. Mech. Appl.*, **371**, 336 (2006).
5. Liang Z.R., Fernandes C.P., Magnani F.S., Philippi P.C., *J. Petrol. Sci. Eng.*, **21**, 273 (1998).
6. Kim K.J., Fane A.G., R. Ben Aim, Liu M.G., Jonsson G., Tessaro I.C., Broek A.P., D.Bargeman, *J. Membran. Sci.*, **81**, 35(1994).
7. Dullien F.A.L., Dhawan G.K., *J. Colloid. Interf. Sci.*, **47**, 337 (1974).
8. Liabastre A.A., Orr C., *J. Colloid. Interf. Sci.*, **64**, 1 (1978).
9. Jena A., Gupta K., *Fluid/particle Sep. J.*, **14**, 227 (2002).
10. Jena A., Gupta K., *J. Filtration Soc.*, **1**, 23 (2001).
11. Jena A., Gupta K., *Int. Nonwovens. J.*, Fall, 45 (2003).

12. Barrett E.P., Joyner L.G., Halenda P.P., *J. Am. Chem. Soc.* **73**, 373 (1951).
13. Calvo J.I., A. Herna Ndez, P. Pra Danos, Martibnez L., Bowen W.R., *J. Colloid. Interf. Sci.*, **176**, 467 (1995).
14. Jena A.K., Gupta K.M., J. Power. Sources, **80**, 46 (1999).
15. Kraus G., Ross J.W., and Girifalco L.A., Phys. Chem., **57**, 330 (1953).
16. Gribble C.M., Matthews G.P., Laudone G.M., Turner A., C.J.Ridgway, Schoelkopf J., Gane P.A.C., *J. Chem. Eng. Sci.*, **66**, 3701 (2011).
17. Deshpande S., Kulkarni A., Sampath S., Herman H., *Surf. Coat. Tech.*, **187**, 6(2004).
18. Tomba E., Facco P., Roso M., Modesti M., Bezzo F., Barolo M., *Ind. Eng. Chem. Res.*, **49**, 2957(2010).
19. Lange D.A., *Cement. Concrete. Res.*, **24**, 841(1994).
20. Masselin I., L. Durand–Bourlier, Laine J.M., Sizaret P.Y., Chasseray X., Lemordant D., *J. Membrane Sci.*, **186**, 85 (2001).
21. Garboczi E.J., Bentz D.P., Martys N.S., *Exp. Meth. Phys. Sci.*, **35**, 1 (1999).
22. Mickel W., Munster S., Jawerth L.M., Vader D.A., Weitz D.A., Sheppard A.P., K.Mecke, Fabry B., G.E. Schroder–Turk, *Biophys. J.*, **95**, 6072 (2008).
23. Roysam B., Lin G., M. Amri Abdul–Karim, O. Al–Kofahi, K. Al–Kofahi, Shain W., Szarowski D.H., Turner J.N., Handbook of Biological Confocal Microscopy, 3rd edition, Springer, New York, 2006.
24. Quiblier J.A., *J. Colloid. Interf. Sci.*, **98**, 84 (1984).
25. Santos L.O.E., Philippi P.C., Damiani M.C., Fernandes C.P., *J. Petrol. Sci. Eng.*, **35**, 109 (2002).
26. Sambaer W., Zatloukal M., Kimmer D., *Chem. Eng. Sci.*, **66**, 613 (2011).
27. Shin C.H., Seo J.M., Bae J.S., *J. Ind. Eng. Chem.*, **15**, 784 (2009).
28. Ye G., K. van Breugel and Fraaij A.L.A., *Cement. Concrete Res.*, **33**, 215 (2003).
29. Holzer L., Münch B., Rizzi M., Wepf R., Marschall P., Graule T., *Appl. Clay. Sci.*, **47**, 330 (2010).
30. Liang Z., Ioannidis M.A., Chatzis I., *J. Colloid. Interf. Sci.*, **221**, 13(2000).
31. R.I. Al–Raoush and Willson C.S., *J. Hydrol.*, **300**, 44 (2005).
32. Fenwick D.H., Blunt M.J., *Adv. Water. Resour.*, **21**, 143 (1998).
33. Santos L.O.E., Philippi P.C., Damiani M.C., Fernandes C.P., *J. Petrol. Sci. Eng.*, **35**,109(2002).
34. Mendoza F., Verboven P., Mebatsion H.K., Kerckhofs, G. M.Wevers and B. Nicolaï, *Planta*, **226**, 559 (2007).
35. Bakke S. and Øren P., *SPE.J.*, **2**, 136 (1997).
36. Yee Ho A.Y., Gao H., Y. Cheong Lam and Rodrıguez I., *Adv. Funct. Mater*, **18**, 2057 (2008).
37. Sakamoto Y., Kim T.W., Ryoo R., Terasaki O., *Angew.Chem*, **116**, 5343 (2004).
38. R.I. Al–Raoush, PhD Thesis, Louisiana State University, 2002.
39. Boissonnat J.D., *ACGraphic M.T.*, **3**, 266 (1984). P
40. Holzer L., Indutnyi F., Ph. Gasser, Münch B., Wegmann M., *J. Microsc.*, **216**, 84 (2004).
41. Pothuaud L., Porion P., Lespessailles E., Benhamou C.L., Levitz P., *J. Microsc.*, **199**, 149 (2000).
42. Yeong C.L.Y., Torquato S., *Phys. Rev. E*, **58**, 224 (1998).
43. Biswal B., Manwart C., Hilfer R., *Physica. A*, **255**, 221 (1998).

44. Desbois G., Urai J.L., Kukla P.A., J.Konstanty and C.Baerle, *J. Petrol. Sci. Eng.*, **78**, 243 (2011).
45. Ulbricht M., *Polymer*, **47**, 2217 (2006).
46. Amendt M.A., PhD Thesis, University Of Minnesota, (2010).
47. P.Szewczykowski, PhD Thesis, University of Denmark, (2009).
48. T.Gullinkala, PhD Thesis, University of Toledo, (2010).
49. S.Naveed and Bhatti I., *J. Res. Sci.*, **17**, 155 (2006).
50. Baker R.W., John Wiley & Sons, England, (2004).
51. Roychowdhury A., MSc Thesis, Louisiana State University, (2007).
52. W.F. Catherina Kools, PhD Thesis, University of Twente, (1998).
53. M.K. Buckley–Smith, PhD Thesis, University of Waikato, (2006).
54. Nunes S.P., Peinemann K.V., Wiley–VCH, Germany, (2001).
55. Li W., PhD Thesis, University of Cincinnati, 2009.
56. Childress A.E., P. Le–Clech, Daugherty J.L., Caifeng Chen and Greg Leslie L., *Desalination*, **180**, 5 (2005).
57. Li L., P.Szewczykowski, Clausen L.D., Hansen K.M., Jonsson G.E., S.Ndoni, *J. Membrane. Sci.*, **384**, 126 (2011).
58. Chaoyiba, PhD thesis, University of Illinois, (2010).
59. Yen C., B.S, PhD thesis, Ohio State University, (2010).
60. Zon X., Kim K., Fang D., Ran S., Hsiao B.S., Chu B., *Polymer*, **43**, 4403 (2002).
61. Ondarçuhu T., Joachim C., Eur. Phys. Lett., **42**, 215(1998).
62. Feng L., Li S., Li Y., Li H., Zhang L., Zhai J., Song Y., Liu B., Jiang L., Zhu D., *Adv. Mater.*, **14**, 1221 (2002).
63. Ma P.X., Zhang R., *J. Biomed. Mater. Res.*, **46**, 60 (1999).
64. Liu G., Ding J., Qiao L., Guo A., Dymov B.P., Gleeson J.T., Hashimoto T.K., *Chem. Eur. J.*, **5**, 2740 (1999).
65. Doshi J., Reneker D.H., *J. Electrostat.*, **35**, 151 (1995).
66. Reneker D.H., Yarin A.L., *Polymer*, **49**, 2387 (2008).
67. Yordem O.S., Papila M., Menceloglu Y.Z., *Mater. Design*, **29**, 34 (2008).
68. Zussman E., Theron A., Yarin A., Appl. Phys. Lett., **82**, 973 (2003).
69. Gibson P.W., H.L. Schreuder–Gibson and Rivin D., *AIChE.J.*, **45**, 190 (1999).
70. Theron A., Zussman E., Yarin A.L., *Nanotechnology*, **12**, 384 (2001).
71. Teo W.E., Inai R., Ramakrishna S., *Sci. Technol. Adv. Mat.*, **12**, 1 (2011).
72. T.Subbiah, Bhat G.S., Tock R.W., Parameswaran S., Ramkumar S.S., *J. Appl. Polym. Sci.*, **96**, 557 (2005).
73. Zong X., Kim K., Fang D., Ran S., Hsiao B.S., Chu B., *Polymer*, **43**, 4403 (2002).
74. Tan S., Huang X., Wu B., *Polym. Int.*, **56**, 1330 (2007).
75. Burger C., Hsiao B.S., Chu B., Annu. *Rev. Mater. Res.*, **36**, 333 (2006).
76. Zhang C., Li Y., Wang W., Zhan N., Xiao N., Wang S., Li Y., Yang Q., *Eur. Polym. J.*, **47**, 2228 (2011).
77. CHOI J., PhD Thesis, Case Western Reserve University, (2010).
78. Reneker D.H., Yarin A.L., Zussman E., Xu H., *Advances In Applied Mechanics*, **41**, 43(2007).
79. Huang Z.M., Zhang Y.Z., Kotaki M., Ramakrishna S., Compos. Sci. Technol., **63**, 2223 (2003).

80. Zander N.E., *Polymers*, **5**, 19 (2013).
81. Bhardwaj N., Kundu S.C., *Biotechnol. Adv.*, **28**, 325 (2010).
82. *Wang N., Burugapalli K., Song W., Halls J., Moussy F., Ray A., Zheng Y., Biomaterials, 34, 888 (2013).*
83. Jung H.R., D.H.Ju, Lee W.J., Zhang X., Kotek R., *Electrochim. Acta.*, **54**, 3630 (2009).
84. Gong Z., Ji G., Zheng M., Chang X., Dai W., Pan L., Shi Y., Zheng Y., *Nanoscale. Res. Lett.*, **4**, 1257 (2009).
85. Wang Y., Zheng M., Lu H., Feng S., Ji G., Cao J., *Nanoscale. Res. Lett.*, **5**, 913 (2010).
86. Yin G.B., *J. Fiber Bioeng. Informatics*, **3**, 137 (2010).
87. Lee J.B., Jeong S.I., Bae M.S., Yang D.H., Heo D.N., Kim C.H., Alsberg E., Kwon I.K., *Tissue. Eng. A*, **17**, 2695 (2011).
88. Taha A.A., Qiao J., Li F., Zhang B., *J. Environ. Sci–China.*, **24**, 610 (2012).
89. Kim G.H., Kim W.D., *J. Biomed. Mater. Res. B: Applied Biomaterials*, **17**, 2695 (2006).
90. Zhang Y.Z., Feng Y., Huang Z.M., Ramakrishna S., Lim C.T., *Nanotechnology*, **17**, 901 (2006).
91. Ramakrishna S., Fujihara K., Teo W.E., Lim T.C., Ma Z., World Scientific Publishing: Singapore, 2005.
92. Burger C., Hsiao B.S., Chu B., *Annu. Rev. Mater. Res*, **36**, 333 (2006).
93. Berkalp O.B., *Fibers. Text. East. Eur.*, **14**, 81 (2006).
94. Choat B., Jansen S., Zwieniecki M.A., Smets E., Holbrook N.M., *J. Exp. Bot.*, **55**, 1569 (2004).
95. Alrawi A.T., Mohammed S.J., *Int. J. Soft. Comput.*, **3**, 1 (2012).
96. Krajewska B., Olech A., *Polym. Gels Networks*, **4**, 33 (1996).
97. Esselburn J.D., MSc Thesis, Wright State University, (2009).
98. Ziabari M., Mottaghitalab V., Haghi A.K., *Korean. J. Chem. Eng.*, **25**, 923 (2008).
99. A.Shrestha, MSc Thesis, University of Colorado, (2012).
100. Borkar N., MSc Thesis, University of Cincinnati, (2010).
101. Cuperus F.P., Smolders C.A., *Adv. Colloid. Interface. Sci.*, **34**, 135 (1991).
102. L. Mart'ınez, F.J. Florido–D'ıaz, Hernández A., Prádanos P., J. Membrane. Sci., **203**, 15(2002).
103. Bloxson J.M., MSc Thesis, Kent State University, (2012).
104. Cao G.Z., Meijerink J., Brinkman H.W., Burggra A.J., *J. Membrane. Sci.*, **83**, 221 (1993).
105. Cuperus F.P., Bargeman D., Smolders C.A., *J. Membrane. Sci.*, **71**, 57 (1992).
106. Fernando J.A., Chuung D.D.L., *J. Porous. Mat.*, **9**, 211 (2002).
107. Shobana K.H., M. Suresh Kumar, Radha K.S., Mohan D., *Sch. J. Eng. Res.*, **1**, 37 (2012).
108. Cañas A., Ariza M.J., Benavente J., *J. Membrane. Sci.*, **183**, 135 (2001).
109. Frey M.W., Li L., *J. Eng. Fiber. Fabr.*, **2**, 31 (2007).
110. J. ˇSirc, Hobzov R., N.Kostina, Munzarov M., M. Jukl'ıˇckov, Lhotka M., S.Kubinov, A.Zaj'ıcov and J. Mich'alek, *J. Nanomater.*, **2012**, 1 (2012).
111. A.B.Venkatarangan, PhD Thesis, University of New York, (2000).
112. Zhou B., PhD Thesis, Massachusetts Institute of Technology, (2006).
113. Manwart C., Aaltosalmi U., Koponen A., Hilfer R., Timonen J., *Phys. Rev. E*, **66**, 016702 (2002).

114. Ekneligoda T.C., Zimmerman R.W., in proceeding of Royal Society A, March 8, 2008, pp. 759–775.
115. L. Ghasemi–Mobarakeh, Semnani D., Morshed M., *J. Appl. Polym. Sci.*, **106**, 2536 (2007).
116. He W., Ma Z., Yong T., Teo W.E., Ramakrishna S., *Biomaterials*, **26**, 7606 (2005).
117. T.Wiederkehr, Klusemann B., Gies D., Müller H., Svendsen B., *Comput. Mater. Sci.*, **47**, 881 (2010).
118. Delerue J.F., Perrie E., Yu Z.Y., Velde B., *Phys. Chem. Earth. A*, **24**, 639 (1999).
119. Diógenes A.N., L.O.E. dos Santos, Fernandes C.P., Moreira A.C., Apolloni C.R., *Therm. Eng.*, **8**, 35 (2009).
120. Faessel M., Delisee C., Bos F., Castera P., *Compos. Sci. Technol.*, **65**, 1931 (2005).

CHAPTER 3

EXPERIMENTAL TECHNIQUES FOR APPLICATION OF RECYCLED POLYMERS IN CONSTRUCTION INDUSTRIES

O. EMAMGHOLIPOUR and A. K. HAGHI

CONTENTS

3.1 INTRODUCTION

This chapter is divided into four sections. In each section we have shown laboratory experiments in order to show the possibilities for converting the different types of polymer waste to wealth. These recycled materials have the required capabilities and potentials to be used in construction industries.

SECTION 1. PRACTICAL HINTS AND UPDATE ON PERFORMANCES OF ORDINARY PORTLAND CEMENT USING RECYCLED GLASS AND RECYCLED TEXTILE FIBERS AS A FRACTION OF AGGREGATES

3.2 EXPERIMENTAL PROCEDURE

Materials used included Ordinary Portland cement type 1, standard sand, silica fume, glass with tow particle size, rice husk ash, tap water and finally fibrillated polypropylene fibers.

The fibers included in this section were monofilament fibers obtained from industrial recycled raw materials that were cut in factory to 6 mm length. Properties of waste Polypropylene fibers are reported in Table 1 and Fig. 1.

TABLE 1 Properties of polypropylene fibers.

Property	Polypropylene
Unit weight [g/cm^3]	0.9–0.91
Reaction with water	Hydrophobic
Tensile strength [ksi]	4.5–6.0
Elongation at break [%]	100–600
Melting point [°C]	175
Thermal conductivity [W/m/K]	0.12

Also the silica fume and rice husk ash contain 91.1% and 92.1% SiO$_2$ with average size of 7.38 μm and 15.83 μm respectively were used. The chemical compositions of all pozzolanic materials containing the reused glass, silica fume and rice husk ash were analyzed using an X-ray microprobe analyzer and listed in Table 2.

FIGURE 1   Polypropylene fiber used in this study.

TABLE 2   Chemical composition of materials.

Oxide	Content (%)		
	Glass C	Silica fume	Rice husk ash
SiO_2	72.5	91.1	92.1
Al_2O_3	1.06	1.55	0.41
Fe_2O_3	0.36	2	0.21
CaO	8	2.24	0.41
MgO	4.18	0.6	0.45
Na_2O	13.1	—	0.08
K_2O	0.26	—	2.31
CL	0.05	—	—
SO_3	0.18	0.45	—
L.O.I	—	2.1	—

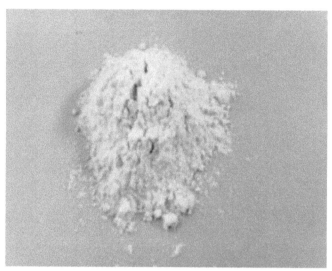

FIGURE 2   Ground waste glass.

To obtain this aim recycled windows clean glass was crushed and grinded in laboratory, and sieved the ground glass to the desired particle size (Fig. 2). To study the particle size effect, two different ground glasses were used, namely:

- Type I: ground glass having particles passing a #80 sieve (180μm);
- Type II: ground glass having particles passing a #200 sieve (75μm).

In addition the particle size distribution for two types of ground glass, silica fume, rice husk ash and ordinary Portland cement were analyzed by laser particle size set, have shown in Fig. 3. As it can be seen in Fig. 3 silica fume has the finest particle size. According to ASTM C618, fine ground glasses under 45μm qualify as a pozzolan due to the fine particle size. Moreover glass type I and II respectively have 42% and 70% fine particles smaller than 45 μm that causes pozzolanic behavior. SEM particle shape of tow kind of glasses is illustrated in Fig. 4.

For the present study, twenty batches were prepared. Control mixes were designed containing standard sand at a ratio of 2.25:1 to the cement in matrix. A partial replacements of cement with pozzolans include ground waste glass (GI, GII), silica fume (SF) and rice husk ash (RH) were used to examine the effects of pozzolanic materials on mechanical properties of PP reinforced mortars at high temperatures. The amount of pozzolans which replaced were 10% by weight of cement which is the rang that is most often used.

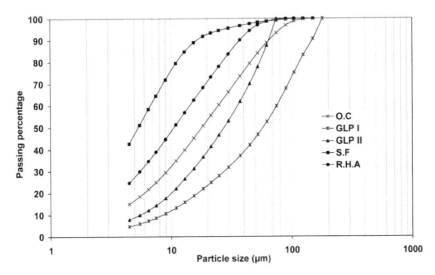

FIGURE 3   Particle size distribution of ground waste glass type I, II, silica fume, rice husk ash and ordinary cement.

FIGURE 4   Particle shape of ground waste glass type I, type II.

TABLE 3 Mixter Properties.

Batch No	Sand/c	w/c	Content (by weight)					PP fibers (by volume)
			O.C	GI	GII	SF	RH	
1	2.25	0.47	100	—	—	—	—	0
2	2.25	0.47	90	10	—	—	—	0
3	2.25	0.47	90	—	10	—	—	0
4	2.25	0.47	90	—	—	10	—	0
5	2.25	0.47	90	—	—	—	10	0
6	2.25	0.6	100	—	—	—	—	0.5
7	2.25	0.6	90	10	—	—	—	0.5
8	2.25	0.6	90	—	10	—	—	0.5
9	2.25	0.6	90	—	—	10	—	0.5
10	2.25	0.6	90	—	—	—	10	0.5

Batch No	Sand/c	w/c	Content (by weight)					PP fibers (by volume)
			O.C	GI	GII	SF	RH	
11	2.25	0.6	100	—	—	—	—	1
12	2.25	0.6	90	10	—	—	—	1
13	2.25	0.6	90	—	10	—	—	1
14	2.25	0.6	90	—	—	10	—	1
15	2.25	0.6	90	—	—	—	10	1
16	2.25	0.6	100	—	—	—	—	1.5
17	2.25	0.6	90	10	—	—	—	1.5
18	2.25	0.6	90	—	10	—	—	1.5
19	2.25	0.6	90	—	—	10	—	1.5
20	2.25	0.6	90	—	—	—	10	1.5

Meanwhile, polypropylene fibers were used as addition by volume fraction of specimens. The reinforced mixtures contained PP fiber with three designated fiber contents of 0.5%, 1% and 1.5% by total volume.

In the plain batches without any fibers, water to cementations ratio of 0.47 was used whereas in modified mixes (with different amount of PP fibers) it changed to 0.6 due to water absorption of fibers. The mix proportions of mortars are given in Table 3.

The strength criteria of mortar specimens and impacts of polypropylene fibers on characteristics of them were evaluated at the age of 60 days.

In our laboratory, the test programme mix conducted as follows:

1. The fibers were placed in the mixer.
2. Three-quarters of the water was added to the fibers while the mixer was running at 60 rpm; mixing continues for one minute.
3. The cement was gradually the cement to mix with the water.
4. The sand and remaining water were added, and the mixer was allowed to run for another two minutes.

After mixing, the samples were casted into the forms 50×50×50 mm for compressive strength and 50×50×20 mm for flexural strength tests. All the molds were coated with mineral oil to facilitate remolding. The samples were placed in two layers. Each layer was tamped 25 times using a hard rubber mallet. The sample surfaces were finished using a metal spatula. After 24 hours, the specimens were demolded and cured in water at 20°C. The suitable propagation of fibers in matrix is illustrated in Fig. 5.

FIGURE 5   Propagation of polypropylene fibers in mortar matrix (Up: 0.5% fiber, Down: 1% fiber).

The heating equipment was an electrically heated set. The specimens were positioned in heater and heated to desire temperature of 300 and 600°C at a rate of 10–12°C/min. After 3 h, heater turned off. It was allowed to cool down before the specimens were removed to prevent thermal shock to the specimens. The rate of cooling was not controlled. The testes to determine the strength were made for all specimens at the age of 60 days. At least three specimens were tested for each variable.

3.3   RESULTS AND DISCUSSION

3.3.1   DENSITY
The initial density of specimens containing polypropylene fibers was less than that of mixes without any fibers. Density of control mixes without any replacement of cement at 23,300 and 600°C are reported in Table 4. According to the results, density decrease of fiber-reinforced specimens was close to that of plain ones. The weight of the melted fibers was negligible. The weight change of mortar was mainly due to the dehydration of cement paste.

TABLE 4   Density of Control Specimens.

Heated at	23°	300°	600°	PP fibers
Density (gr/cm³)	2.57	2.45	2.45	0%
Density (gr/cm³)	2.50	2.36	2.36	0.5%
Density (gr/cm³)	2.44	2.28	2.27	1%
Density (gr/cm³)	2.41	2.23	2.21	1.5%

3.3.2   COMPRESSIVE STRENGTH
In order to asses the effect of elevated temperatures on mortar mixes under investigation, measurements of mechanical properties of test specimens were made shortly before and after heating, when specimens were cooled down to room temperature. Compressive strength of reference specimen and heated ones at the age of 60 days are illustrated in Fig. 6.

According to the results, by increasing the amount of polypropylene fibers in matrix the compressive strength of specimens reduced. Also, it's clear that the compressive strength of specimens were decreased by increasing the temperature to 300 and 600°C, respectively, as supported by previous literatures.

The rate of strength reduction in fiber-reinforced specimens is more than the plain samples and by rising the temperature it goes up.

The basic factor of strength reduction in plain specimens is related to matrix structural properties exposed to elevated temperature, but this factor for fiber-reinforced specimens is related to properties of fibers. Fibers melt at temperature higher than 190°C and generate lots of holes in the matrix. These holes are the most important reasons of strength reduction for fiber-reinforced specimens.

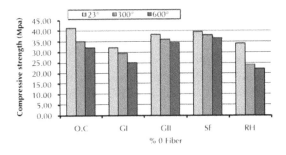

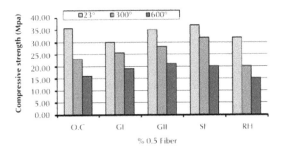

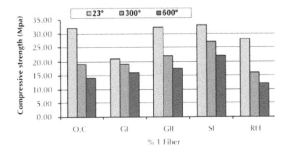

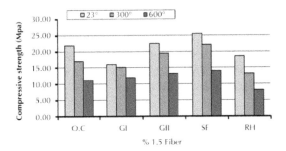

FIGURE 6   Compressive strength of samples at different temperatures.

Also, results indicate that silica fume and glass type II have an appropriate potential to apply as a partial replacement of cement due to their respective pozzolanic activity index values (according to ASTM C618 and C989, and Table2).

3.3.3 FLEXURAL STRENGTH

The specimens were used for flexural testes were 50×50×200 mm. The results of plain specimens and samples containing 1.5% fibers are shown in Table 5. The heat resistance of the flexural strength appeared to decrease when polypropylene fibers were incorporated into mortar. This is probably due to the additional porosity and small channels created in the matrix of mortar by the fibers melting like compressive strength. However the effect of the pozzolans on flexural strength is not clear but it seems that silica fume and glass types II have better impact on strength in compare with control specimens than rice husk ash and glass type I.

TABLE 5   Flexural strength of control samples with 0% and 1.5% fibers.

Batch	Flexural Strength (Mpa)			PP fibers
No	23°	300°	600°	(by volume)
1	4	3.1	2.8	0
2	3	2.4	2.1	0
3	3.7	3.2	2.7	0
4	3.9	3.4	2.7	0
5	3.3	2.8	2.6	0
16	2.2	1	0.7	1.5
17	1.6	0.8	0.6	1.5
18	2	1.1	0.8	1.5
19	2.4	1.3	0.8	1.5
20	1.8	1	0.6	1.5

SECTION 2. PRACTICAL HINTS AND UPDATE ON PERFORMANCES OF ORDINARY PORTLAND CEMENT USING RECYCLED TIRE FIBERS AND CHIPS AS A FRACTION OF AGGREGATES

3.4 EXPERIMENTAL PROCEDURE

It should be noted that concrete strength is greatly affected by the properties of its constituents and the mixture design parameters. In this section, the raw materials used included Portland cement Type 1, mixture of aggregates (coarse and medium), and sand, water and tire fibers. Also silica fume and rice husk ash as high reactive pozzolans were used in this study. Tires were cut by hand and a cuter in laboratory in form of fiber and chips. They were cut into strips of 30 mm × 5 mm × 5 mm, 60 mm × 5 mm × 5 mm as fibers and 10 mm × 10 mm × 5 mm and 20 mm × 20 mm × 5 mm as chips that illustrated in Fig. 1. Also the chemical composition of ordinary cement, silica fumes and rice husk ash are reported in Table 1.

TABLE 1   Chemical composition of materials.

Materials	Ordinary cement	Silica fume	Rice husk ash
SiO_2	21.24	91.1	92.1
Al_2O_3	5.97	1.55	0.41
Fe_2O_3	3.34	2	0.21
CaO	62.72	2.24	0.41
MgO	2.36	0.6	0.45
Na_2O	0.13	—	0.08
K_2O	0.81	—	2.31
CL	—	—	—
SO_3	1.97	0.45	—
L.O.I.	1.46	2.1	—

FIGURE 1   Rubber in form of Fibers and chips used in experiments.

The concrete mix was rated at 40 Mpa (compression strength). A control mix was designed using ACI Standard 211.1 mix design methods. The modified batches designed to be compared with. In the modified batches, 15% by volume of the coarse aggregates was replaced by tires. The mix ratio by weight for control concrete was cement: water: gravel: sand: = 1: 0.50: 3.50: 1.88. The mix ratio by weight for rubberized concrete was cement: water: gravel: waste tire: sand: = 1: 0.50: 3.40: 0.10: 1.88.

Also all rubberized batches treated with pozzolanic materials as the additional part of cement. In this study 10% and 20% of cement content by weight are selected for addition of pozzolan to the rubberized concrete mixtures.

After the concrete was mixed, it was placed in a container to set for 24 hours after that, the specimens were demolded and cured in water at 20°C.

Twenty-one batches of 15 cm radius by 30 cm height cylinders were prepared according to ACI specifications. One batch was made without waste tires and pozzolanic addition to be the control while four batches were prepared using waste tire fibers, without any pozzolanic addition.

Remain specimens prepared with waste tire and additional cementations part.

Waste tire measurement of main batches is reported in Table 2. According to this table rubberized modified batches include silica fume and rice husk ash prepared as followed.

TABLE 2   The size of tire in each batch.

Batch number	Waste tire shape	Length (mm)	Width (mm)	Height (mm)
1	—	—	—	—
2	Tires Fiber	30	5	5
3		60	5	5
4	Tire Chips	10	10	5
5		20	20	5

3.4.1   TESTING METHODS

3.4.1.1   COMPRESSIVE AND TENSILE STRENGTH

ASTM C 39 Standard was used in conducting compressive tests and ASTM C496–86 Standard was used for the split tensile strength tests. Three specimens of each mixture were tested to determine the average strength.

3.4.1.2   FLEXURAL TOUGHNESS

Flexural toughness determine according to ASTM C1018.

3.4.1.3   SLUMP

Slump tests were also conducted to measure the workability of concrete. The slump should never exceed 15 cm. Slump test was performed according to ASTM C143.

3.5   RESULTS AND DISCUSSIONS

3.5.1   COMPRESSIVE AND TENSILE STRENGTH OF CONCRETE

The results of the tests for strength performed on the samples in the experiments are shown in Fig. 2.

It can be seen from Fig. 2, that there was a significant and almost consistent decrease in the compressive strength of the rubberized concrete (*Rubcrete*) batches. Of all the batches tested, the control batch had the highest compressive strength. There was approximately 40% decrease in the compressive strength with the addition of the waste tire fibers. Also batch 4 had the highest compressive strength of all the modified samples.

Also Fig. 2 shows that the control samples had the highest split tensile strength. Batch 3 which consisted of waste tires fiber had the highest split tensile strength of all the rubberized concrete (*Rubcrete*) samples.

Totally results indicate that the size, proportions and surface texture of rubber particles noticeably affect compressive strength of Rubcrete mixtures.

Also its obvious that concrete mixtures with tire chips rubber aggregates exhibited higher compressive and lower splitting tensile strengths than other modified batches, but never has passed the regular Portland cement concrete specimens.

Generally there was approximately 40% reduction in compressive strength and 30% reduction in splitting tensile strength when 15% by volume of coarse aggregates were replaced with rubber fibers and chips.

All of modified mixtures demonstrated a ductile failure and had the ability to absorb a large amount of energy under compressive and tensile loads.

The beneficial effect of silica fume and rice husk ash as high reactive pozzolans on mechanical characteristics of rubberized concrete respect to pozzolans is exhibit in Fig. 2.

It is clear that by increasing the amount of pozzolans in Rubcrete, compressive strengths rise up. The rate of strength increase ranged from 5% to 40% for rubberized concrete, depending on the variation in size of rubber fibers and chips, amount of silica fume and rice husk ash and the fine particles in pozzolans, in general pozzolan activity index.

Results indicate that silica fume had grater impact on strength of specimens than rice husk ash. It is relevant to pozzolanic behavior and activity index.

3.5.2 MODULUS OF ELASTICITY

Static modulus of elasticity test results for reference samples as a function of rubber shape are depicted in Fig. 3.

Bar chart in Fig. 3 show that the static elastic modulus decreased by replacement of rubber whit course aggregated, similar to that observed in both compressive and splitting tensile strengths. It shows that the control samples without any rubber particle possessed the highest modulus of elasticity. Also batch 4 indicates the highest module of elasticity in compare with modified batches. In addition it's very close to batch 1 as control.

The general deviation in values for the rubberized samples was rather small. The volume and modulus of the aggregate are the factors that are mainly responsible for the modulus of elasticity of concrete. Therefore, small additions of tire fiber would not be able to significantly change the modulus of the composite, especially according to recent studies replacement up to 10% has no considerable effect on module of elasticity.

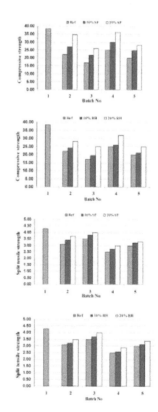

FIGURE 2   Compressive and split tensile strength of Rubbercrete (MPa).

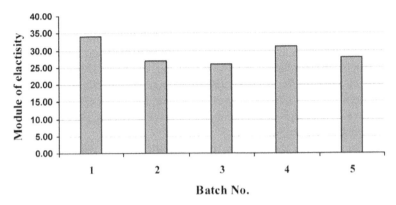

FIGURE 3   Variation of modulus of elasticity (Mpa).

3.5.3   WORKABILITY OF CONCRETE

One of the concerns when adding waste tires to the concrete was whether the workability of the concrete would be negatively affected. Workability refers to the ability of the concrete to be easily molded.

From Fig. 4 it was seen that there was a variation of approximately +/– 0.6 cm between the slum of the reference samples and the rubberized concrete (*Rubcrete*).

Results imply that the workability of the concrete was not adversely affected by the addition of waste tires.

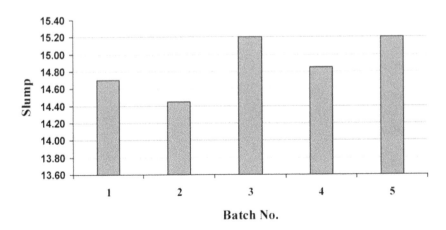

FIGURE 4   Variation of slump (mm) of concrete (MPa).

3.5.4 EFFECT OF WASTE TIRE ON TOUGHNESS OF CONCRETE

Figure 4 shows the applied load – displacement curves for split tensile testing of a control sample and a rubberized sample from batch 2. Toughness is defined as the ability of materials to sustain load after initial cracking and is measured as the total strain experienced at failure. Upon cracking, tier fibers are able to bridge the initial crack and hold the crack together and sustain the load until the fibers either pullout from the matrix (in early age) or fracture that say flexural toughness. According to Fig. 5 it can be observed that after crack initiation, fibers can still carry load and absorb the energy.

For plain concrete, the behavior was in a brittle manner. When the strain energy was high enough to cause the crack to self-propagate, fracture occurred almost while the peak load was reached (this is due to the tremendous amount of energy being released). According to Fig. 5, the tire fiber bridging effect helped to control the rate of energy release significantly. Thus, fibers still can carry load even after the peak. With the effect of fibers bridging across the crack surface, fibers were able to maintain the load carrying ability even after the concrete had been cracked. These are accordance to ASTM C1018, in which toughness or energy absorption defined as the area under the load-deflection curve from crack point to 1/150 of span.

Our laboratory results indicate that, the area beneath the curve for the concrete without rubber is very small compared to the area beneath the curve for the concrete with rubber. This implies that concrete with rubber is much tougher than concrete without rubber.

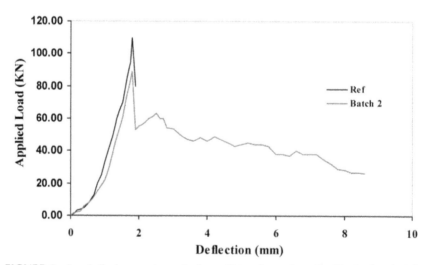

FIGURE 5 Load displacement results for split tensile testing of rubberized and plain concrete.

SECTION 3. PRACTICAL HINTS AND UPDATE ON PERFORMANCES OF ORDINARY PORTLAND CEMENT USING RECYCLED MELAMINE AS A FRACTION OF AGGREGATES

3.6 EXPERIMENTAL PROCEDURE

The materials used in present study are as follows:

Ordinary Portland Cement: Type I Portland cement conforming to ASTM C150–94.

Sand: Fine aggregate is taken from natural sand. Therefore, it was used after separating by sieve in accordance with the grading requirement for fine aggregate (ASTM C33–92). Table 1 presents the properties of the sand and its gradation is presented in Fig. 1.

TABLE 1 Properties of sand and melamine aggregates.

Properties	Sand	Melamine
Density (g/cm³)	2.60	1.574
Bulk density (g/cm³) (Melamine)	—	0.3–0.6
Water absorption (%)	1.64	7.2
Max size (mm)	4.75	1.77
Min size (mm)	—	0.45
Sieve 200 (%)	0.24	—
Flammability (Melamine)	Nonflammable	Nonflammable
Decomposition (Melamine)	—	At > 280°C formation of NH_3

Thermosetting plastic: Melamine is a widely used type of thermosetting plastic. Therefore, in the present work has been selected for application in the mixed design of composite. The mechanical and physical properties of melamine are shown in Table 1. The melamine waste was ground with a grinding machine. The ground melamine waste was separated under sieve analysis. Scanning electron micrographs (SEM) of melamine aggregates is shown in Fig. 2. Those have irregular shape and rough surface texture. The grain size distributions were then plotted as shown in Fig. 1. It was observed that the gradation curve of the combination of sand and plastic aggregates after sieve number 16 meets most of the requirements of ASTM C33–92.

The melamine aggregates were been saturated surface dry. Therefore, melamine aggregates immerse in water at approximately 21°C for 24 h and removing surface moisture by warm air bopping.

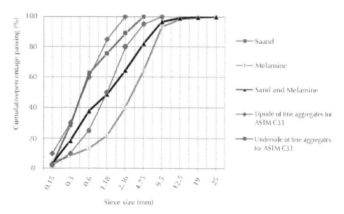

FIGURE 1   Gap-grading analysis of sand and melamine aggregates according to ASTM C33–92.

FIGURE 2   SEM photographs of materials: (up) melamine aggregates and (down) silica fume.

Aluminum powder: In the present study, aluminum powder was selected as an agent to produce hydrogen gas (air entrainment) in the cement. This type of lightweight concrete is then called aerated concrete. The following are possible chemical reactions of aluminum with water:

$$2Al + 6H_2O \ 2Al(OH)_3 + 3H_2 \tag{1}$$

$$2Al + 4H_2O \ 2AlO(OH) + 3H_2 \tag{2}$$

$$2Al + 3H_2O \ Al_2O_3 + 3H_2 \tag{3}$$

The first reaction forms the aluminum hydroxide bayerite ($Al(OH)_3$) and hydrogen, the second reaction forms the aluminum hydroxide boehmite ($AlO(OH)$) and hydrogen, and the third reaction forms aluminum oxide and hydrogen. All these reactions are thermodynamically favorable from room temperature past the melting point of aluminum (660°C). All are also highly exothermic. From room temperature to 280°C, $Al(OH)_3$ is the most stable product, while from 280–480°C, $AlO(OH)$ is most stable. Above 480°C, Al_2O_3 is the most stable product. The following equation illustrates the combined effect of hydrolysis and hydration on tricalcium silicate.

$$3CaO.SiO_2 + water \ xCaO.ySiO_2(aq.) + Ca(OH)_2 \tag{4}$$

In considering the hydration of Portland cement it is demonstrate that the more basic calcium silicates are hydrolyzed to less basic silicates with the formation of calcium hydroxide or 'slaked lime' as a by-product. It is this lime, which reacts with the aluminum powder to form hydrogen in the making of aerated concrete from Portland cement:

$$2Al + 3Ca(OH)_2 + 6H_2O \ 3CaO.Al_2O_3.6H_2O + 3H_2 \tag{5}$$

Hydrogen gas creates many small air (hydrogen gas) bubbles in the cement paste. The density of concrete becomes lower than the normal weight concrete due to this air entrainment.

Silica fume: In the present work, Silica fume has been used. Its chemical compositions and physical properties are being given in Tables 2 and 3, respectively. Scanning electron micrographs (SEM) of silica fume is shown in Fig. 2.

Super plasticizer: Premia 196 with a density of 1.055 ± 0.010 kg/m³ was used. It was based on modified polycarboxylate.

TABLE 2   Chemical composition of Silica fume.

Chemical composition	Silica fume
SiO_2 (%)	86–94
Al_2O_3 (%)	0.2–2
Fe_2O_3 (%)	0.2–2.5
C (%)	0.4–1.3
Na_2O (%)	0.2–1.5
K_2O (%)	0.5–3
MgO (%)	0.3–3.5
S (%)	0.1–0.3
CaO (%)	0.1–0.7
Mn (%)	0.1–0.2
SiC (%)	0.1–0.8

TABLE 3   Physical properties of Silica fume.

Items	Silica fume
Specific gravity (gr/cm³)	2.2–2.3
Particle size (µm)	< 1
Specific surface area (m²/gr)	15–30
Melting point (°C)	1230
Structure	Amorphous

3.6.1 MIX DESIGN

To determine the suitable composition of each material, the mixing proportions were tested in the laboratory, as shown in Table 4. In this section, the mix proportions were separated for five experimental sets. For each set, the cement and Aluminum powder contents was specified as a constant proportion. The proportion of each of the remaining materials, that is, sand, water, silica fume, aluminum powder, and melamine, was varied for each mix design.

3.6.2 EXPERIMENTAL TECHNIQUES

Mortar was mixed in a standard mixer and placed in the standard mold of 50 × 50 × 50 mm according to ASTM C109-02. In the pouring process of mortar, an expansion of volume due to the aluminum powder reaction had to be considered. The expanded portion of mortar was removed until finishing. The fresh mortar was tested for slump according to ASTM C143-03. The specimens were cured by wet curing at normal room temperature. The hardened mortar was tested for dry density, compressive strength, water absorption and voids for the curing age of 7 and 28 days. The test results for melamine, sand and water contents were reported for 7 days curing age for mix nos. 1–3, because these were very close to the results of 28 days. When silica fume was added in the latter mix nos. 4 and 5, the test results were presented for 28 days. This is because the presence of silica fume increases the duration for completion of the chemical reaction. The testing procedures of dry density, water absorption and voids were performed according to ASTM C642-97 and compressive strength was performed according to ASTM C109-02.

TABLE 4 Mix proportions of melamine lightweight composites (by weight).

Mix no.	Cement	Aluminum powder	Sand	Water	Silica fume	Melamine	Super plasti-cizer
1. Determination of melamine content (1st trial mix design)	1.0	0.004	1.0	0.35	—	1.0	—
						1.5	
						2.0	
						2.5	
						3.0	
2. Determination of sand content	1.0	0.004	1.0	0.35	—	1.0	—
			1.2				
			1.4				
			1.6				
			1.8				

TABLE 4 *(Continued)*

3. Determination of water content or water–cement ratio (w/c)	1.0	0.004	1.4	0.30	—	1.0	—
				0.35			
				0.40			
				0.45			
				0.50			
				0.55			
4. Determination of silica fume content	1.0	0.004	1.4	0.35	0.10	1.0	0.005
					0.15		0.007
					0.20		0.009
					0.25		0.012
					0.30		0.015
					0.35		0.020
5. Determination of melamine content (final mix design)	1.0	0.004	1.4	0.35	0.25	1.0	0.012
						1.2	0.011
						1.4	0.010
						1.6	0.009
						1.8	0.008
						2.0	0.007
						2.2	0.006

3.7 RESULTS AND DISCUSSION

3.7.1 MIX NUMBER 1 (DETERMINATION OF MELAMINE CONTENT FOR THE FIRST TRIAL MIX DESIGN)

Figure 3 present the variations in the compressive strength and dry density for 7 days age of mortars as a function of the value of melamine substitutes used. It can initially be seen, to increased melamine, the compressive strength and dry density of composites decreased. The reduction in the compressive strength is due to the addition of melamine aggregates or could be due to either a poor bond between the cement paste and the melamine aggregates or to the low strength that is characteristic of plastic aggregates.

TABLE 5   Specification of non-load-bearing lightweight concrete (ASTM C129).

Type	Compressive strength (MPa)	Density (kg/m³)
	Average of three unit Individual unit	
II	4.1–3.5	< 1680

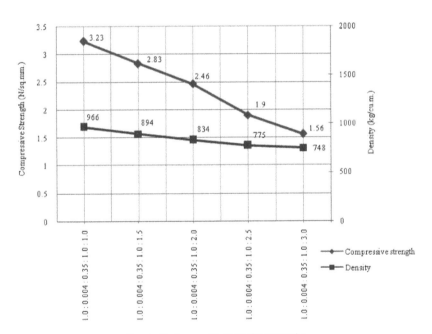

Cement : Aluminum powder : Water : Sand : Melamine

FIGURE 3   Compressive strength and density for varying melamine content (curing for 7 days).

FIGURE 4   Absorption after immersion for varying melamine content (curing for 7 days).

The absorption is an indirect parameter to examine the inside porosity of mortar. The results showed that the absorption after immersion and voids of mortar increased as the melamine content increased (see Figs. 4 and 5). Therefore, to increased melamine plastic, the inside porosity of mortar increased. This might be other reason for the reduction in the compressive strength and density.

FIGURE 5   Optical photographs of samples containing varying melamine, right to left containing 1.0, 1.5, 2.0, 2.5 and 3.0 the weight percentage of melamine

3.7.2 MIX NUMBER 2 (DETERMINATION OF SAND CONTENT)

The results of compressive strength and dry density for 7 days age are shown in Fig. 6. It can be seen that a reduction of sand leads to a reduction in the strength and dry density. The compressive strength and dry density for sand content equal to or greater than 1.4 exactly satisfy the standard value.

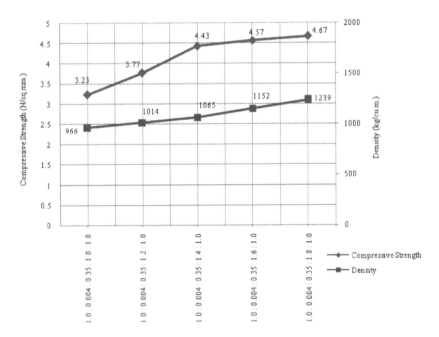

Cement : Aluminum powder : Water : Sand : Melamine

FIGURE 6   Compressive strength and density for the determination of the optimum sand content (curing for 7 days).

Figure 7 present the variations in the absorption after immersion and voids as a function of the value of sand substitutes used. The results showed that the absorption after immersion and voids of mortar decreased as the sand content increased.

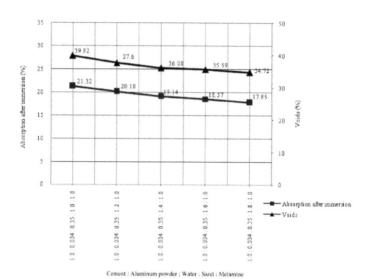

Cement : Aluminum powder : Water : Sand : Melamine

FIGURE 7   Absorption after immersion and voids for varying sand content (curing for 7 days).

3.7.3 MIX NUMBER 3 (DETERMINATION OF WATER CONTENT)

The results of compressive strength and dry density for 7 days age are shown in Fig. 8. The results showed that the compressive strength and dry density of mortar decreased as the water content increased.

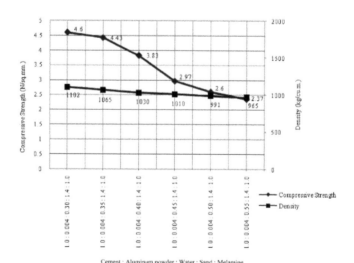

Cement : Aluminum powder : Water : Sand : Melamine

FIGURE 8   Compressive strength and density for the determination of the optimum water content (curing for 7 days).

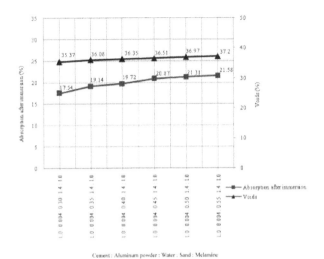

Cement : Aluminum powder : Water . Sand : Melamine

FIGURE 9 Absorption after immersion and voids for varying water content (curing for 7 days).

3.7.4 MIX NUMBER 4 (DETERMINATION OF SILICA FUME CONTENT)

Figure 10 present the variations in the compressive strength and dry density for 28 days age of mortars as a function of the value of silica fume substitutes used. It was found that the results of compressive strength for seven days age do not increase when compared with those without silica fume.

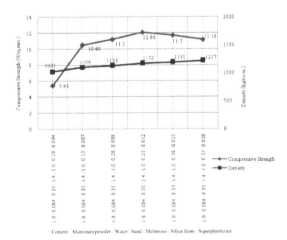

Cement : Aluminum powder : Water : Sand : Melamine : Silica fume : Superplasticizer

FIGURE 10 Compressive strength and density for the determination of the optimum silica fume content (curing for 28 days).

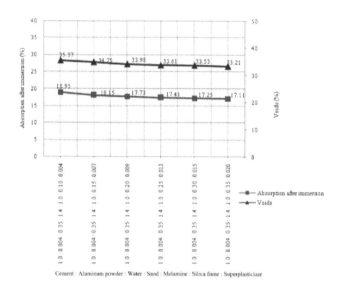

Cement : Alumimum powder : Water : Sand : Melamine : Silica fume : Superplasticizer

FIGURE 11   Absorption after immersion and voids for varying silica fume content (curing for 28 days).

3.7.5 MIX NUMBER 5 (DETERMINATION OF THE FINAL MELAMINE PLASTIC CONTENT)

Figure 12 presents the variations in the compressive strength and dry density for 28 days age of mortars as a function of the value of melamine substitutes used. It was found that the presence of melamine caused a reduction in the dry density and compressive strength of concretes as discussed previously. Figure 13 shows that the scanning electron microscopy analysis of composites reveals that cement paste-melamine aggregates adhesion is imperfect and weak. Therefore, the problem of bonding between plastic particles and cement paste is main reason to decrease of compressive strength. An optimum melamine content of 2.0 was selected. The results of compressive strength and dry density, which are 7.06 MPa and 887 kg /m3, are according to ASTM C129 Type II standard.

The results showed that the absorption after immersion and voids of mortar increased as the melamine content increased (see Fig. 14). Also, the structure analysis of mortars by scanning electron microscopy has revealed a low level of compactness in mortars when the value of melamine plastic increased (see Fig. 15). It was confirmed that to increased melamine plastic, the inside porosity of mortar increased.

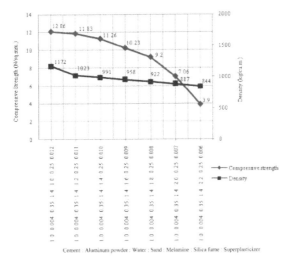

FIGURE 12   Compressive strength and density for the determination of the optimum melamine plastic content (curing for 28 days).

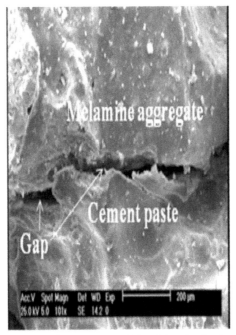

FIGURE 13   Microstructure of concrete containing 1.6 melamine by weight of cement, as obtained using SEM (enlargement: 101×).

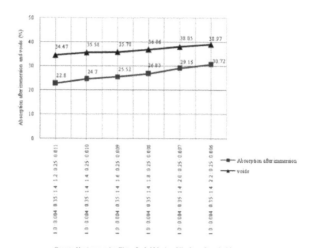

Cement : Aluminum powder : Water : Sand : Melamine : Silica fume : Superplasticizer

FIGURE 14   Absorption after immersion and voids for varying melamine plastic content (curing for 28 days).

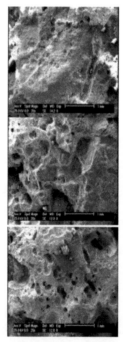

FIGURE 15   Scanning electron micrographs of various mortars containing melamine plastic aggregates. (a) 1.2 by weight of cement (enlargement: 25×). (b) 1.6 by weight of cement (enlargement: 25×). (c) 2.0 by weight of cement (enlargement: 25×).

3.7.6 COMPARISON RESEARCH FINDINGS ON THE USE OF WASTE PLASTIC IN CONCRETE

The results of this study are in a perfectly agreement with the other research findings on the use of waste plastic in concrete (see Table 6).

SECTION 4. PRACTICAL HINTS AND UPDATE ON PERFORMANCES OF CEMENT-STABILIZED SOIL USING RECYCLED TEXTILE FIBERS AS A FRACTION OF AGGREGATES

3.8 EXPERIMENTAL PROCEDURE

The soil samples used in the present study were obtained from the northern region sites of Iran. The soil is classified as clayey sand (SC) according to unified soil classification system. The general characteristics of this type of soil could be seen in Table 1. The standard penetration test was used to obtain the soil information at site. The test is carried out according to ASTM D1586 and the derived number commonly called SPT numbers of the tested samples can be seen in Table 1.

TABLE 1   Characteristics of clayey sand.

Specific Gravity (gr/cm³)	SPT number	% Finer than 0.002 mm	Plastic Limit	Liquid Limit
1.45–1.65	15–30	15–30%	20–24	37–45

The cement used in this study is ordinary Portland cement that its physical and chemical properties are given in Table 2 and the waste fibers are Polyamide and Acrylic, which their characteristics are given in Table 3. The waste fibers are cut such that the lengths are 8 mm (0.315 inch).

TABLE 2   Physical and chemical properties of the cement.

Physical properties	Cement
Fineness	3.12
Chemical composition	
Silica (SiO_2)	20.44%
Alumina (Al_2O_3)	5.5%
Calcium oxide (CaO)	64.86%
Potash (K_2O)	22.31%
Magnesia (MgO)	1.59%
Loss on ignition	1.51%

TABLE 2   *(Continued)*

Physical properties	Cement
PH	12.06
3CaO. SiO$_2$	66.48%
2CaO. SiO$_2$	10.12%
4CaO. Al$_2$O$_3$. Fe$_2$O$_3$	9.43%
Free lime	1.65%
3CaO. Al2O3	8.06%

TABLE 3   Mechanical and physical properties of polyamide and Acrylic.

Fiber	Water Absorption	Shear Modulus	Tensile Strength	Flexural Strength	Compressive Yield Strength
Acrylic	0.3–2%	203 ksi	7,980–12,300 psi	11,700–20,000 psi	14,500–17,000 psi
Polyamide	2.1–4%	8,560 psi	2,180–12,300 psi	13,100–15,200 psi	2,470 psi

The specimens were prepared according to the standard definitions with three different cement contents (5%, 7% and 9%) and two kinds of fiber polymers (Polyamide and Acrylic). The reinforced samples were prepared with three different fiber contents of 0.1%, 0.2% and 0.3%. The specimens are prepared with maximum dry density and optimum moisture content. Testing was performed on specimens with fibers distributed uniformly in space and with as much as possible uniform distribution of fiber orientation in all directions. Such a distribution requires an elaborate technique for preparation of the specimens. This technique included a five-layer specimen construction. As the fibers tend to assume a close-to-horizontal orientation during mixing with dry soil, a specially designed tool was used for reorienting the fibers in the specimen before compaction.

For each test and period of curing, three specimens were prepared. The specimens were tested after curing in a moist chamber.

3.9   EXPERIMENTAL TESTS

In this study, compaction, unconfined compression, indirect tensile, flexure, direct shear and durability tests were carried out on reinforced and non-reinforced samples. The testing procedures used are described in table 4. Flexure tests were carried with the strain rate of 0.2mm/min on the rectangular specimens with dimension of 51×51×150 mm.

TABLE 4   Testing procedures.

Test type	Test procedure
Compaction test	ASTM D558–96
Unconfined compressive strength	ASTM D1633
Indirect tensile test	ASTM C496
Flexure tests	ASTM D1635–95
Direct shear tests	ASTM D3080–90
Durability tests	ASTM D 559–93

3.10   RESULTS AND DISCUSSIONS

3.10.1   THE EFFECT OF ADDING RECYCLED TEXTILE FIBERS ON COMPACTION

The maximum dry density (MDD) and optimum water content (OWC) vs. cement and fiber contents curves obtained in compaction tests are given in Figs. 1 and 2.

As shown in Fig. 1, the percent of optimum water content increases as the cement and fiber content increases. The increase of the cement content will result in an increment in the fine-grained content of the soil. Therefore, the specific surface of the soil would increase, and consequently much more water is needed to reach to the maximum dry density.

On the other hand, added polymer fibers to the soil-cement mixtures, will absorb a portion of the water content and correspond to the increase of fiber polymers percentage, the amount of water absorption will increase. Consequently, greater amount of water is needed to provide the OWC of the mixture.

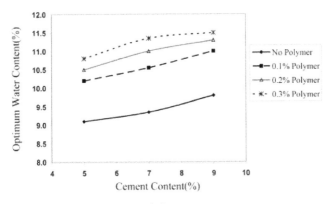

FIGURE 1   Optimum water content variation.

From Fig. 2 it can be concluded that in the specimens stabilized with 5% cement and reinforced with fibers, those specimens with 0.1% fibers, have the maximum dry density. Increasing the fibers content to 0.2% and 0.3%, the maximum dry density decreases. The maximum dry density in the reinforced specimens with 0.3% of fibers is less than those of non-reinforced ones. Moreover, in the stabilized specimens with 7% and 9% of cement, correspond to the increases of fibers up to 0.2%, maximum dry density increases. The greater amount of fibers content of 0.2%, the maximum dry density decreases. Furthermore, It is deduced from the Fig. 2 that the amount of maximum dry density in the stabilized specimens with 7% and 9% cement and reinforced with 0.3% fibers, have greater values in comparison with those of non-reinforced.

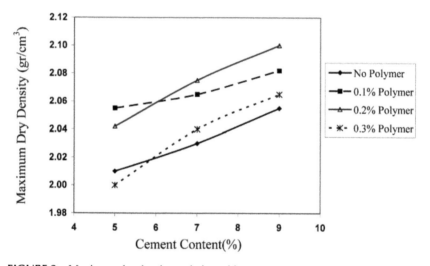

FIGURE 2   Maximum dry density variation with cement content.

3.10.2   THE EFFECT OF ADDING FIBERS ON UCS

Figure 3 shows the variations of maximum UCS (q_u) of the stabilized clayey sand-cement which were reinforced with fibers, and that of non-reinforced ones cured after 7 and 28 days. The experimental evident indicate that, in the stabilized specimens with 5% cement and reinforced with fiber polymers, the maximum value of q_u occurs for the reinforced specimens with 0.1% of fibers. In the stabilized specimens with 7% and 9% of cement, reinforced with fiber polymers, the q_u, in both of polymers, occurs in the specimens, which have 0.2% fibers. Therefore, in the compressive test, when the percent of cement changes, the percent of optimum fibers changes. By considering the following diagrams and the effect of polyamide and Acrylic on q_u, it can be seen that there is no significant difference between

q_u of the soil-cement mixtures reinforced with polyamide and acrylic. Because the non-uniform distribution of the fibers in the soil-cement mixture, the existing differences can be the result of the distribution quality of the fibers. Studies carried out by many researchers described that the variation of q_u, with increase of cement is linear. In the case of soil-cement specimens, not only this behavior can be observed but also all reinforced soil-cement specimens with polymers were showed such behavior (Fig. 3).

Based on the results of unconfined compression tests shown in Figs. 2 and 3, it can be concluded that the increment of UCS by using fiber reinforcement on specimens containing less than 5% of cement content is more considerable. The reason for this phenomenon can be explained by the fact that the UCS value of specimens containing more than 5% of cement content is more affected by cement than by fibers, while the fiber-reinforcement effect on improving UCS value of specimens containing less than 5% of cement is more significant. Comparing the results of compressive strength and maximum dry density of the reinforced and non-reinforced specimens showed that there is a direct relation between maximum UCS and maximum dry density. The stabilized and reinforced specimens with 5% of cement, the highest compressive strength and maximum dry density refers to the reinforced specimens which have 0.1% of fibers, while in the stabilized and reinforced specimens with 7 and 9% of cement, the highest compressive strength and maximum dry density refers to the specimens which have 0.2% of fibers.

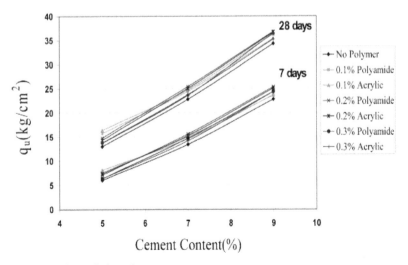

FIGURE 3   The variation of qu versus cement percentage.

3.10.3 THE EFFECT OF ADDING RECYCLED TEXTILE FIBERS ON AXIAL STRESS – STRAIN BEHAVIOR

Before performing the test, surfaces of specimens are covered with a thin layer of kaolinite powder (specimen capping) so that a smooth surface is obtained for uniform distribution of stress. In most of curves an initial reduction is evident after which curves takes the normal shape, which is due to kaolinite layer compression that lead to specimen strain. Stress-strain characteristics of cement-stabilized soil are very important to determine the behavior of these material used in pavement construction under repeated loading. Stress-strain behavior of cement-stabilized soil is nonlinear but for low stresses and also for limited loading unloading, this behavior may be idealized linear. Thus, Stress-strain curves must be corrected to determine their behavior. To do this, linear section must be extended until it intersects the strain curve. All strains are then subtracted as much as origin strain. Resulting curve is named modified stress-strain curve. Figs. 4 and 5 show the real and corrected path for a series of tests.

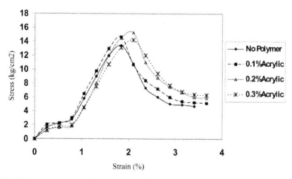

FIGURE 4   Real (non-corrected) stress-strain curve.

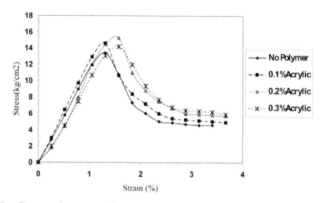

FIGURE 5   Corrected stress-strain curve.

The corrected curve of axial stress-strain behavior of stabilized and reinforced soil with different percents of cement and fibbers after 7 and 28 days' curing are shown in Figs. 6 to 11. The results indicate that adding 0.1% of fiber to the stabilized soil increases maximum compressive strength. The strain related to the maximum stress of the fiber-reinforced specimens will grow by using greater amount of polymer content, which increases the ductility of the specimens. This subject confirms the last research in the field of stabilizing and reinforcing soil by cement and polymer, and shows the effect of fibers in making soil-cement ductile.

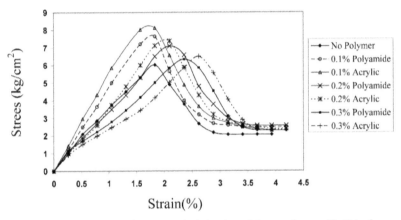

FIGURE 6   The variation of stress-strain behavior of the specimen with 5% of cement and different percents of polymer after 7 days' curing.

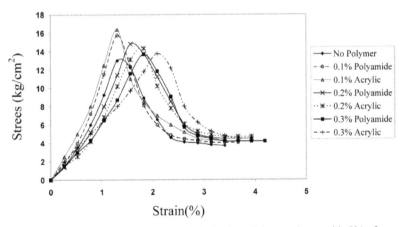

FIGURE 7   The variation of stress-strain behavior of the specimen with 5% of cement and different percents of polymer after 28 days' curing.

Evaluating the experimental evident, the residual strength in all reinforced specimens is more than that of non-reinforced specimens. The only exception is about the stabilized specimens with 5% of cement that are reinforced with 0.2% of fibers. In this case, the amount of residual strength decreases with the increase of fibers content.

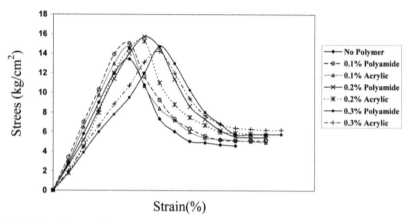

FIGURE 8   The variation of stress-strain behavior of the specimen with 7% of cement and different percents of polymer after 7 days' curing.

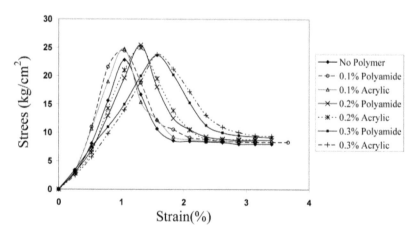

FIGURE 9   The variation of stress-strain behavior of the specimen with 7% of cement and different percents of polymer after 28 days' curing.

However, in the stabilized specimens with 7% and 9% of cement, the amount of maximum residual strength belongs to the reinforced specimens with 0.3% fibers. Generally, the reason of greater residual strength of the reinforced specimens with fibers in comparison with that of non-reinforced is that existing fibers in the soil-cement mixture prohibit the complete failure of the composite structure after reaching to maximum bearable stress of soil. In addition, adding fibers improve the load bearing capacity of the specimens. In the other word, the reinforced specimen undergoes more stress before failure in comparison with the non-reinforced specimens.

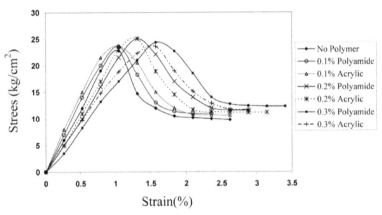

FIGURE 10   The variation of stress-strain behavior of the specimen with 9% of cement and different percents of polymer after 7 days' curing.

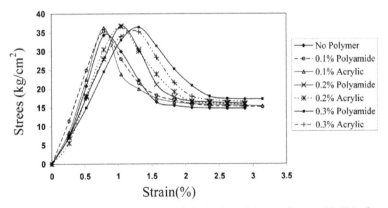

FIGURE 11   The variation of stress-strain behavior of the specimen with 9% of cement and different percents of polymer after 28 days' curing.

3.10.4 THE EFFECT OF RECYCLED TEXTILE FIBERS ON MODULUS OF ELASTICITY

Figures 6–11 show the stress-strain properties of the stabilized and reinforced specimens with different percents of cement and fibers. Secant modulus can be used as the value of elastic modulus of geomaterials and concrete. The calculating method of secant modulus at 50% of compressive strength (E_{50}) has been used to calculate modulus of elasticity. Secant modulus (E_{50}) of the stabilized specimens with different percent of cement and fibers were calculated and shown in Figs. 12 and 13. The results showed that using more cement content and curing period will increase the E_{50} value of the mixtures. The E_{50} value of reinforced soil-cement mixes will increase by using up to 0.1% of fiber content in comparison with non-reinforced specimens, while using more than 0.1% of fiber content will decrease the E_{50} to an amount less than non-reinforced specimens.

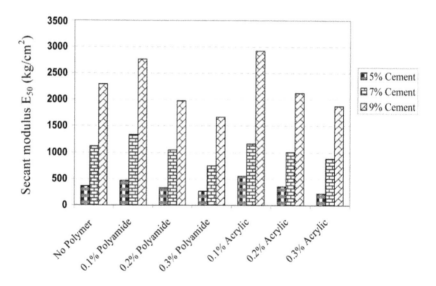

FIGURE 12   The secant modulus of the specimens after 7 days' curing.

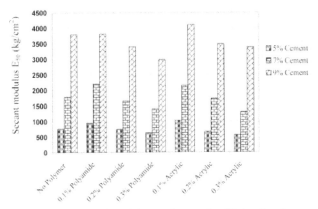

FIGURE 13   The secant modulus of the specimens after 28 days' curing.

The variations of secant modulus at 50% of compressive strength (E_{50}) based on the maximum UCS of the reinforced and non-reinforced cement-stabilized soils are shown by a power function in Fig. 14. Following numerical E–q equations were drawn for different mixtures:

E_{50} (non-reinforced) = 29.96 $q_u^{1.35}$
E_{50} (Reinforced with 0.1% polymer) = 31.467 $q_u^{1.349}$
E_{50} (Reinforced with 0.2%, 0.3% polymer) = 18.836 $q_u^{1.4463}$

Using the above equations it can be seen that adding 0.1% of fibers to the soil-cement, increase the secant modulus (E50) between 2.1% and 5.6%, while adding 0.2% and 0.3% of fibers decrease the secant modulus (E50) between 11.5% and 26.8%.

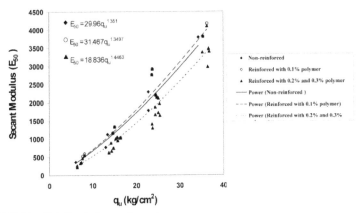

FIGURE 14   Relation between E_{50} and q_u.

3.10.5 THE EFFECT OF RECYCLED TEXTILE FIBERS ON INDIRECT TENSILE STRENGTH

The results of fiber-reinforcement on indirect tensile strength (ITS) of the specimens are shown in Figs. 15 and 16. It can be concluded from these figures that, first, tensile strength increase with the increase of cement percentage in the mixture and then the tensile strength in all of the fiber-reinforced specimens is more then that of non-reinforced specimens. In addition, with the increase of fiber polymers, the tensile strength increases by a constant percent of cement. From the Figs. 15 and 16, it is clear that the specimens with 0.3% of fibers have the greatest tensile strength.

Generally, these results show that, these two kinds of polymers are suitable to improve tensile strength. The different effects of polyamide and acrylic on the tensile strength of soil-cement mixture are, also, related to the distributing way of fibers in the mixture.

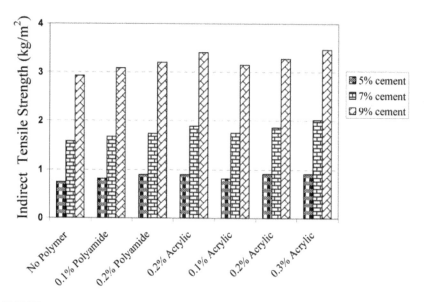

FIGURE 15 Comparison between tensile strength for reinforced and non-reinforced specimens after 7 days' curing.

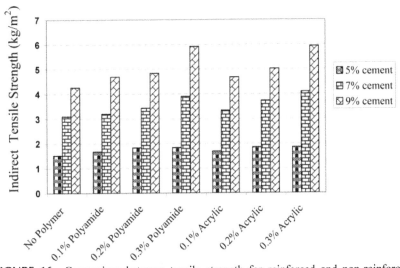

FIGURE 16 Comparison between tensile strength for reinforced and non-reinforced specimens after 28 days' curing.

Relation between compressive and tensile strengths for stabilized reinforced and non-reinforced soil specimens are shown in Figs. 17 (a and b). Many researchers proposed that tensile strength of cement-stabilized embankment is about 8% of unconfined compressive strength (Fig. 17a). As it can be seen the ratio of tensile strength to compressive strength for non-reinforced clayey sand specimens is about 12% but this ratio is approximately 15% for polymeric fiber mixed specimens. In this research it can be concluded that effect of fiber on tensile strength is greater compared to compressive strength.

The pore size distribution in a fibrous material has significant impact on the mixture transport process. These distributions are clearly shown in Fig. 18. Pore size distribution can influence the spontaneous uptake of liquids (i.e., moisture content) and therefore the strength keeps on increasing. It should be noted that the amount of pores within the fibers could have significant effects on the strength improvement as well. The greater the pores, the more moist aggregates the fibers can hold.

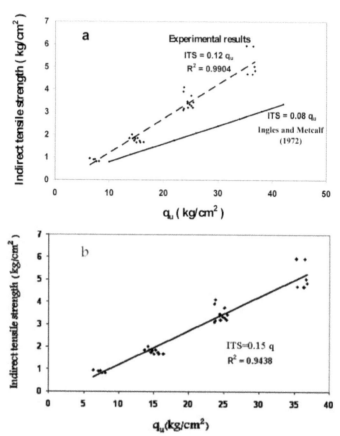

FIGURE 17 (a) Relation between compressive strength and tensile strength of cement-stabilized soil specimens, non-reinforced. (b) Relation between compressive strength and tensile strength of cement-stabilized soil specimens-reinforced.

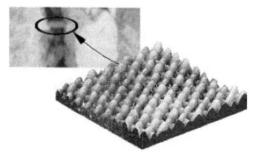

FIGURE 18 Pore size distribution within the polyamide fibers and tensile failure mode in an indirect tensile test.

Figure 19 shows the crack distribution adjacent to the polyamide fibers. It is notable that the cracks did not spread out in the direction where the fibers are located.

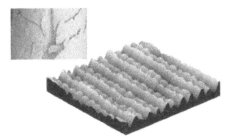

FIGURE 19   Optical micrograph of cut-polyamide fibers.

3.10.6   THE EFFECT OF RECYCLED FIBERS ON MODULUS OF RUPTURE (M_R)

The result of the flexure tests, for the specimens after 7 and 28 days' curing, are shown respectively in Figs. 20 and 21. As it is expected, the MR increases with the increase of fibers content. In addition, it is revealed that, first, in a constant percent of cement; all fiber-reinforced specimens exhibit greater MR value than non-reinforced ones. Second, by the increase of fiber's percentage MR increases. As it is illustrated in theses figures, the reinforced specimens with 0.3% of fibers possess the highest MR and tensile strength; it should be noted that adding more than a specific amount of fibers content increase the porosity of the soil. As a result, it causes MR and tensile strength to decrease. Therefore, it is possible that, increasing more percents of fibers for preparation of specimens causes MR and tensile strength to decrease in comparison with non-reinforced ones.

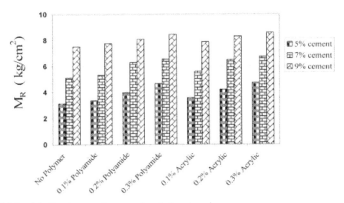

FIGURE 20 M_R for the specimens after 7 days' curing.

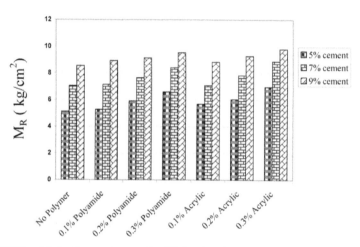

FIGURE 21 M_R for the specimens after 28 days' curing.

3.10.7 THE EFFECT OF FIBERS ON BEHAVIOR FLEXURAL LOAD-DEFORMATION

Flexural behavior of the load-displacement with different percents of cement, for the cured specimens after 7 and 28 days' curing is shown in Fig. 22 from (a) to (c) and Fig. 24 from (a) to (c). It can be seen that for all of the specimens, before the first cracking, an approximate linear relation creates within the load-displacement that coincides with the maximum load bearing of the specimens. On the basis of the experimental results it could be concluded that all of the reinforced specimens exhibit more displacement and will not exactly rupture after undergoing maximum load bearing in spite of non-reinforced specimens. In other word, deviation is seen in the direction of axial displacement. The presence of fibers in the soil-cement mixture increases the flexural strength of the soil-cement-fiber compound in comparison with the non-reinforced specimens. The main effect of fibers appears as the soil reaches to its own maximum strength. Presence of fibers will delay the rupture of the specimen. Also, the increase of displacement with the applied load is resultant of the polymer's tension, and finally the rupture of the specimen due to the maximum bearable stress is the result of the polymer rupture. This could be a result of the bridge effect of fibers used in the specimens. Generally, the presence of fibers in the soil-cement mixture causes mixture's ductility to increase and its behavior changes from fragile state to flexible behavior.

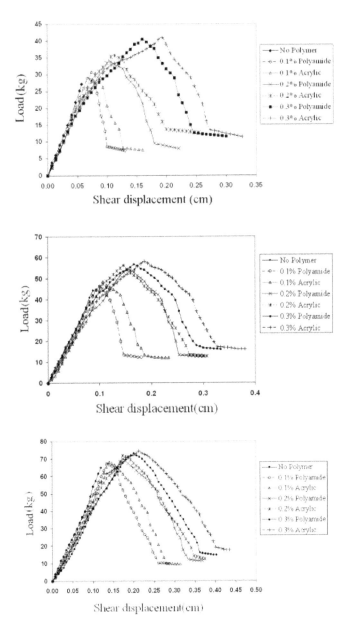

FIGURE 22 (a) Flexural load-displacement of the specimens with 5% cement after 7 days' curing. (b) Flexural load-displacement of the specimens with 7% cement after 7 days' curing. (c) Flexural load-displacement of the specimens with 9% cement after 7 days' curing.

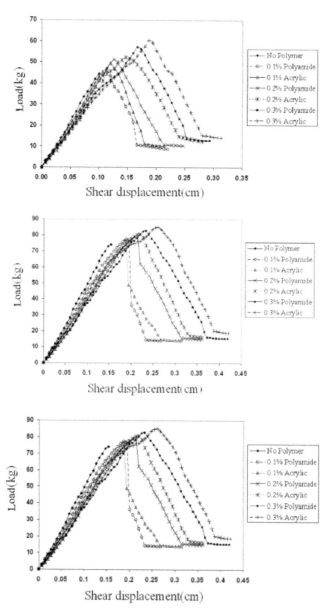

FIGURE 23   (a) Flexural load-displacement of the specimens with 5% cement after 28 days' curing. (b) Flexural load-displacement of the specimens with 7% cement after 28 days' curing. (c) Flexural load-displacement of the specimens with 9% cement after 28 days' curing.

3.10.8   THE EFFECT OF TEXTILE FIBERS ON FRACTURE ENERGY

The area under the load-deformation curves up to failure (approximately point of zero load) divided by the specimen's cross sectional area, is a measure of the energy absorption capacity or the toughness of the composite, and it is sometimes termed the fracture energy. The average fracture energy of specimens after 7 and 28 days curing are shown in Figs. 24 and 25. It can be concluded that the amount of fracture energy increases with the increase of cement percentage and, for a constant percent of cement, with the increase of fibers, it increases as well.

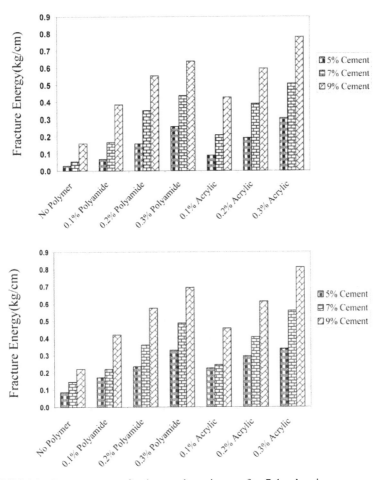

FIGURE 24   Fracture energy for the cured specimens after 7 days' curing.

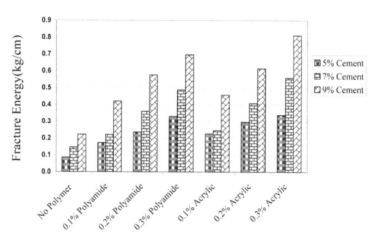

FIGURE 25   Fracture energy for the cured specimens after 28 days' curing.

On the basis of experimental results for the reinforced and non-reinforced soil-cement specimens containing different amount of cement and fibers with two different curing period, fracture energy for the specimens containing 5% of cement, by the effect of fibers, is more than that in the specimens which contain 7% of cement.

The effect of fibers in the specimens containing 7% of cement is more significant than that of the specimens containing 9% of cement. This phenomenon indicates the superiority of fiber polymers in the low percent of cement.

3.10.9 THE EFFECT OF RECYCLED TEXTILE FIBERS ON SHEAR STRENGTH

Shear strength test have been conducted on reinforced and non-reinforced soil-cement specimens containing different percents of fibers content. The applied perpendicular stress 1kg/cm^2 (100 KN/m^2) has been chosen in all tests. The shear stress-strain diagram for the specimens that contain 5% and 9% of cement are shown in Figs. 26 and 27, respectively.

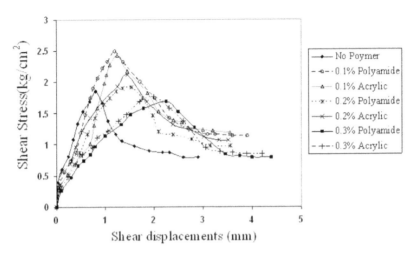

FIGURE 26   Shear stress-strain behavior of the cured specimen with 5% cement in 7 days' curing.

Figure 26 shows that, cement-stabilized specimens with 5% cement and reinforced with 0.1% of fibers have the highest shear strength, and by increasing the fibers content, shear strength decreases. This occurs in a way that the shear strength in the stabilized specimens with 5% of cements and reinforced with 0.1% and 0.2% of fibers is more than that of non-reinforced specimens. On the other hand, the shear strength of specimens containing 0.3% of fibers, is less than that of non-reinforced ones. The residual strength of reinforced specimens is more than that of non-reinforced specimens. It is readily observed from Figs. 26 and 27 that the overall soil behavior is significantly influenced by the investigated variables. Peak strength, stress-strain behavior and residual response are changes as s consequence of either the separate or the joined effects of fiber and cement inclusions. Shear strength and stress-strain behavior (stiffness) are dramatically increased by increasing the cement content for each percent of fiber contents. By using more fiber content, a moderate increase in shear strength is accompanied by a rigidity loss; the residual strength is increased and the volumetric response becomes more compressive in the early stages of loading and less expansive afterwards. For the specimens containing 0.1% of fibers the residual strength have a greater value in comparison with other specimens. Adding fibers increase the strain related to the maximum shear strength of the specimens, which results in a more ductile specimen. This ductility increases with the increase percent of fibers.

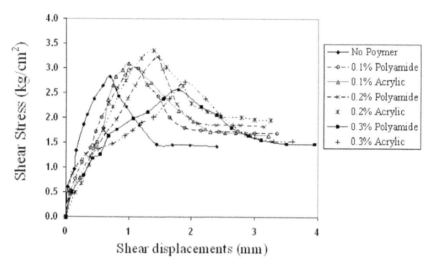

FIGURE 27   Shear stress-strain behavior of the cured specimen with 9% cement after 7 days' curing.

It can be drawn from Fig. 27 that, in the stabilized specimens with 9% of cement, adding fibers up to 0.2% of fibers content increases the shear strength to its maximum value while adding more fibers decrease the shear strength. Therefore, the shear strength of the stabilized specimens with 9% of cements and reinforced with 0.3% of fibers is less than that of non-reinforced specimens.

Also, in the stabilized specimens with 5% and 9% of cement and reinforced with fibers, the residual strength in all reinforced specimens is more than that of non-reinforced.

3.10.10   THE EFFECT OF RECYCLED TEXTILE FIBERS ON DILATION

The mechanical properties of stabilized samples in shear box (for condition of unsaturated and at optimum moisture) are similar to soft rocks and practically their dilation is a consequence of contact of teeth generated over the shear surface of samples. It is important that the numerical rates of dilation depend on amounts of the normal loads, used in direct shear tests. Dilation in direct shear test distinguishes by monitoring of the vertical displacement, among the horizontal displacement of soil shear samples. In this study, all of the vertical displacements were positive, namely volume of all samples was increased. The effect of fibers on dilation of the cured specimens after 7 days' curing is shown in Fig. 28.

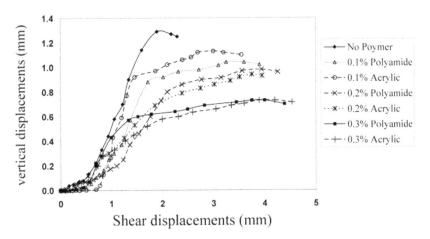

FIGURE 28   Influence of adding fibers on dilation of shear samples content of 5% cement after 7 days' curing.

It can be seen in this figure that the dilation in reinforced specimens is less than that of non-reinforced specimens and, their amount dilation decrease with the increase of fibers. This phenomenon could be the result of this matter that using fibers make a smoother shear surface in comparison with non-reinforced specimens.

3.10.11   EFFECT ON DURABILITY

The durability of stabilized soil is primarily evaluated from the degree to which the engineering properties are retained. The durability concern may arise from the exposure to wetting and drying, freezing and thawing, sulphate attack, etc. Although dominating exposure condition(s) may vary geographically, such exposures are common and, should therefore, be considered as a part of the design procedure. During construction, soil cement is compacted to a high density. As the cement hydrates, the soil cement mixture hardens in this dense state and becomes a slab-like structural material. Soil cement can bridge over small, local weak subgrade areas. If the appropriate freeze-thaw and wet-dry soil cement design criteria have been satisfied, the soil cement will not rut or shove during spring thaws and will be minimally affected by water or, freezing and thawing. Compatibility is related to the pore structure of a specimen. When a stabilized soil specimen is placed in water or dried in an oven, water penetrates into or leaves the specimen through the interconnected pore structure. It is also known that the surface tension of water is high enough to generate substantial capillary pressure to cause cracking of the surrounding matrix, especially for stabilized soils with very little tensile strength. However, to add fiber polymers, as reinforced factor, prohibits the small crack to develop and cause durability of the sol-cement specimens to increase.

Tables 5 and 6 show the rates of decreasing weight of the specimens after wet and dry cycles of durability test. The decreasing weight of specimens with 5% of cement curing after 7 days do not satisfy the criterion of the Table 4, but as it is, first, the existing fiber causes the weight loss of specimens to decrease and, second, the decreasing weight of cured specimens satisfies the criteria mentioned in Table 4. Generally, the existing fibers improve the properties of the durability of specimens and decrease the weight loss of specimens.

TABLE 5 The rate of decreasing weight after cycles of durability test for cured specimens in 7 and 28 days' curing.

	5%	7%	9%	5%	7%	9%
	7 days			28 days		
No Polymer	27.64%	4.68%	1.87%	6.90%	2.18%	0.66%
0.1% Polyamide	19.37%	4.00%	1.61%	6.50%	1.96%	0.60%
0.2% Polyamide	21.99%	3.35%	1.44%	6.65%	1.84%	0.54%
0.3% Polyamide	24.51%	4.20%	1.56%	6.83%	2.02%	0.62%
0.1% Acrylic	18.84%	4.27%	1.60%	6.56%	1.89%	0.58%
0.2% Acrylic	20.86%	3.54%	1.51%	6.78%	1.79%	0.55%
0.3% Acrylic	23.37%	4.29%	1.67%	6.87%	2.05%	0.62%

TABLE 6 The rate of decreasing weight after cycles of durability test for cured specimens in 90 and 180 days' curing.

	5%	7%	9%	5%	7%	9%
	90 Days			180 Days		
No Polymer	4.23%	1.20%	0.46%	2.62%	0.78%	0.26%
0.1% Polyamide	3.83%	1.18%	0.41%	2.53%	0.70%	0.21%
0.2% Polyamide	3.95%	0.94%	0.39%	2.35%	0.64%	0.19%
0.3% Polyamide	4.19%	1.16%	0.44%	2.59%	0.76%	0.24%
0.1% Acrylic	3.95%	1.16%	0.42%	2.55%	0.75%	0.22%
0.2% Acrylic	4.08%	1.00%	0.40%	2.28%	0.68%	0.20%
0.3% Acrylic	4.20%	1.18%	0.45%	2.40%	0.76%	0.25%

3.11 CONCLUDING REMARKS

The introduction and development of advanced composite material opened the door to new and innovative application in civil and structural engineering [1–11]. Key points of this investigation are to evaluate the application of recycled materials in concrete composites and:

- To convert waste into useful product.
- To consume wastes; this would otherwise go to landfill.
- To protect the environment from being heavily contaminated.

KEYWORDS

- **composite materials**
- **construction industries**
- **environment**
- **landfill**
- **recycled polymers**
- **waste conversion**

REFERENCES

1. Albano, C., Camacho, N., Hernandez, M., Matheus, A., Gutierrez. A., 2009. Influence of content and particle size of waste pet bottles on concrete behavior at different w/c ratios. Waste Management 29, 2707–2716.
2. Al-Salem, S.M., Lettieri, P., Baeyens, J., 2010. The valorization of plastic solid waste (PSW) by primary to quaternary routes: From re-use to energy and chemicals. Progress in Energy and Combustion Science 36, 103–129.
3. Chan, Y.N.S, Ji, X, 1999. Comparative study of the initial surface absorption and chloride diffusion of high performance zeolite, silica fume and PFA concretes. Cement and Concrete Composites 21 (1999) 293–300.
4. Choi, Y.W., Moon, D.J., Kim, Y.J., Lachemi, M., 2009. Characteristics of mortar and concrete containing fine aggregate manufactured from recycled waste polyethylene terephthalate bottles. Construction and Building Materials 23, 2829–2835.
5. Choi, Y.W., Moon, D.J., Chung, J.S., Cho, S.K., 2005. Effects of waste PET bottles aggregate on the properties of concrete. Cement and Concrete Research 35, 776–781.
6. Ismail, Z.Z., AL-Hashmi, E.A., 2008. Use of waste plastic in concrete mixture as aggregate replacement. Waste Management 28, 2041–2047.
7. Marzouk, O.Y., Dheilly, R.M., Queneudec, M., 2007. Valorization of post-consumer waste plastic in cementations concrete composites. Waste Management 27, 310–318.
8. Naik, T.R., Singh, S.S., Huber, C.O., Brodersen, B.S., 1996. Use of postconsumer waste plastics in cement-based composites. Cement and Concrete Research 26 (10), 1489–1492.

9. Panyakapo, P., Panyakapo M., 2008. Reuse of thermosetting plastic waste for lightweight concrete. Waste Management 28, 1581–1588.

10. Rao, G.A., 2003. Investigations on the performance of silica fume incorporated cement pastes and mortars. Cement and Concrete Research 33 (2003) 1765–1770.

11. Siddique, R., Khatib, J., Kaur, I., 2008. Use of recycled plastic in concrete: A review. Waste Management 28, 1835–1852.

CHAPTER 4

NANOCOMPOSITES POLYETHYLENE/ORGANOCLAY ON SUPRASEGMENTAL LEVEL

G. V. KOZLOV, K.S. DIBIROVA, G. M. MAGOMEDOV, and G. E. ZAIKOV

CONTENTS

4.1   INTRODUCTION

The structural mechanism of polymer nanocomposites filled with organoclay on suprasegmental level was offered. Within the frameworks of this mechanism nanocomposites elasticity modulus is defined by local order domains (nanoclusters) sizes similarly to natural nanocomposites (polymers). Densely packed interfacial regions formation in nanocomposites at nanofiller introduction is the physical basis of nanoclusters size decreasing.

Very often a filler (nanofiller) is introduced in polymers with the purpose of the latter stiffness increase. This effect is called polymer composites (nanocomposites) reinforcement and it is characterized by reinforcement degree E_c/E_m (E_n/E_m), where E_c, E_n and E_m are elasticity moduli of composite, nanocomposite and matrix polymer, accordingly. The indicated effect significance results to a large number of quantitative models development, describing reinforcement degree: micromechanical [1], percolation [2] and fractal [3] ones. The principal distinction of the indicated models is the circumstance, that the first ones take into consideration the filler (nanofiller) elasticity modulus and the last two – don't. The percolation [2] and fractal [3] models of reinforcement assume, that the filler (nanofiller) role comes to modification and fixation of matrix polymer structure. Such approach is obvious enough, if to account for the difference of elasticity modulus of filler (nanofiller) and matrix polymer. So, for the considered in the present paper nanocomposites low density polyethylene/Na$^+$-montmorillonite the matrix polymer elasticity modulus makes up 0.2 GPa [4] and nanofiller – 400–420 GPa [5], that is, the difference makes up more than three orders. It is obvious, that at such conditions organoclay strain is equal practically to zero and nanocomposites behavior in mechanical tests is defined by polymer matrix behavior.

Lately it was offered to consider polymers amorphous state structure as a natural nanocomposite [6]. Within the frameworks of cluster model of polymers amorphous state structure it is supposed, that the indicated structure consists of local order domains (clusters), immersed in loosely-packed matrix, in which the entire polymer free volume is concentrated [7, 8]. In its turn, clusters consist of several collinear densely packed statistical segments of different macromolecules, i.e. they are an amorphous analogue of crystallites with stretched chains. It has been shown [9], that clusters are nanoworld objects (true nanoparticles-nanoclusters) and in case of polymers representation as natural nanocomposites they play nanofiller role and loosely packed matrix-nanocomposite matrix role. It is significant that the nanoclusters dimensional effect is identical to the indicated effect for particulate filler in polymer nanocomposites – sizes decrease of both nanoclusters [10] and disperse particles [11] results to sharp enhancement of nanocomposite reinforcement degree (elasticity modulus). In connection with the indicated observations the question arises: how organoclay introduction in polymer matrix influences on nanoclusters size and how the variation of the latter influences on nanocomposite elasticity modulus value. The purpose of the present paper is these two problems solution on the example of nanocomposite linear low-density polyethylene/Na$^+$-montmorillonite [4].

4.2   EXPERIMENTAL

Linear low density polyethylene (LLDPE) of mark Dowlex-2032, having melt flow index 2.0 g/10 min and density 926 kg/m^3, that corresponds to crystallinity degree of 0.49, used as a matrix polymer. Modified Na$^+$-montmorillonite (MMT), obtained by cation exchange reaction between MMT and quaternary ammonium ions, was used as nanofiller MMT contents makes up 1–7 mass % [4].

Nanocomposites linear low-density polyethylene/Na$^+$-montmorillonite (LLDPE/MMT) were prepared by components blending in melt using Haake twin-screw extruder at temperature 473 K [4].

Tensile specimens were prepared by injection molding on Arburg Allounder 305-210-700 molding machine at temperature 463 K and pressure 35 MPa. Tensile tests were performed by using tester Instron of the model 1137 with direct digital data acquisition at temperature 293 K and strain rate ~3.35 × 10^{-3} s^{-1}. The average error of elasticity modulus determination makes up 7%, yield stress – 2% [4].

4.3   RESULTS AND DISCUSSION

For the solution of the first from the indicated problems the statistical segments number in one nanocluster n_{cl} and its variation at nanofiller contents change should be estimated. The parameter n_{cl} calculation consistency includes the following stages. At first the nanocomposite structure fractal dimension d_f is calculated according to the equation [12]:

$$d_f = (d-1)(1+v),\qquad (1)$$

where d is dimension of Euclidean space, in which a fractal is considered (it is obvious, that in our case $d=3$), v is Poisson's ratio, which is estimated according to mechanical tests results with the aid of the relationship [13]:

$$\frac{\sigma_Y}{E_n} = \frac{1-2v}{6(1+v)},\qquad (2)$$

where σ_Y and E_n are yield stress and elasticity modulus of nanocomposite, accordingly.

Then nanoclusters relative fraction φ_{cl} can be calculated by using the following equation [8]:

$$d_f = 3 - 6\left(\frac{\phi_{cl}}{C_\infty S}\right)^{1/2}, \tag{3}$$

where C_∞ is characteristic ratio, which is a polymer chain statistical flexibility indicator [14], S is macromolecule cross-sectional area.

The value C_∞ is a function of d_f according to the relationship [8]:

$$\tilde{N}_\infty = \frac{2d_f}{d(d-1)(d-d_f)} + \frac{4}{3} \tag{4}$$

The value S for low-density polyethylenes is accepted equal to 14.9 Å² [15]. Macromolecular entanglements cluster network density v_{cl} can be estimated as follows [8]:

$$v_{cl} = \frac{\phi_{cl}}{C_\infty l_0 S}, \tag{5}$$

where l_0 is the main chain skeletal bond length, which for polyethylenes is equal to 0.154 nm [16].

Then the molecular weight of the chain part between nanoclusters M_{cl} was determined according to the equation [8]:

$$M_{cl} = \frac{\rho_p N_A}{v_{cl}}, \tag{6}$$

where ρ_p is polymer density, which for the studied polyethylenes is equal to ~ 930 kg/m³, N_A is Avogadro number.

And at last, the value n_{cl} is determined as follows [8]:

$$n_{cl} = \frac{2M_e}{M_{cl}}, \tag{7}$$

where M_e is molecular weight of a chain part between entanglements traditional nodes ("binary hookings"), which is equal to 1390 g/mole for low-density polyethylenes [17].

In Fig. 1, the dependence of nanocomposite elasticity modulus E_n on value n_{cl} is adduced, from which E_n enhancement at n_{cl} decreasing follows. Such behavior of nanocomposites LLDPE/MMT is completely identical to the behavior of both particulate-filled [11] and natural [10] nanocomposites.

In Ref. [18], the theoretical dependences of E_n as a function of cluster model parameters for natural nanocomposites was obtained:

$$E_n = c\left(\frac{\phi_{cl}V_{cl}}{n_{cl}}\right), \tag{8}$$

where c is constant, accepted equal to 5.9×10^{-26} m^3 for LLDPE.

In Fig. 1, the theoretical dependence $E_n(n_{cl})$, calculated according to the equation (8), for the studied nanocomposites is adduced, which shows a good enough correspondence with the experiment (the average discrepancy of theory and experiment makes up 11.6%, that is comparable with mechanical tests experimental error). Therefore, at organoclay mass contents W_n increasing within the range of 0–7 mass % n_{cl} value reduces from 8.40 up to 3.17 that are accompanied by nanocomposites LLDPE/MMT elasticity modulus growth from 206 up to 569 MPa.

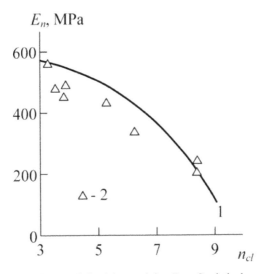

FIGURE 1   The dependences of elasticity modulus E_n on Statistical segments number per one nanocluster n_{cl} for nanocomposites LLDPE/MMT. 1 – calculation according to the Eq. (8), 2 – the experimental data.

Let us consider the physical foundations of n_{cl} reduction at W_n growth. The main equation of the reinforcement percolation model is the following one [9]:

$$\frac{E_n}{E_m} = 1 + 11\left(\phi_n + \phi_{if}\right)^{1.7},$$ (9)

where ϕ_n and ϕ_{if} are relative volume fractions of nanofiller and interfacial regions, accordingly.

The value ϕ_n can be determined according to the equation [5]:

$$\phi_n = \frac{W_n}{\rho_n},$$ (10)

where ρ_n is nanofiller density, which is equal to ~1700 kg/m³ for Na⁺-montmorillonite [5].

Further the Eq. (9) allows to estimate the value ϕ_{if}. In Fig. 2, the dependence $n_{cl}(\phi_{if})$ for nanocomposites LLDPE/MMT is adduced. As one can see, n_{cl} reduction at ϕ_{if} increasing is observed, i.e. formed on organoclay surface densely packed (and, possibly, subjecting to epitaxial crystallization [9]) interfacial regions as if pull apart nanoclusters, reducing statistical segments number in them. As it follows from the Eqs. (8) and (9), these processes have the same direction, namely, nanocomposite elasticity modulus increase.

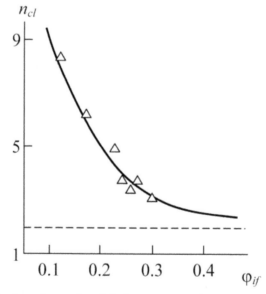

FIGURE 2   The dependence of statistical segments number per one nanocluster n_{cl} on interfacial regions relative fraction ϕ_{if} for nanocomposites LLDPE/MMT. Horizontal shaded line indicates the minimum value $n_{cl}=2$.

4.4   CONCLUSIONS

Hence, the obtained in the present paper results demonstrated common reinforcement mechanism of natural and artificial (filled with inorganic nanofiller) polymer nanocomposites. The statistical segments number per one nanocluster reduction at nanofiller contents growth is such a mechanism on suprasegmental level. The indicated effect physical foundation is the densely packed interfacial regions formation in artificial nanocomposites.

KEYWORDS

- elasticity modulus
- interfacial regions
- nanoclusters
- nanocomposite
- organoclay
- polyethylene

REFERENCES

1. Ahmed S., Jones F.R. J. Mater. Sci., 1990, v. 25, № 12, 4933–4942.
2. Bobryshev A.N., Kozomazov V.N., Babin L.O., Solomatov V.I. Synergetics of Composite Materials. Lipetsk, NPO ORIUS, 1994, 154 p.
3. Kozlov G.V., Yanovskii Yu.G., Zaikov G.E. Structure and Properties of Particulate–Filled Polymer Composites: the Fractal Analysis. New York, Nova Science Publishers, Inc., 2010, 282 p.
4. Hotta S., Paul D.R. Polymer, 2004, v. 45, № 21, 7639–7654.
5. Sheng N., Boyce M.C., Parks D.M., Rutledge G.C., Abes J.I., Cohen R.E. Polymer, 2004, v. 45, № 2, 487–506.
6. Bashorov M.T., Kozlov G.V., Mikitaev A.K. Materialovedenie, 2009, № 9, 39–51.
7. Kozlov G.V., Novikov V.U. Uspekhi Fizicheskikh Nauk, 2001, v. 171, № 7, 717–764.
8. Kozlov G.V., Zaikov G.E. Structure of the Polymer Amorphous State. Utretch, Boston, Brill Academic Publishers, 2004, 465 p.
9. Mikitaev A.K., Kozlov G.V., Zaikov G.E. Polymer Nanocomposites: Variety of Structural Forms and Applications. New York, Nova Science Publishers, Inc., 2008, 319 p.
10. Kozlov G.V., Mikitaev A.K. Polymers as Natural Nanocomposites: Unrealized Potential. Saarbrücken, Lambert Academic Publishing, 2010, 323 p.
11. Edwards D.C. J. Mater. Sci., 1990, v. 25, № 12, 4175–4185.
12. Balankin A.S. Synergetics of Deformable Body. Moscow, Publishers of Ministry Defence SSSR, 1991, 404 p.
13. Kozlov G.V., Sanditov D.S. Anharmonic Effects and Physical–Mechanical Properties of Polymers. Novosibirsk, Nauka, 1994, 261 p.

14. Budtov V.P. Physical Chemistry of Polymer Solutions. Sankt–Peterburg, Khimiya, 1992, 384 p.
15. Aharoni S.M. Macromolecules, 1985, v. 18, № 12, 2624–2630.
16. Aharoni S.M. Macromolecules, 1983, v. 16, № 9, 1722–1728.
17. Wu S. J. Polymer Sci.: Part B: Polymer Phys., 1989, v. 27, № 4, 723–741.
18. Kozlov G.V. Recent Patents on Chemical Engineering, 2011, v. 4, № 1, p. 53–77.

CHAPTER 5

QUANTUM AND WAVE CHARACTERISTICS OF SPATIAL-ENERGY INTERACTIONS

G. A. KORABLEV and G. E. ZAIKOV

CONTENTS

5.1   INTRODUCTION

It is demonstrated that for two-particle interactions the principle of adding reciprocals of energy characteristics of subsystems is performed for processes flowing by the potential gradient, and the principle of their algebraic addition – for the processes against the potential gradient.

The equation of the dependence of spatial-energy parameter of free atoms on their wave, spectral and frequency characteristics has been obtained.

Quantum conceptualizations on the composition of atoms and molecules make the foundation of modern natural science theories. Thus, the electronic angular momentum in stationary condition equals the integral multiple from Planck's constant. This main quantum number and three other combined explicitly characterize the state of any atom. The repetition factors of atomic quantum characteristics are also expressed in spectral data for simple and complex structures.

It is known that any periodic processes of complex shape can be shown as separate simple harmonic waves. "By Fourier theory, oscillations of any shape with period T can be shown as the total of harmonic oscillations with periods T_1, T_2, T_3, T_4, etc. Knowing the periodic function shape, we can calculate the amplitude and phases of sinusoids, with this function as their total" [1].

Therefore, many regularities in intermolecular interactions, complex formation and nanothermodynamics are explained with the application of functional divisible quantum or wave energy characteristics of structural interactions.

In this research we tried to apply the conceptualizations on spatial-energy parameter (P-parameter) for this.

5.2   ON TWO PRINCIPLES OF ADDING ENERGY CHARACTERISTICS OF INTERACTIONS

The analysis of the kinetics of various physic-chemical processes demonstrates that in many cases the reciprocals of velocities, kinetic or energy characteristics of the corresponding interactions are added.

Here are some examples: ambipolar diffusion, total rate of topochemical reaction, change in the light velocity when transiting from vacuum into the given medium, effective permeability of biomembranes.

In particular, such assumption is confirmed by the formula of electron transport probability (W_∞) due to the overlapping of wave functions 1 and 2 (in stationary state) during electron-conformation interactions:

$$W_\infty = \frac{1}{2}\frac{W_1 W_2}{W_1 + W_2} \tag{1}$$

Equation (1) is applied when evaluating the characteristics of diffusion processes accompanied with non-radiating electron transport in proteins [2].

Also: "It is known from the traditional mechanics that the relative motion of two particles with the interaction energy U(r) is the same as the motion of a material point with the reduced mass μ :

$$\frac{1}{\mu} = \frac{1}{m_1} + \frac{1}{m_2}$$

(2)

in the field of central force U(r), and total translational motion – as the free motion of the material point with the mass:

$$m = m_1 + m_2$$

(3)

Such situation can be also found in quantum mechanics" [3].

The problem of two-particle interactions flowing by the bond line was solved in the time of Newton and Lagrange:

$$\mathring{A} = \frac{m_1 v_1^2}{2} + \frac{m_2 v_2^2}{2} + U\left(\bar{r}_2 - \bar{r}_1\right),$$

(4)

where E – system total energy, first and second components – kinetic energies of the particles, third – potential energy between particles 1 and 2, vectors $\bar{r}_2$ and $\bar{r}_1$ characterize the distance between the particles in final and initial states.

For moving thermodynamic systems the first law of thermodynamics can be shown as follows [4]:

$$\delta\mathring{A} = d\left(U + \frac{mv^2}{2}\right) \pm \delta A,$$

(5)

where: δE – amount of energy transferred to the system; component $d\left(U + \frac{mv^2}{2}\right)$ characterizes changes in internal and kinetic energies of the system; $+\delta A$ – work performed by the system; $-\delta A$ – work performed on the system.

Since the work numerically equals the change in the potential energy, then:

$$+\delta A = -\Delta U \text{ и } -\delta A = +\Delta U$$

(6 and 7)

Probably not only the value of potential energy but its changes are important in thermodynamic and also in many other processes in the dynamics of interactions of moving particles. Therefore, by the analogy with Eq. (4) the following should be fulfilled for two-particle interactions:

$$\delta E = d\left(\frac{m_1 v_1^2}{2} + \frac{m_2 v_2^2}{2}\right) \pm \Delta U$$

(8)

Here $$\Delta U = U_2 - U_1,$$ (9)

where U_2 and U_1 – potential energies of the system in final and initial states.

At the same time, the total energy (E) and kinetic energy $\left(\dfrac{mv^2}{2} \right)$ can be calculated from their zero value. In this case only the last component is modified in the equation (4).

The character of the changes in the potential energy value (ΔU) was analyzed by its index for different potential fields as given in Table 1.

From the table it is seen that the values of $-\Delta U$ and consequently $+\delta A$ (positive work) correspond to the interactions taking place by the potential gradient, and ΔU and $-\delta A$ (negative work) take place during the interactions against the potential gradient.

The solution of two-particle problem of the interaction of two material points with masses m_1 and m_2 obtained under the condition of no external forces available corresponds to the interactions taking place by the gradient, the positive work is performed by the system (similar to attraction process in the gravitation field).

The solution for this equation through the reduced mass (μ) [5] is Lagrangian equation for the relative motion of the isolated system of two interacting material points with masses m_1 and m_2, in coordinate x it looks as follows:

$$\mu \cdot x'' = -\frac{\partial U}{\partial x}; \frac{1}{\mu} = \frac{1}{m_1} + \frac{1}{m_2}.$$

Here: U – mutual potential energy of material points; μ – reduced mass. At the same time $x'' = a$ (characteristic of system acceleration). For elementary regions of interactions Δx can be taken as follows:

$$\frac{\partial U}{\partial x} \approx \frac{\Delta U}{\Delta x}$$ That is: $\mu a \Delta x = -\Delta U$. Then:

$$\frac{1}{1/(a\Delta x)}\frac{1}{1/m_1 + 1/m_2} \approx -\Delta U; \frac{1}{a/(m_1 a \Delta x)} \approx -\Delta U$$

or: $$\frac{1}{\Delta U} \approx \frac{1}{\Delta U_1} + \frac{1}{\Delta U_2}$$ (10)

where ΔU_1 and ΔU_2 – potential energies of material points on the elementary region of interactions, ΔU – resulting (mutual) potential energy of these interactions.

TABLE 1 Directedness of interaction processes.

No	Systems	Potential field type	Process	U	r_2/r_1 (x_2/x_1)	U_2/U_1	Index ΔU	Index δA	Process directedness in the potential field
1	Opposite electric charges	Electrostatic	Attraction	$-k\dfrac{q_1q_2}{r}$	$r_2<r_1$	$U_2>U_1$	—	+	By gradient
			Repulsion	$-k\dfrac{q_1q_2}{r}$	$r_2>r_1$	$U_2<U_1$	+	—	Against gradient
2	Same electric charges	Electrostatic	Attraction	$k\dfrac{q_1q_2}{r}$	$r_2<r_1$	$U_2>U_1$	+	—	Against gradient
			Repulsion	$k\dfrac{q_1q_2}{r}$	$r_2>r_1$	$U_2<U_1$	—	+	By gradient
3	Elementary masses m_1 and m_2	Gravitational	Attraction	$-\gamma\dfrac{m_1m_2}{r}$	$r_2<r_1$	$U_2>U_1$	—	+	By gradient
			Repulsion	$-\gamma\dfrac{m_1m_2}{r}$	$r_2>r_1$	$U_2<U_1$	+	—	Against gradient
4	Spring deformation	Field of spring forces	Compression	$k\dfrac{\Delta x^2}{2}$	$x_2<x_1$	$U_2>U_1$	+	—	Against gradient
			Stretching	$k\dfrac{\Delta x^2}{2}$	$x_2>x_1$	$U_2>U_1$	+	—	Against gradient
5	Photoeffect	Electrostatic	Repulsion	$k\dfrac{q_1q_2}{r}$	$r_2>r_1$	$U_2<U_1$	—	+	By gradient

Thus:
1. In systems in which the interaction takes place by the potential gradient (positive work), the resultant potential energy is found by the principle of adding the reciprocals of the corresponding energies of subsystems [6]. The reduced mass for the relative motion of isolated system of two particles is calculated in the same way.

2. In systems in which the interaction takes place against the potential gradient (negative work), their masses and corresponding energies of subsystems (similar to Hamiltonian) are added algebraically.

5.3 INITIAL CRITERIA

From the Eq. (10) it is seen that the resultant energy characteristic of the system of interaction of two material points is found by the principle of adding the reciprocals of initial energies of interacting subsystems.

"Electron with the mass m moving near the proton with the mass M is equivalent to the particle with the mass $m_r = \dfrac{mM}{m+M}$ " [7].

Therefore modifying the Eq. (10), we can assume that the energy of atom valence orbitals (responsible for interatomic interactions) can be calculated [6] by the principle of adding the reciprocals of some initial energy components based on the equations:

$$\frac{1}{q^2/r_i} + \frac{1}{W_i n_i} = \frac{1}{P_E} \qquad (11)$$

or

$$\frac{1}{P_0} = \frac{1}{q^2} + \frac{1}{(Wrn)_i}; \qquad (12)$$

$$P_E = P_0/r_i \qquad (13)$$

where: W_i – orbital energy of electrons [8]; r_i – orbital radius of i orbital [9]; $q=Z^*/n^*$ – by [10,11], n_i – number of electrons of the given orbital, Z^* and n^* – nucleus effective charge and effective main quantum number, r – bond dimensional characteristics.

P_0 is called a spatial-energy parameter (SEP), and P_E – effective P-parameter (effective SEP). Effective SEP has a physical sense of some averaged energy of valence orbitals in the atom and is measured in energy units, for example, in electron-volts (eV).

The values of P_0 parameter are tabulated constants for electrons of the given atom orbital.

For SEP dimensionality:

$$[P_0] = [q] = [E] \cdot [r] = [h] \cdot [v] = \frac{kgm^3}{s^2} = Jm,$$

where [E], [h] and [υ] – dimensionalities of energy, Plank's constant and velocity.

The introduction of P-parameter should be considered as further development of quasi-classical concepts with quantum-mechanical data on atom structure to obtain the criteria of phase-formation energy conditions. For the systems of similarly charged (e.g., orbitals in the given atom) homogeneous systems the principle of algebraic addition of such parameters is preserved:

$$\Sigma P_E = \Sigma \left(P_0 / r_i \right); \qquad (14)$$

$$\Sigma P_E = \frac{\Sigma P_0}{r} \qquad (15)$$

or:

$$\Sigma P_0 = P_0' + P_0'' + P_0''' + ...; \qquad (16)$$

$$r\Sigma P_E = \Sigma P_0 \qquad (17)$$

Here P-parameters are summed up by all atom valence orbitals.

To calculate the values of P_E-parameter at the given distance from the nucleus either the atomic radius (R) or ionic radius (r_1) can be used instead of r depending on the bond type.

Let us briefly explain the reliability of such an approach. As the calculations demonstrated the values of P_E-parameters equal numerically (in the range of 2%) the total energy of valence electrons (U) by the atom statistic model. Using the known correlation between the electron density (β) and intra-atomic potential by the atom statistic model [12], we can obtain the direct dependence of P_E-parameter on the electron density at the distance r_i from the nucleus.

The rationality of such technique was proved by the calculation of electron density using wave functions by Clementi [13] and comparing it with the value of electron density calculated through the value of P_E-parameter.

5.4 WAVE EQUATION OF P-PARAMETER

To characterize atom spatial-energy properties two types of P-parameters are introduced. The bond between them is a simple one:

$$P_E = \frac{P_0}{R}$$

where R – atom dimensional characteristic. Taking into account additional quantum characteristics of sublevels in the atom, this equation can be written down in coordinate x as follows:

$$\Delta P_E \approx \frac{\Delta P_0}{\Delta x} \text{ or } \partial P_E = \frac{\partial P_0}{\partial x}$$

where the value ΔP equals the difference between P_0-parameter of i orbital and P_{CD} – countdown parameter (parameter of main state at the given set of quantum numbers).

According to the established [6] rule of adding P-parameters of similarly charged or homogeneous systems for two orbitals in the given atom with different quantum characteristics and according to the energy conservation rule we have:

$$\Delta P''_E - \Delta P'_E = P_{E,\lambda}$$

where $P_{E,\lambda}$ – spatial-energy parameter of quantum transition.

Taking for the dimensional characteristic of the interaction $\Delta\lambda = \Delta x$, we have:

$$\frac{\Delta P''_0}{\Delta\lambda} - \frac{\Delta P'_0}{\Delta\lambda} = \frac{P_0}{\Delta\lambda} \text{ or: } \frac{\Delta P'_0}{\Delta\lambda} - \frac{\Delta P''_0}{\Delta\lambda} = -\frac{P_0\lambda}{\Delta\lambda}$$

Let us again divide by $\Delta\lambda$ term by term:

where:

$$\left(\frac{\Delta P'_0}{\Delta\lambda} - \frac{\Delta P''_0}{\Delta\lambda}\right) \bigg/ \Delta\lambda = -\frac{P_0}{\Delta\lambda^2},$$

That is,

$$\left(\frac{\Delta P'_0}{\Delta\lambda} - \frac{\Delta P''_0}{\Delta\lambda}\right) \bigg/ \Delta\lambda \sim \frac{d^2 P_0}{d\lambda^2},$$

$$\frac{d^2 P_0}{d\lambda^2} + \frac{P_0}{\Delta\lambda^2} \approx 0$$

Taking into account only those interactions when $2\pi\Delta x = \Delta\lambda$ (closed oscillator), we have the following equation:

$$\frac{d^2 P_0}{dx^2} + 4\pi^2 \frac{P_0}{\Delta\lambda^2} \approx 0$$

Since $\Delta\lambda = \dfrac{h}{mv}$,

then: $\dfrac{d^2 P_0}{dx^2} + 4\pi^2 \dfrac{P_0}{h^2} m^2 v^2 \approx 0$

or $\qquad\qquad \dfrac{d^2 P_0}{dx^2} + \dfrac{8\pi^2 m}{h^2} P_0 E_k = 0$ \hfill (18)

where $E_k = \dfrac{mV^2}{2}$ – electron kinetic energy.

Schrodinger equation for the stationery state in coordinate x:

$$\frac{d^2 \psi}{dx^2} + \frac{8\pi^2 m}{h^2} \psi E_k = 0$$

When comparing these two equations we see that P_0-parameter numerically correlates with the value of Ψ-function:

$$P_0 \approx \Psi \; ,$$

and is generally proportional to it: $P_0 \sim \Psi$. Taking into account the broad practical opportunities of applying the P-parameter methodology, we can consider this criterion as the materialized analogue of Ψ-function [14, 15].

Since P_0-parameters like Ψ-function have wave properties, the superposition principles should be fulfilled for them, defining the linear character of the equations of adding and changing P-parameter.

5.5 QUANTUM PROPERTIES OF P-PARAMETER

According to Planck, the oscillator energy (E) can have only discrete values equaled to the whole number of energy elementary portions-quants:

$$nE = hv = hc/\lambda \hfill (19)$$

where h – Planck's constant, v – electromagnetic wave frequency, c – its velocity, λ – wavelength, $n = 0, 1, 2, 3\ldots$

Planck's equation also produces a strictly definite bond between the two ways of describing the nature phenomena – corpuscular and wave.

P_0-parameter as an initial energy characteristic of structural interactions, similarly to the Eq. (19), can have a simple dependence from the frequency of quantum transitions:

$$P_0 \sim \hbar(\lambda v_0) \qquad (20)$$

where: λ – quantum transition wavelength [16]; $\hbar = h/(2\pi)$; v_0 – kayser, the unit of wave number equaled to $2.9979\cdot10^{10}$ Hz.

In accordance with Rydberg equation, the product of the right part of this equation by the value $(1/n^2 - 1/m^2)$, where n and m – main quantum numbers – should result in the constant.

Therefore the following equation should be fulfilled:

$$P_0(1/n^2_1 - 1/m^2_1) = N\hbar(\lambda v_0)(1/n^2 - 1/m^2) \qquad (21)$$

where the constant N has a physical sense of wave number and for hydrogen atom equals $2\times10^2\text{Å}^{-1}$.

The corresponding calculations are demonstrated in Table 2. There: $r_i' = 0.5292$ Å – orbital radius of 1S-orbital and $r_i'' = 2^2\times0.5292 = 2.118$ Å – the value approximately equaled to the orbital radius of 2S-orbital.

The value of P_0-parameter is obtained from the equation (12), for example, for 1S-2P transition:

$$1/P_0 = 1/(13.595 \times 0.5292) + 1/14.394 \rightarrow P_0 = 4.7985 \text{ eVÅ}$$

The value q^2 is taken from Refs. [10, 11], for the electron in hydrogen atom it numerically equals the product of rest energy by the classical radius.

The accuracy of the correlations obtained is in the range of percentage error 0.06 (%), that is, the Eq. (21) is in the accuracy range of the initial data.

In the Eq. (21) there is the link between the quantum characteristics of structural interactions of particles and frequencies of the corresponding electromagnetic waves.

But in this case there is the dependence between the spatial parameters distributed along the coordinate. Thus in P_0-parameter the effective energy is multiplied by the dimensional characteristic of interactions, and in the right part of the Eq. (21) the kayser value is multiplied by the wavelength of quantum transition.

In Table 2 you can see the possibility of applying the Eq. (21) and for electron Compton wavelength ($\lambda_к = 2.4261 \times 10^{-12}$ m), which in this case is as follows:

$$P_0 = 10^7 \hbar (\lambda_\kappa v_0) \tag{22}$$

(with the relative error of 0.25%).

Integral-valued decimal values are found when analyzing the correlations in the system "proton-electron" given in Table 3:

1. Proton in the nucleus, energies of three quarks $5 + 5 + 7 \approx 17$ (MeV) $\rightarrow$ $P_p \approx 17$ MeV $\times 0.856 \times 10^{-15}$ m $\approx 14.552 \times 10^{-9}$ eVm. Similarly for the electron $P_e = 0.511$ (MeV) $\times 2.8179 \times 10^{-15}$ m (electron classic radius) $\rightarrow$ $P_e = 1.440 \times 10^{-9}$ eVm.

Therefore:

$$P_p \approx 10\ P_E \tag{23}$$

2. Free proton $P_n = 938.3$ (MeV) $\times 0.856 \times 10^{-15}$ (m) $= 8.0318 \times 10^{-7}$ eVm. For electron in the atom $P_a = 0.511$ (MeV) $\times 0.5292 \times 10^{-5}$ (m) $= 2.7057 \times 10^{-5}$ eVm.

Then:

$$3P_a \approx 10^2 P_n \tag{24}$$

The relative error of the calculations by these equations is found in the range of the accuracy of initial data for the proton ($\delta \approx 1\%$).

From Tables 2 and 3 we can see that the wave number N is quantized by the decimal principle:

$$N = n10^Z,$$

where n and Z – whole numbers.

Other examples of electrodynamics equations should be pointed out in which there are integral-valued decimal functions, for example, in the formula:

$$4\pi\varepsilon_0 c^2 = 10^7,$$

where ε_0 – electric constant.

In [17] the expression of the dependence of constants of electromagnetic interactions from the values of electron P_e-parameter was obtained:

$$k\mu_0 c = k/(\varepsilon_0 c) = P_e^{1/2} c^2 \approx 10/\alpha \tag{25}$$

where: $k = 2\pi/\sqrt{3}$; μ_0 – magnetic constant; c – electromagnetic constant; α – fine structure constant.

All the above conclusions are based on the application of rather accurate formulas in the accuracy range of initial data.

TABLE 2   Quantum properties of hydrogen atom parameters.

Orbitals	W_i (eV)	r_i (Å)	q_i^2 (eVÅ)	P_0 (eVÅ)	$P_0(1/n_i^2-1/m_i^2)$ (eVÅ)	N (Å^{-1})	λ (Å)	Quantum transition	$Nh\lambda\nu_0$ (eVÅ)	$Nh\lambda\nu_0 \times(1/n^2-1/m^2)$ (eVÅ)
1S	13.595	0.5292	14.394	4.7985	3.5989	2×10^2	1215	1S-2P	4.7951	3.5963
1S						2×10^2	1025	1S-3P	4.0452	3.5954
1S						2×10^2	912	1S-nP	3.5990	3.5990
2S	3.3988	2.118	14.394	4.7985	3.5990	2×10^2	6562	2S-3P		3.5967
2S						2×10^2	4861	2S-4P		3.5971
2S						2×10^2	3646	2S-nP		3.5973
1S	13.595	0.5292	14.394	4.7985		10^7	2.4263×10^{-2}	–	4.7878	

TABLE 3   Quantum ratios of proton and electron parameters.

Particle	E (eV)	r (Å)	$P = Er$ (eVÅ)	Ratio
Free proton	938.3×10^6	0.856×10^{-5}	$8.038 \times 10^3 = P_n$	$3P_a/P_n \approx 10^2$
Electron in an atom	0.511×10^6	0.5292	$2.7042 \times 10^5 = P_a$	
Proton in atom nuclei	$(5 + 5 + 7) \times 10^6 = 17 \times 10^6$	0.856×10^{-5}	$145.52 = P_p$	$P_p/P_e \approx 10$
Electron	0.511×10^6	2.8179×10^{-5}	$14.399 = P_e$	

5.6   CONCLUSIONS

1. Two principles of adding interaction energy characteristics are functionally defined by the direction of interaction by potential gradient (positive work) or against potential gradient (negative work).
2. Equation of the dependence of spatial-energy parameter on spectral and frequency characteristics in hydrogen atom has been obtained.

KEYWORDS

- frequency characteristics
- interaction energy
- negative work
- quantum
- spatial-energy
- spectral characteristics

REFERENCES

1. Gribov L.A., Prokofyeva N.I. Basics of physics. M.: Vysshaya shkola, 1992, 430 p.
2. Rubin A.B. Biophysics. Book 1. Theoretical biophysics. M.: Vysshaya shkola, 1997, 319 p.
3. Blokhintsev D.I. Basics of quantum mechanics. M.: Vysshaya shkola, 1991, 512 p.
4. Yavorsky B.M., Detlaf A.A. Reference-book in physics. M.: Nauka, 1998, 939 p.
5. Christy R.W., Pytte A. The structure of matter: an introduction to odern physics. Translated from English. M.: Nauka, 1999, 596 p.
6. Korablev G.A. Spatial–Energy Principles of Complex Structures Formation, Netherlands, Brill Academic Publishers and VSP, 2005, 426p. (Monograph).
7. Eyring G., Walter J., Kimball G. Quantum chemistry. M., F. L., 1998, 528 p.
8. Fischer C.F. Atomic Data, 1992, № 4, 301–399.
9. Waber J.T., Cromer D.T. J. Chem. Phys, 1965, vol 42, № 12, 4116–4123.
10. Clementi E., Raimondi D.L. Atomic Screening constants from S.C.F. Functions, J. Chem. Phys., 1993, v.38, №11, 2686–2689.
11. Clementi E., Raimondi D.L. J. Chem. Phys., 1997, v. 47, № 4, 1300–1307.
12. Gombash P. Statistic theory of an atom and its applications. M.: I.L., 1991, 398 p.
13. Clementi E. J.B.M. S. Res. Develop. Suppl., 1995, v. 9, № 2, 76.
14. Korablev G.A., Zaikov G.E. J. Appl. Polym. Sci., USA, 2006, v. 101, № 3, 2101–2107.
15. Korablev G.A., Zaikov G.E. Progress on Chemistry and Biochemistry, Nova Science Publishers, Inc. New York, 2009, 355–376.
16. Allen K.W. Astrophysical values. M.: Mir, 1997, 446 p.
17. Korablev G.A. Exchange spatial-energy interactions. Izhevsk. Publishing house "Udmurt University", 2010, 530 p. (Monograph).

CHAPTER 6

KEY ELEMENTS ON NANOPOLYMERS—FROM NANOTUBES TO NANOFIBERS

A. K. HAGHI and G. E. ZAIKOV

CONTENTS

6.1 INTRODUCTION

The appearance of "nanoscience" and "nanotechnology" stimulated the burst of terms with "nano" prefix. Historically the term "nanotechnology" appeared before and it was connected with the appearance of possibilities to determine measurable values up to 10^{-9} of known parameters: 10^{-9}m-nm (nanometer), 10^{-9}s-ns (nanosecond), 10^{-9} degree (nanodegree, shift condition). Nanotechnology and molecular nanotechnology comprise the set of technologies connected with transport of atoms and other chemical particles (ions, molecules) at distances contributing the interactions between them with the formation of nanostructures with different nature. Although Nobel laureate Richard Feyman (1959) showed the possibility to develop technologies on nanometer level, Eric Drexler is considered to be the founder and ideologist of nanotechnology. When scanning tunnel microscope was invented by Nobel laureates Rorer and Binig (1981) there was an opportunity to influence atoms of a substance thus stimulating the work in the field of probe technology, which resulted in substantiation and practical application of nanotechnological methods in 1994. With the help of this technique it is possible to handle single atoms and collect molecules or aggregates of molecules, construct various structures from atoms on a certain substrate (base). Naturally, such a possibility cannot be implemented without preliminary computer designing of so-called "nanostructures architecture." Nanostructures architecture assumes a certain given location of atoms and molecules in space that can be designed on computer and afterwards transferred into technological programme of nanotechnological facility.

The notion "science of nanomaterials" assumes scientific knowledge for obtaining, composition, properties and possibilities to apply nanostructures, nanosystems and nanomaterials. A simplified definition of this term can be as follows: material science dealing with materials comprising particles and phases with nanometer dimensions. To determine the existence area for nanostructures and nanosystems it is advisable to find out the difference of these formations from analogous material objects.

From the analysis of literature the following can be summarized: the existence area of nanosystems and nanoparticles with any structure is between the particles of molecular and atomic level determined in picometers and aggregates of molecules or per molecular formations over micron units. Here, it should be mentioned that in polymer chemistry particles with nanometer dimensions belong to the class of per molecular structures, such as globules and fibrils by one of parameters, for example, by diameter or thickness. In chemistry of complex compounds clusters with nanometer dimensions are also known.

In 1991, Japanese researcher Idzhima was studying the sediments formed at the cathode during the spray of graphite in an electric arc. His attention was attracted by the unusual structure of the sediment consisting of microscopic fibers and filaments. Measurements made with an electron microscope showed that the diameter of these filaments does not exceed a few nanometers and a length of one to several microns.

Having managed to cut a thin tube along the longitudinal axis, the researchers found that it consists of one or more layers, each representing a hexagonal grid of graphite, which is based on hexagon with vertices located at the corners of the carbon atoms. In all cases, the distance between the layers is equal to 0.34 nm that is the same as that between the layers in crystalline graphite.

Typically, the upper ends of tubes are closed by multilayer hemispherical caps, each layer is composed of hexagons and pentagons, reminiscent of the structure of half a fullerene molecule.

The extended structure consisting of rolled hexagonal grids with carbon atoms at the nodes are called nanotubes.

Lattice structure of diamond and graphite are shown in Fig. 1. Graphite crystals are built of planes parallel to each other, in which carbon atoms are arranged at the corners of regular hexagons. Each intermediate plane is shifted somewhat toward the neighboring planes, as shown in Fig. 1.

The elementary cell of the diamond crystal represents a tetrahedron, with carbon atoms in its center and four vertices. Atoms located at the vertices of a tetrahedron form a center of the new tetrahedron, and thus, are also surrounded by four atoms each, etc. All the carbon atoms in the crystal lattice are located at equal distance (0.154 nm) from each other.

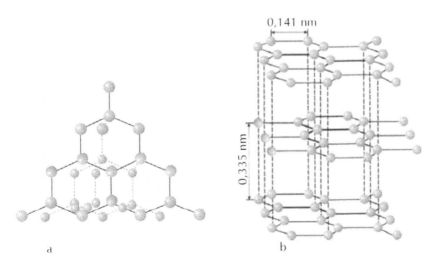

FIGURE 1   The structure of (a) the diamond lattice and (b) graphite.

Nanotubes are rolled into a cylinder (hollow tube) graphite plane, which is lined with regular hexagons with carbon atoms at the vertices of a diameter of several nanometers. Nanotubes can consist of one layer of atoms – single-wall nanotubes SWNT and represent a number of "nested" one into another layer pipes – multi-walled nanotubes – MWNT.

Nanostructures can be built not only from individual atoms or single molecules, but the molecular blocks. Such blocks or elements to create nanostructures are graphene, carbon nanotubes and fullerenes.

6.2 GRAPHENE

Graphene is a single flat sheet, consisting of carbon atoms linked together and forming a grid, each cell is like a bee's honeycombs (Fig. 2). The distance between adjacent carbon atoms in graphene is about 0.14 nm.

Graphite, from which slates of usual pencils are made, is a pile of graphene sheets (Fig. 3). Graphenes in graphite is very poorly connected and can slide relative to each other. So, if you conduct the graphite on paper, then after separating graphene from sheet the graphite remains on paper. This explains why graphite can write.

FIGURE 2   Schematic illustration of the graphene. Light balls – the carbon atoms, and the rods between them – the connections that hold the atoms in the graphene sheet.

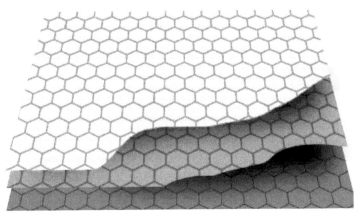

FIGURE 3 Schematic illustration of the three sheets of graphene, which are one above the other in graphite.

6.3 CARBON NANOTUBES

Many perspective directions in nanotechnology are associated with carbon nanotubes.

Carbon nanotubes are a carcass structure or a giant molecule consisting only of carbon atoms.

Carbon nanotube is easy to imagine, if we imagine that we fold up one of the molecular layers of graphite – graphene (Fig. 4).

FIGURE 4 Carbon nanotubes.

The way of folding nanotubes – the angle between the direction of nanotube axis relative to the axis of symmetry of graphene (the folding angle) – largely determines its properties.

Of course, no one produces nanotubes, folding it from a graphite sheet. Nanotubes formed themselves, for example, on the surface of carbon electrodes during arc discharge between them. At discharge, the carbon atoms evaporate from the surface, and connect with each other to form nanotubes of all kinds – single, multi-layered and with different angles of twist (Fig. 5).

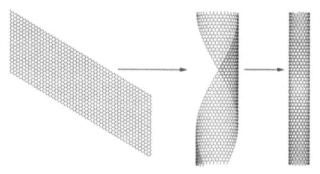

FIGURE 5   One way of imaginary making nanotube (right) from the molecular layer of graphite (left).

The diameter of nanotubes is usually about 1 nm and their length is a thousand times more, amounting to about 40 microns. They grow on the cathode in perpendicular direction to surface of the butt. The so-called self-assembly of carbon nanotubes from carbon atoms occurs (Fig. 6). Depending on the angle of folding of the nanotube they can have conductivity as high as that of metals, and they can have properties of semiconductors.

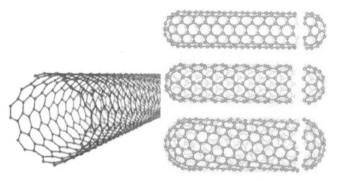

FIGURE 6   Left – schematic representation of a single-layer carbon nanotubes, on the right (top to bottom) – two-ply, straight and spiral nanotubes.

Carbon nanotubes are stronger than graphite, although made of the same carbon atoms, because carbon atoms in graphite are located in the sheets. And everyone knows that sheet of paper folded into a tube is much more difficult to bend and break than a regular sheet. That's why carbon nanotubes are strong. Nanotubes can be used as a very strong microscopic rods and filaments, as Young's modulus of single-walled nanotube reaches values of the order of 1–5 TPa, which is much more than steel! Therefore, the thread made of nanotubes, the thickness of a human hair is capable to hold down hundreds of kilos of cargo.

It is true that at present the maximum length of nanotubes is usually about a hundred microns – which is certainly too small for everyday use. However, the length of the nanotubes obtained in the laboratory is gradually increasing – now scientists have come close to the millimeter border. So there is every reason to hope that in the near future, scientists will learn how to grow a nanotube length in centimeters and even meters.

6.4 FULLERENES

The carbon atoms, evaporated from a heated graphite surface, connecting with each other, can form not only nanotube, but also other molecules, which are closed convex polyhedral, for example, in the form of a sphere or ellipsoid. In these molecules, the carbon atoms are located at the vertices of regular hexagons and pentagons, which make up the surface of a sphere or ellipsoid.

The molecules of the symmetrical and the most studied fullerene consisting of 60 carbon atoms (C_{60}), form a polyhedron consisting of 20 hexagons and 12 pentagons and resembles a soccer ball (Fig. 7). The diameter of the fullerene C_{60} is about 1 nm.

FIGURE 7   Schematic representation of the fullerene C_{60}.

6.5 CLASSIFICATION OF NANOTUBES

The main classification of nanotubes is conducted by the number of constituent layers.

Single-walled nanotubes – the simplest form of nanotubes. Most of them have a diameter of about 1 nm in length, which can be many thousands of times more. The structure of the nanotubes can be represented as a "wrap" a hexagonal network of graphite (graphene), which is based on hexagon with vertices located at the corners of the carbon atoms in a seamless cylinder. The upper ends of the tubes are closed by hemispherical caps, each layer is composed of hexa- and pentagons, reminiscent of the structure of half of a fullerene molecule. The distance d between adjacent carbon atoms in the nanotube is approximately equal to $d = 0.15$ nm.

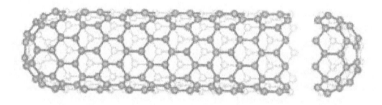

FIGURE 8   Graphical representation of single-walled nanotube.

Multi-walled nanotubes consist of several layers of graphene stacked in the shape of the tube. The distance between the layers is equal to 0.34 nm, which is the same as that between the layers in crystalline graphite.

Carbon nanotubes could find applications in numerous areas:

- additives in polymers;
- catalysts (autoelectronic emission for cathode ray lighting elements, planar panel of displays, gas discharge tubes in telecom networks);
- absorption and screening of electromagnetic waves;
- transformation of energy;
- anodes in lithium batteries;
- keeping of hydrogen;
- composites (filler or coating);
- nanosondes;
- sensors;
- strengthening of composites;
- supercapacitors.

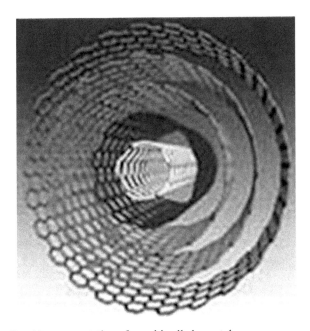

FIGURE 9   Graphic representation of a multiwalled nanotube.

More than a decade, carbon nanotubes, despite their impressive performance characteristics have been used, in most cases, for scientific research.

To date, the most developed production of nanotubes has Asia, the production capacity, which is 2–3 times higher than in North America and Europe combined. Is dominated by Japan, which is a leader in the production of MWNT. Manufacturing North America, mainly focused on the SWNT. Growing at an accelerated rate production in China and South Korea. In the coming years, China will surpass the level of production of the U.S. and Japan, and by 2013, a major supplier of all types of nanotubes, according to experts, could be South Korea.

6.6   CHIRALITY

Chirality – a set of two integer positive indices (n, m), which determines how the graphite plane folds and how many elementary cells of graphite at the same time fold to obtain the nanotube.

From the value of parameters (n, m) are distinguished direct (achiral) high-symmetry carbon nanotubes

- armchair $n = m$
- zigzag $m = 0$ or $n = 0$
- helical (chiral) nanotube

In Fig. 10a is shown a schematic image of the atomic structure of graphite plane – graphene, and shown how a nanotube can be obtained the from it. The nanotube is fold up with the vector connecting two atoms on a graphite sheet. The cylinder is obtained by folding this sheet so that were combined the beginning and end of the vector. That is, to obtain a carbon nanotube from a graphene sheet, it should turn so that the lattice vector $\bar{R}$ has a circumference of the nanotube in Fig. 10b. This vector can be expressed in terms of the basis vectors of the elementary cell graphene sheet $\vec{R} = n\vec{r_1} + m\vec{r_2}$. Vector $\bar{R}$, which is often referred to simply by a pair of indices (n, m), called the chiral vector. It is assumed that $n > m$. Each pair of numbers (n, m) represents the possible structure of the nanotube.

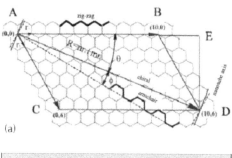

(a)

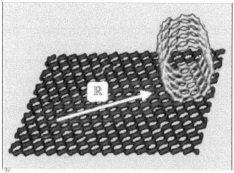

FIGURE 10   Schematic representation of the atomic structure of graphite plane.

In other words the chirality of the nanotubes (n, m) indicates the coordinates of the hexagon, which as a result of folding the plane has to be coincide with a hexagon, located at the beginning of coordinates (Fig. 11).

Many of the properties of nanotubes (for example, zonal structure or space group of symmetry) strongly depend on the value of the chiral vector. Chirality indicates what property has a nanotube – a semiconductor or metallicheskm. For example, a nanotube (10,10) in the elementary cell contains 40 atoms and is the type of metal, whereas the nanotube (10,9) has already in 1084 and is a semicon-ductor (Fig. 12).

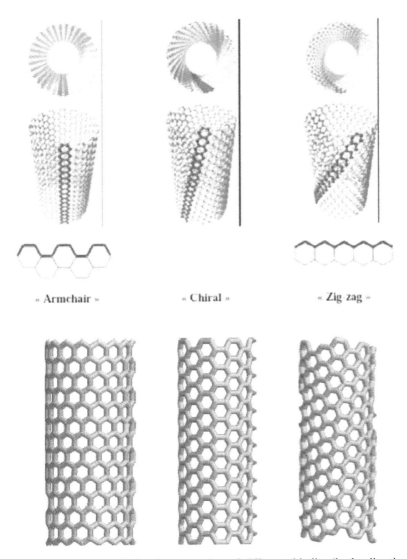

FIGURE 11   Single-walled carbon nanotubes of different chirality (in the direction of convolution). Left to right: the zigzag (16,0), armchair (8,8) and chiral (10,6) carbon nanotubes.

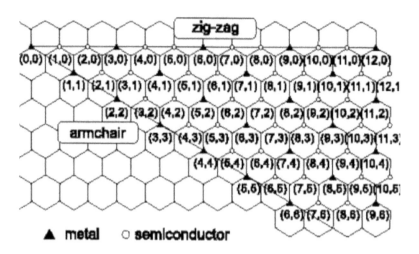

FIGURE 12 The scheme of indices (n,m) of lattice vector $\overline{R}$ tubes having semiconductor and metallic properties.

If the difference $n - m$ is divisible by 3, then these CNTs have metallic properties. Semimetals are all achiral tubes such as "chair". In other cases, the CNTs show semiconducting properties. Just type chair CNTs $(n = m)$ are strictly metal.

6.7 EFFECTIVE PARAMETERS

In addition to size effects that occur in micro and nano, we should note the following factors that determine the processes in low-dimensional systems:

- surface roughness (resistance to flow effects, interactions with the particles, etc.);
- dissolved gases (formation of bubbles, sticking to the surface, etc.);
- chemical surface properties (chemical reactions, etc.);
- hydrophobic – hydrophilic of surface;
- contaminants;
- heating due to uncontrollable processes;
- electrical properties of the surface (double layer, the change of surface and volume charge, charge transfer, etc.).

Viscous forces in the fluid can lead to large dispersion flow along the axis of motion. They have a significant impact, both on the scale of individual molecules, and the scale of microflows – near the borders of the liquid-solid (beyond a few molecular layers), during the motion on a complex and heterogeneous borders.

Influence of the effect of boundary regions on the particles and fluxes have been observed experimentally in the range of molecular thicknesses up to hundreds of nanometers. If the surface has a superhydrophobic properties, this range can extend to the micron thickness. *Molecular theory can predict the effect of hydrophobic surfaces in the system only up to tens of nanometers.*

Fluids, the flow of liquid or gas, have properties that vary continuously under the action of external forces. In the presence of fluid shear forces are small in magnitude, leads large changes in the relative position of the element of fluid. In contrast, changes in the relative positions of atoms in solids remain small under the action of any small external force. Termination of action of the external forces on the fluid does not necessarily lead to the restoration of its initial form.

6.7.1 CAPILLARY EFFECTS

To observe the capillary effects, you must open the nanotube, that is, remove the upper part – lids. Fortunately, this operation is quite simple.

The first study of capillary phenomena have shown that there is a relationship between the magnitude of surface tension and the possibility of its being drawn into the channel of the nanotube. It was found that the liquid penetrates into the channel of the nanotube, if its surface tension is not higher than 200 mN/m. Therefore, for the entry of any substance into the nanotube using solvents having a low surface tension. For example concentrated nitric acid with surface tension of 43 mN/m is used to inject certain metals into the channel of a nanotube. Then annealing is conducted at 4000°C for 4 h in an atmosphere of hydrogen, which leads to the recovery of the metal. Thus, the obtained nanotubes containing nickel, cobalt and iron.

Along with the metals carbon nanotubes can be filled with gaseous substances, such as hydrogen in molecular form. This ability is of great practical importance, since opening the ability to safely store hydrogen, which can be used as a clean fuel in internal combustion engines.

6.7.2 SPECIFIC ELECTRICAL RESISTANCE OF CARBON NANOTUBES

Due to small size of carbon nanotubes only in 1996 they succeeded to directly measure their electrical resistivity ρ. The results of direct measurements showed that the resistivity of the nanotubes can be varied within wide limits to 0.8 ohm/cm. The minimum value is lower than that of graphite. Most of the nanotubes have metallic conductivity, and the smaller shows properties of a semiconductor with a band gap of 0.1 to 0.3 eV.

The resistance of single-walled nanotube is independent of its length, because of this it is convenient to use for the connection of logic elements in microelectronic devices. The permissible current density in carbon nanotubes is much greater than in metallic wires of the same cross section, and one hundred times better achievement for superconductors.

The results of the study of emission properties of the material, where the nano-tubes were oriented perpendicular to the substrate, have been very interesting for practical use. Attained values of the emission current density of the order of 0.5 mA/mm^2. The value obtained is in good agreement with the Fowler–Nordheim expression.

The most effective and common way to control microflow substances are *electrokinetic* and *hydraulic*. At the same time the most technologically advanced and automated considered electrokinetic.

Charges transfer in mixtures occurs as a result of the directed motion of charge carriers – ions. There are different mechanisms of such transfer, but usually are *convection, migration and diffusion.*

Convection is called mass transfer the macroscopic flow. *Migration* – the movement of charged particles by electrostatic fields. The velocity of the ions depends on field strength. In microfluidics a special role is played *electrokinetic processes* that can be divided into four types: *electro-osmosis, electrophoresis, streaming potential and sedimentation potential.* These processes can be qualitatively described as follows:

1. *Electro-osmosis* – the movement of the fluid volume in response to the applied electric field in the channel of the electrical double layers on its wetted surfaces.
2. *Electrophoresis* – the forced motion of charged particles or molecules, in mixture with the acting electric field.
3. *Streamy potential* – the electric potential, which is distributed through a channel with charged walls, in the case when the fluid moves under the action of pressure forces. Joule electric current associated with the effect of charge transfer is flowing stream.
4. *The potential of sedimentation* – an electric potential is created when charged particles are in motion relative to a constant fluid. The driving force for this effect – usually gravity.

In general, for the microchannel cross-section S amount of introduced probe (when entering electrokinetic method) depends on the applied voltage U, time t during which the received power, and mobility of the sample components μ:

$$Q = \frac{\mu S U t}{L} \cdot c$$

where: c – probe concentration in the mixture, L – the channel length.

Amount of injected substance is determined by the electrophoretic and total electro-osmotic mobilities μ.

In the hydrodynamic mode of entry by the pressure difference in the channel or capillary of circular cross section, the volume of injected probe V_c:

$$V_c = \frac{4}{128} \cdot \frac{\Delta p \pi dt}{\eta L}$$

where: Δp – pressure differential, d – diameter of the channel, η – viscosity.

In the simulation of processes in micron-sized systems the following basic principles are fundamental:

1. hypothesis of *laminar* flow (sometimes is taken for granted when it comes to microfluidics);
2. continuum hypothesis (detection limits of applicability);
3. laws of formation of the velocity profile, mass transfer, the distribution of electric and thermal fields;
4. boundary conditions associated with the geometry of structural elements (walls of channels, mixers zone flows, etc.).

Since we consider the physical and chemical transport processes of matter and energy, mathematical models, most of them have the form of systems of differential equations of second order partial derivatives. Methods for solving such equations are analytical (Fourier and its modifications, such as the method of Greenberg, Galerkin, in some cases, the method of d'Alembert and the Green's functions, the Laplace operator method, etc.) or numerical (explicit or, more effectively, implicit finite difference schemes) – traditional. The development involves, basically, numerical methods and follows the path of saving computing resources, and increasing the speed of modern computers.

Laminar flow – a condition in which the particle velocity in the liquid flow is not a random function of time. The small size of the microchannels (typical dimensions of 5 to 300 microns) and low surface roughness create good conditions for the establishment of laminar flow. Traditionally, the image of the nature of the flow gives the dimensionless characteristic numbers: the Reynolds number and Darcy's friction factor.

In the motion of fluids in channels the turbulent regime is rarely achieved. At the same time, the movement of gases is usually turbulent.

Although the liquid – are quantized in the length scale of intermolecular distances (about 0.3 nm to 3 nm in liquids and for gases), they are assumed to be continuous in most cases, microfluidics. Continuum hypothesis (continuity, continuum) suggests that the macroscopic properties of fluids consisting of molecules, the same as if the fluid was completely continuous (structurally homogeneous). Physical characteristics: mass, momentum and energy associated with the volume of fluid containing a sufficiently large number of molecules must be taken as the sum of all the relevant characteristics of the molecules.

Continuum hypothesis leads to the concept of fluid particles. In contrast to the ideal of a point particle in ordinary mechanics, in fluid mechanics, particle in the fluid has a finite size.

At the atomic scale we would see large fluctuations due to the molecular structure of fluids, but if we the increase the sample size, we reach a level where it is possible to obtain stable measurements. This volume of probe must contain a sufficiently large number of molecules to obtain reliable reproducible signal with small statistical fluctuations. For example, if we determine the required volume as a cube with sides of 10 nm, this volume contains some of the molecules and determines the level of fluctuations of the order of 0.5%.

The most important position in need of verification is to analyze the admissibility of mass transfer on the basis of the continuum model that can be used instead of the concentration dependence of the statistical analysis of the ensemble of individual particles. The position of the continuum model is considered as a necessary condition for microfluidics.

The applicability of the hypothesis is based on comparison of free path length of a particle λ in a liquid with a characteristic geometric size d. The ratio of these lengths – the Knudsen number: $Kn = \lambda / d$. Based on estimates of the Knudsen number defined two important statements:

1. $Kn < 10^{-3}$ – justified hypothesis of a continuous medium; and
2. $Kn < 10^{-1}$ – allowed the use of adhesion of particles to the solid walls of the channel.

Wording of the last condition can also be varied: both in form $U = 0$ and in a more complex form, associated with shear stresses. The calculation of λ can be carried out as

$$\lambda \approx \sqrt[3]{\bar{V} / Na} ,$$

where: $\bar{V}$ – molar volume, Na – Avogadro's number.

Under certain geometrical approximations of the particles of substance free path length can be calculated as $\lambda \approx 1 / \left(\sqrt{2} \pi r_S^2 Na \right)$, if used instead r_S Stokes radius, as a consequence of the spherical approximation of the particle. On the other hand, for a rigid model of the molecule r_S should be replaced by the characteristic size of the particles R_g – the radius of inertia, calculated as $R_g = n_i \cdot \delta_l / \sqrt{6}$. Here δ_l – the length of a fragment of the chain (link), n_i – the number of links.

Of course, the continuum hypothesis is not acceptable when the system under consideration is close to the molecular scale. This happens in nanoliquid, such as liquid transport through nanopores in cell membranes or artificially made nanochannels.

In contrast to the continuum hypothesis, the essence of modeling the molecular dynamics method is as follows. We consider a large ensemble of particles, which simulate atoms or molecules, i.e., all atoms are material points. It is believed that the particles interact with each other and, moreover, may be subject to external influence. Interatomic forces are represented in the form of the classical potential force (the gradient of the potential energy of the system).

The interaction between atoms is described by means of van der Waals forces (intermolecular forces), mathematically expressed by the Lennard–Jones potential:

$$V(r) = \frac{Ae^{-\sigma r}}{r} - \frac{C_6}{r^6}$$

where: A and C_6 – some coefficients depending on the structure of the atom or molecule, σ – the smallest possible distance between the molecules.

In the case of two isolated molecules at a distance of r_0 the interaction force is zero, that is, the repulsive forces balance attractive forces. When $r > r_0$ the resultant force is the force of gravity, which increases in magnitude, reaching a maximum at $r = r_m$ and then decreases. When $r < r_0$ – a repulsive force. Molecule in the field of these forces has potential energy $V(r)$, which is connected with the force of $f(r)$ by the differential equation

$$dV = -f(r)dr$$

At the point $r = r_0$, $f(r) = 0$, $V(r)$ reaches an extremum (minimum).

The chart of such a potential is shown below in Fig. 13. The upper (positive) half-axis r corresponds to the repulsion of the molecules, the lower (negative) half-plane – their attraction. We can say simply: at short distances the molecules mainly repel each, on the long – draw each other. Based on this hypothesis, and now an obvious fact, the van der Waals received his equation of state for real gases.

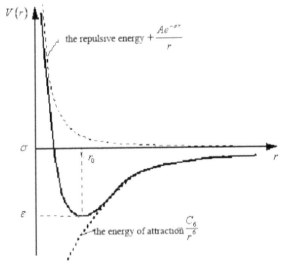

FIGURE 13 The chart of the potential energy of intermolecular interaction.

The exponential summand in the expression for the potential describing the repulsion of the molecules at small distances, often approximated as

$$\frac{Ae^{-\sigma r}}{r} \approx \frac{C_{12}}{r^{12}}$$

In this case we obtain the Lennard–Jones potential:

$$V(r) = \frac{C_{12}}{r^{12}} - \frac{C_6}{r^6} \tag{1}$$

The interaction between carbon atoms is described by the potential

$$V_{CC}(r) = K(r-b)^2,$$

where: K – constant tension (compression) connection, $b = 1,4A$ – the equilibrium length of connection, r – current length of the connection.

The interaction between the carbon atom and hydrogen molecule is described by the Lennard–Jones

$$V(r) = 4\varepsilon\left[\left(\frac{\sigma}{r}\right)^{12} - \left(\frac{\sigma}{r}\right)^6\right]$$

For all particles (Fig. 14) the equations of motion are written:

$$m\frac{d^2\overline{r_i}}{dt^2} = \overline{F}_{T-H_2}(\overline{r_i}) + \sum_{j\neq i}\overline{F}_{H_2-H_2}(\overline{r_i} - \overline{r_j}),$$

where: $\overline{F}_{T-H_2}(\overline{r})$ – force, acting by the CNT, $\overline{F}_{H_2-H_2}(\overline{r_i} - \overline{r_j})$ – force acting on the i-th molecule from the j-th molecule.

The coordinates of the molecules are distributed regularly in the space, the velocity of the molecules are distributed according to the Maxwell equilibrium distribution function according to the temperature of the system:

$$f(u,v,w) = \frac{\beta^3}{\pi^{3/2}}\exp\left(-\beta^2\left(u^2 + v^2 + w^2\right)\right) \quad \beta = \frac{1}{\sqrt{2RT}}$$

The macroscopic flow parameters are calculated from the distribution of positions and velocities of the molecules:

$$\overline{V} = \left\langle \overline{v_i} \right\rangle = \frac{1}{n}\sum_i \overline{v_i},$$

$$\frac{3}{2}RT = \frac{1}{2}\left\langle \left|\overline{v_i'}\right|^2 \right\rangle, \quad \overline{v_i'} = \overline{v_i} - \overline{V},$$

$$\rho = \frac{m}{V_0},$$

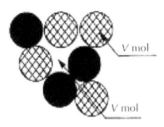

FIGURE 15   Position of particles.

The resulting system of equations is solved numerically. However, the molecular dynamics method has limitations of applicability:

1. the de Broglie wavelength h/mv (where h – Planck's constant, m – the mass of the particle, v – velocity) of the atom must be much smaller than the interatomic distance;
2. Classical molecular dynamics cannot be applied for modeling systems consisting of light atoms such as helium or hydrogen;
3. at low temperatures, quantum effects become decisive for the consideration of such systems must use quantum chemical methods;
4. necessary that the time at which we consider the behavior of the system were more than the relaxation time of the physical quantities.

In 1873, Van der Waals proposed an equation of state is qualitatively good description of liquid and gaseous systems. It is for one mole (one mole) is:

$$\left(p + \frac{a}{v^2}\right)(v - b) = RT \tag{2}$$

Note that at $p > \dfrac{a}{v^2}$ and $v \gg b$ this equation becomes the equation of state of ideal gas

$$pv = RT \tag{3}$$

Van der Waals equation can be obtained from the Clapeyron equation of Mendeleev by an amendment to the magnitude of the pressure a/v^2 and the amendment

b to the volume, both constant a and b independent of T and v but dependent on the nature of the gas.

The amendment b takes into account:

1. the volume occupied by the molecules of real gas (in an ideal gas molecules are taken as material points, not occupying any volume);
2. so-called "dead space", in which can not penetrate the molecules of real gas during motion, i.e. volume of gaps between the molecules in their dense packing.

Thus, $b = v_{мол.} + v_{заз.}$ (Fig. 16). The amendment a/v^2 takes into account the interaction force between the molecules of real gases. It is the internal pressure, which is determined from the following simple considerations. Two adjacent elements of the gas will react with a force proportional to the product of the quantities of substances enclosed in these elementary volumes.

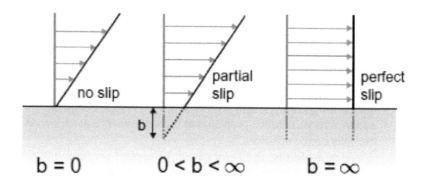

FIGURE 16   Location scheme of molecules in a real gas.

Therefore, the internal pressure P_{BH} is proportional to the square of the concentration n:

$$P_{BH} \sim n^2 \sim \rho^2 \sim \frac{1}{v^2},$$

where: ρ – the gas density.

Thus, the total pressure consists of internal and external pressures:

$$p + P_{BH} = p + \frac{a}{v^2}$$

Equation (3) is the most common for an ideal gas. Under normal physical conditions and from Eq. (3) we obtain:

$$R = \frac{R\mu}{\mu} = \frac{8314}{\mu}$$

Knowing $R\mu$ we can find the gas constant for any gas with the help of the value of its molecular mass μ (Table 1):

TABLE 1 The molecular weight of some gases.

Gas	N	Ar	H_2	O_2	CO	CO_2	Ammonia	Air
μ	28	40	2	32	28	44	17	29

For gas mixture with mass M state equation has the form:

$$pv = MR_{CM}T = \frac{8314MT}{\mu_{CM}} \tag{4}$$

where: R_{CM} – gas constant of the mixture.

The gas mixture can be given by the mass proportions g_i, voluminous r_i or mole fractions n_i respectively, which are defined as the ratio of mass m_i, volume v_i or number of moles N_i of i gas to total mass M, volume v or number of moles N of gas mixture. Mass fraction of component is $g_i = \frac{m_i}{M}$, where $i = 1, n$. It is obvious that $M = \sum_{i=1}^{n} m_i$ and $\sum_{i=1}^{n} g_i = 1$. The volume fraction is $r_i = \frac{v_i}{v_{CM}}$, where v_i – partial volume of component mixtures.

Similarly, we have $\sum_{i=1}^{n} v_i = v_{CM}, \sum_{i=1}^{n} r_i = 1$.

Depending on specificity of tasks the gas constant of the mixture may be determined as follows:

$$R_{CM} = \sum_{i=1}^{n} g_i R_i; \quad R_{CM} = \frac{1}{\sum_{i=1}^{n} r_i R_i^{-1}}$$

If we know the gas constant R_{CM} , the seeming molecular weight of the mixture is equal to

$$\mu_{CM} = \frac{8314}{R_{CM}} = \frac{8314}{\sum\limits_{i=1}^{n} g_i R_i} = 8314 \sum\limits_{i=1}^{n} r_i R_i^{-1}$$

The pressure of the gas mixture p is equal to the sum of the partial pressures of individual components in the mixture p_i:

$$p = \sum\limits_{i=1}^{n} p_i \tag{5}$$

Partial pressure p_i – pressure that has gas, if it is one at the same temperature fills the whole volume of the mixture ($p_i v_{CM} = RT$).

With various methods of setting the gas mixture partial pressures

$$p_i = pr_i; \; p_i = \frac{p g_i \mu_{CM}}{\mu_i} \tag{6}$$

From the Eq. (6) we see that for the calculation of the partial pressures p_i necessary to know the pressure of the gas mixture, the volume or mass fraction i of the gas component, as well as the molecular weight of the gas mixture μ and the molecular weight of i of gas μ_i.

The relationship between mass and volume fractions are written as follows:

$$g_i = \frac{m_i}{m_{CM}} = \frac{\rho_i v_i}{\rho_{CM} v_{CM}} = \frac{R_{CM}}{R_i} r_i = \frac{\mu_i}{\mu_{CM}} r_i$$

We rewrite Eq. (2) as

$$v^3 - \left(b + \frac{RT}{p}\right)v^2 + \frac{a}{p}v - \frac{ab}{p} = 0 \tag{7}$$

When $p = p_k$ and $T = T_k$, where p_k and T_k – critical pressure and temperature, all three roots of Eq. (7) are equal to the critical volume v_k

$$v^3 - \left(b + \frac{RT_k}{p_k}\right)v^2 + \frac{a}{p_k}v - \frac{ab}{p_k} = 0 \tag{8}$$

Because the $v_1 = v_2 = v_3 = v_k$, then Eq. (8) must be identical to the equation

$$(v - v_1)(v - v_2)(v - v_3) = (v - v_k)^3 = v^3 - 3v^2 v_k + 3vv_k^2 - v_k^3 = 0 \qquad (9)$$

Comparing the coefficients at the equal powers of v in both equations leads to the equalities

$$b + \frac{RT_k}{p_k} = 3v_k; \quad \frac{a}{p_k} = 3v_k^2; \quad \frac{ab}{p_k} = v_k^3 \qquad (10)$$

Hence

$$a = 3v_k^2 p_k; \quad b = \frac{v_k}{3} \qquad (11)$$

Considering Eq. (10) as equations for the unknowns p_k, v_k, T_k, we obtain

$$p_k = \frac{a}{27b^2}; \quad v_k = 3b; \quad T_k = \frac{8a}{27bR} \qquad (12)$$

From Eqs. (10) and (11) or (12) we can find the relation

$$\frac{RT_k}{p_k v_k} = \frac{8}{3} \qquad (13)$$

Instead of the variables p, v, T let's introduce the relationship of these variables to their critical values (leaden dimensionless parameters)

$$\pi = \frac{p}{p_k}; \quad \omega = \frac{v}{V_k}; \quad \tau = \frac{T}{T_k} \qquad (14)$$

Substituting Eqs. (12) and (14) in Eq. (7) and using Eq. (13), we obtain

$$\left(\pi p_k + \frac{3v_k^2 p_k}{\omega^2 v_k^2} \right)\left(\omega v_k - \frac{v_k}{3} \right) = RT_k \tau,$$

$$\left(\pi + \frac{3}{\omega^2} \right)(3\omega - 1) = 3 \frac{RT_k}{p_k v_k} \tau,$$

$$\left(\pi + \frac{3}{\omega^2}\right)(3\omega - 1) = 8\tau \tag{15}$$

6.8 SLIPPAGE OF THE FLUID PARTICLES NEAR THE WALL

Features of the simulation results of Poiseuille flow in the microtubules, when the molecules at the solid wall and the wall atoms at finite temperature of the wall make chaotic motion lies in the fact that in the intermediate range of Knudsen numbers there is slippage of the fluid particles near the wall.

Researchers describe three possible cases:

1. the liquid can be stable (no slippage),
2. slides relative to the wall (with slippage flow),
3. the flow profile is realized; this is when the friction of the wall is completely absent (complete slippage).

In the framework of classical continuum fluid dynamics, according to the Navier boundary condition the velocity slip is proportional to fluid velocity gradient at the wall:

$$v\big|_{y=0} = L_S \, dv / dy\big|_{y=0} \tag{16}$$

Here and in Fig. 17, L_S represents the "slip length" and has a dimension of length.

Because of the slippage, the average velocity in the channel $\langle v_{pdf} \rangle$ increases.

In a rectangular channel (of width >> height h and viscosity of the fluid η) due to an applied pressure gradient – dp/dx the authors of that article obtained:

$$\langle v_{pdf} \rangle = \frac{h^2}{12\eta}\left(-\frac{dp}{dx}\right)\left(1 + \frac{6L_S}{h}\right) \tag{17}$$

The results of molecular dynamics simulation for nanosystems with liquid, with characteristic dimensions of the order of the size of the fluid particles, show that a large slippage lengths (of the order of microns) should occur in the carbon nanotubes of nanometer diameter and, consequently, can increase the flow rate by three orders of ($6L_S/h > 1000$). Thus, the flow with slippage is becoming more and more important for hydrodynamic systems of small size.

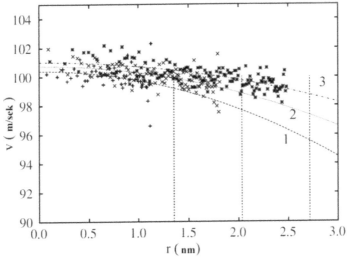

FIGURE 17   Three cases of slip flow past a stationary surface. The slip length b is indicated.

The results of molecular dynamics simulation of unsteady flow of mixtures of water – water vapor, water and nitrogen in a carbon nanotube are reported by many researchers.

Carbon nanotubes have been considered "zigzag" with chiral vectors (20, 20), (30, 30) and (40, 40), corresponding to pipe diameters of 2,712, 4,068 and 5,424 nm, respectively.

Knowing the value of the flow rate and the system pressure, which varies in the range of 600–800 bars, are high enough to ensure complete filling of the tubes. This pressure can be achieved by the total number of water molecules 736, 904, and 1694, respectively.

The effects of slippage of various liquids on the surface of the nanotube were studied in detail.

The length of slip, can be calculated using the current flow velocity profiles of liquid, shown in Fig. 18, were 11, 13, and 15 nm for the pipes of 2,712, 4,068 and 5,424 nm respectively. The dotted line marked by theoretical modeling data. The vertical lines indicate the position of the surface of carbon nanotubes.

It was found out that as the diameter decreases, the speed of slippage of particles on the wall of nanotube also decreases. The authors attribute this to the increase of the surface friction.

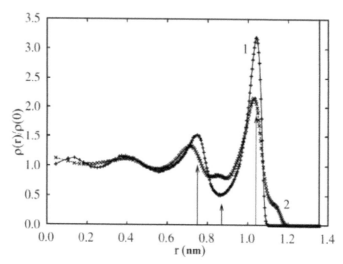

FIGURE 18   Time average streaming velocity profiles of water in a carbon nanotubes of different diameter: 2.712 nm, curve 1; 4.068 nm, curve 2; 5.424 nm, curve 3.

Experiments with various pressure drops in nanotubes demonstrated slippage of fluid in micro- and nanosystems. The most remarkable were the two recent experiments, which were conducted to improve the flow characteristics of carbon nanotubes with the diameters of 2 and 7 nm respectively. In the membranes in which the carbon nanotubes were arranged in parallel, there was a slip of the liquid in the micrometer range. This led to a significant increase in flow rate – up to three – four orders of magnitude.

In the experiments for the water moving in microchannels on smooth hydrophobic surfaces, there are slidings at about 20 nm. If the wall of the channel is not smooth but twisty or rough, and at the same time, hydrophobic, such a structure would lead to an accumulation of air in the cavities and become superhydrophobic (with contact angle greater than 160°). It is believed that this leads to creation of contiguous areas with high and low slippage, which can be described as "effective slip length". This effective length of the slip occurring on the rough surface can be several tens of microns, which was indeed experimentally confirmed by many researchers.

Another possible problem is filling of the hydrophobic systems with liquid. Filling of micron size hydrophobic capillaries is not a big problem, because pressure of less than 1 atm is sufficient. Capillary pressure, however, is inversely proportional to the diameter of the channel, and filling for nanochannels can be very difficult.

6.9　THE DENSITY OF THE LIQUID LAYER NEAR A WALL OF CARBON NANOTUBE

Researchers showed radial density profiles of oxygen averaged in time and hydrogen atoms in the "zigzag" carbon nanotube with chiral vector (20, 20) and a radius $R = 1,356$ nm (Fig. 19). The distribution of molecules in the area near the wall of the carbon nanotube indicated a high density layer near the wall of the carbon nanotube. Such a pattern indicates the presence of structural heterogeneity of the liquid in the flow of the nanotube. In Fig. (19) $\rho^*(r) = \rho(r)/\rho(0)$, 2,712 nm diameter pipe is completely filled with water molecules at $300^0 K$. The overall density $r(0) = 1000 \kappa s / M^3$. The arrows denote the location of distinguishable layers of the water molecules and the vertical line the position of the CNT wall.

The distribution of molecules in the region of $0.95 \leq r \leq R$ nm indicates a high density layer near the wall carbon nanotube. Such a pattern indicates the presence of structural heterogeneity of the liquid in the flow of the nanotube.

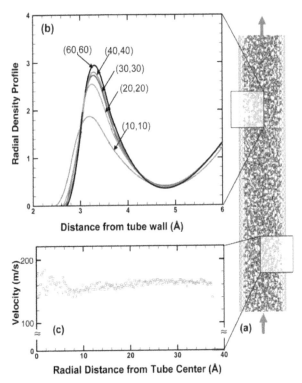

FIGURE 19　Radial density profiles of oxygen (curve 1) and hydrogen (curve 2) atoms for averaged in the time interval.

Many researchers obtained similar results for the flow of the water in the carbon nanotubes using molecular dynamic simulations.

Figure 20 shows the scheme of the initial structure and movement of water molecules in the model carbon nanotube (a), the radial density profile of water molecules inside nanotubes with different radii (b) and the velocity profile of water molecules in a nanotube chirality (60,60) (c).

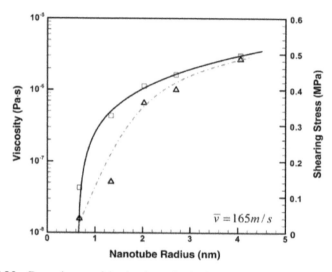

FIGURE 20 Dependences of the density and velocity profiles of water molecules inside the nanotubes with different radii (Xi Chen et al., 2008).
Schematic of the initial structure and transport of water molecules in a model CNT.
The radial density profile (RDP) of water molecules inside CNTs with different radii.
The representative radial velocity profile (RVP) of water molecules inside a (60,60) nanotube.

It is mentioned that the area available for molecules of the liquid is less than the area bounded by a solid wall, primarily due to the van der Waals interactions.

6.10   THE EFFECTIVE VISCOSITY OF THE LIQUID IN A NANOTUBE

Scientists showed a significant increase in the effective viscosity of the fluid in the nanovolumes compared to its macroscopic value. It was shown that the effective viscosity of the liquid in a nanotube depends on the diameter of the nanotube.

The effective viscosity of the liquid in a nanotube is defined as follows.

Let's establish a conformity nanotubes filled with liquid, possibly containing crystallites with the same size tube filled with liquid, considered as a homogeneous medium (i.e., without considering the crystallite structure), in which Poiseuille flow is realized at the same pressure drop and consumption rate. The viscosity of a homogeneous fluid, which ensures the coincidence of these parameters, will be called the effective viscosity of the flow in the nanotube.

Researchers showed that while flowing in the narrow channels of width less than 2 nm, water behaves like a viscous liquid. In the vertical direction water behaves as a rigid body, and in a horizontal direction it maintains its fluidity.

It is known that at large distances the van der Waals interaction has a magnetting tendency and occurs between any molecules like polar as well as nonpolar. At small distances it is compensated by repulsion of electron shells. Van der Waals interaction decreases rapidly with distance. Mutual convergence of the particles under the influence magnetting forces continues until these forces are balanced with the increasing forces of repulsion.

Knowing the deceleration of the flow (Fig. 20) of water a, the authors of the cited article calculates the effective shear stress between the wall of the pipe length l and water molecules by the formula

$$\tau = Nma / (2\pi Rl) \tag{18}$$

Here, the shear stress is a function of tube radius and flow velocity $\bar{v}$, m – mass of water molecules, the average speed is related to volumetric flow $\bar{v} = Q/(\pi R^2)$.

Denoting n_0 the density of water molecules number, we can calculate the shear stress in the form of:

$$\tau\big|_{r=R} = n_0 mRa / 2 \tag{19}$$

Figure 21 shows the results of calculations of the authors of the cited article – influence of the size of the tube R_0 on the effective viscosity (squares) and shear stress τ (triangles), when the flow rate is approximately 165 m/sec.

According to classical mechanics of liquid flow at different pressure drops Δp along the tube length l is given by Poiseuille formula

$$Q_P = \frac{\pi R^4 \Delta p}{8\eta l} \qquad Q_P = \pi R^2 \bar{v} \tag{20}$$

Therefore,

$$\tau = \frac{\Delta p R}{2l} \tag{21}$$

and the effective viscosity of the fluid can be estimated as $\eta = \tau \cdot R / (4\bar{v})$.

The change in the value of shear stress directly causes the dependence of the effective viscosity of the fluid from the pipe size and flow rate. In this case the effective viscosity of the transported fluid can be determined from Eqs. (20) and (21) as

$$\eta = \frac{\tau \cdot R}{4\bar{v}} \tag{22}$$

With increasing radius τ according to Eq. (21) increases.

Calculations showed that the magnitude of the shear stress τ is relatively small in the range of pipe sizes considered. This indicates that the surface of carbon nanotubes is very smooth and the water molecules can easily slide through it.

In fact, shear stress is primarily due to van der Waals interaction between the solid wall and the water molecules. It is noted that the characteristic distance between the near-wall layer of fluid and pipe wall depends on the equilibrium distance between atoms O and C, the distribution of the atoms of the solid wall and bend of the pipe.

From Fig. 21 we can see, that the effective viscosity η increases by two orders of magnitude when R_0 changes from 0.67 to 5.4 nm. The value of the calculated viscosity of water in the tube (10,10) is 8.5×10^{-8} Pa/s, roughly four orders of magnitude lower than the viscosity of a large mass of water.

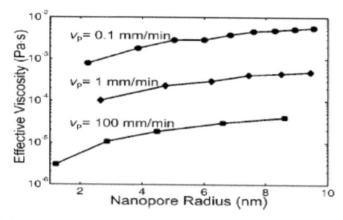

FIGURE 21   Size effect of shearing stress (triangle) and viscosity (square), with $\bar{v}=165$ M/C.

According to Eqs. (20)–(22), the effective viscosity can be calculated as $\eta = \frac{\pi R^4 \Delta p}{8QL}$. The results of calculations (Xi Chen et al., 2008) are shown in Fig. 22.

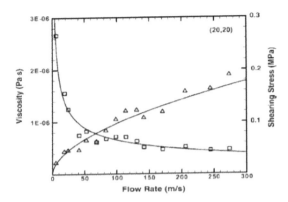

FIGURE 22   Effective viscosity as a function of the nanopore radius and the loading rate.

The dependence of the shear stress on the flow rate is illustrated in Fig. 23. From the example of the tube (20,20) it's clear that τ increases with v. The growth rate slowed down at higher values $\bar{v}$.

At high speeds $\bar{v}$, while water molecules are moving along the surface of the pipe, the liquid molecules do not have enough time to fully adjust their positions to minimize the free energy of the system. Therefore, the distance between adjacent carbon atoms and water molecules may be less than the equilibrium van der Waals distances. This leads to an increase in van der Waals forces of repulsion and leads to higher shear stress.

Scientists showed that, even though the equation for viscosity is based on the theory of the continuum, it can be extended to a complex flow to determine the effective viscosity of the nanotube.

Figure 23 also shows a dependence η on $\bar{v}$ on the inside of the nanotube (20,20). It is seen that η decreases sharply with increasing flow rate and begins to asymptotically approaches a definite value when $\bar{v} > 150$ m/sec. For the current pipe size and flow rate ranges $\eta \sim 1/\sqrt{v}$, this trend is the result of addiction $\tau - \bar{v}$, contained the same Fig. 23. According to Fig. 22 high-speed effects are negligible.

One can easily see that the dependence of viscosity on the size and speed is consistent qualitatively with the results of molecular dynamic simulations. In all studied cases, the viscosity is much smaller than its macroscopic analogy. When the radius of the pores varies from about 1 nm to 10 nm, the value of the effective viscosity increases by an order of magnitude respectively. A more significant change occurs when the speed increases from 0.1 mm/min up to 100 mm/min. This results in a change in the value of viscosity η, respectively, by 3–4 orders. The discrepancy between simulation and test data can be associated with differences in the structure of the nanopores and liquid phase.

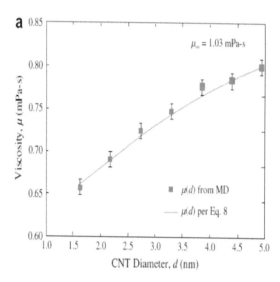

FIGURE 23   Flow rate effect of shearing stress (triangle) and viscosity (square), with R_0 =1.336 HM.

Figure 24 shows the viscosity dependence of water, calculated by the method of DM, the diameter of the CNT. The viscosity of water, as shown in the figure, increases monotonically with increasing diameter of the CNT.

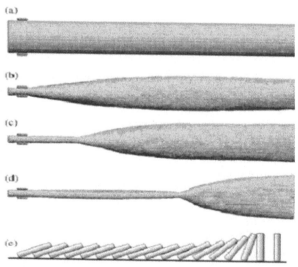

FIGURE 24   Variation of water viscosity with CNT diameter.

6.11 ENERGY RELEASE DUE TO THE COLLAPSE OF THE NANOTUBE

Scientists theoretically predicted the existence of a "domino effect" in single-walled carbon nanotube.

Squashing it at one end by two rigid moving to each other by narrow graphene planes (about 0.8 nm in width and 8.5 nm in length), one can observe it rapidly (at a rate exceeding 1 km/s) release its stored energy by collapsing along its length like a row of dominoes. The effect resembles a tube of toothpaste squeezing itself (Fig. 25).

The structure of a single-walled carbon nanotube has two possible stable states: circular or collapsed. Chang realized that for nanotubes wider than 3.5 nanometers, the circular state stores more potential energy than the collapsed state as a result of van der Waal's forces. He performed molecular dynamics simulations to find out what would happen if one end of a nanotube was rapidly collapsed by clamping it between two graphene bars.

This phenomenon occurs with the release of energy, and thus allows for the first time to talk about carbon nanotubes as energy sources. This effect can also be used as a accelerator of molecules.

The tube does not collapse over its entire length at the same time, but sequentially, one after the other carbon ring, starting from the end, which is tightened (Fig. 25). It happens just like a domino collapses, arranged in a row (this is known as the "domino effect"). Only here the role of bone dominoes is performed by a ring of carbon atoms forming the nanotube, and the nature of this phenomenon is quite different.

Recent studies have shown that nanotubes with diameters ranging from 2 to 6 nm, there are two stable equilibrium states – cylindrical (tube no collapses) and compressed (imploded tube) – with difference values of potential energy, the difference between which and can be used as an energy source.

Researchers found that switching between these two states with the subsequent release of energy occurs in the form of arising domino effect wave. The scientists have shown that such switching is not carried out in carbon nanotubes with diameters of 2 nm and not more, as evident from previous studies, but with a little more, starting from 3.5 nm.

A theoretical study of the "domino effect" was conducted using a special method of classical molecular dynamics, in which the interaction between carbon atoms was described by van der Waals forces (intermolecular forces), mathematically expressed by the Lennard–Jones potential.

The main reason for the observed effect, in author's opinion, is the competition of the potential energy of the van der Waals interactions, which "collapses" the nanotube with the energy of elastic deformation, which seeks to preserve the geometry of carbon atoms, which eventually leads to a bistable (collapsed and no collapsed) configuration of the carbon nanotube.

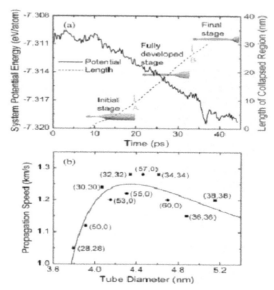

FIGURE 25 "Domino effect" in a carbon nanotube. (a) The initial form of carbon nanotubes – cylindrical. (b) One end of the tube is squeezed. (c), (d) Propagation of domino waves – the configuration of the nanotube 15 and 25 picoseconds after the compression of its end. (e) Schematic illustration of the "domino effect" under the influence of gravity.

For small diameter tubes the dominant is the energy of elastic deformation, the cylindrical shape of such a nanotube is stable. For nanotubes with sufficiently large diameter the van der Waals interaction energy is dominant. This means more stability and less compressed nanotube stability, or, as physicists say, metastability (i.e., apparent stability) of its cylindrical shape.

Thus, "domino effect" wave can be produced in a carbon nanotube with a relatively large diameter (more than 3.5 nm, as the author's calculations), because only in such a system the potential energy of collapsing structures may be less than the potential energy of the "normal" nanotube. In other words, the cylindrical and collapsing structure of a nanotubes with large diameters are, respectively, its metastable and stable states.

Change of the potential energy of a carbon nanotube with a propagating "domino effect" wave in it with time is represented as a graph in Fig. 26a.

This chart shows three sections of features in the change of potential energy. The first (from 0 ps to 10 ps) are composed of elastic strain energy, which appears due to changes in the curvature of the walls of the nanotubes in the process of collapsing, the energy change of van der Waals interactions occurring between the opposite walls of the nanotubes, as well as the interaction between the tube walls and graphene planes, compressing her end.

The second region (from 10 ps to 35 ps) corresponds to the "domino effect" – "domino effect" wave spreads along the surface of the carbon nanotube. Energetically, it looks like this: at every moment, when carbon ring collapses, some of the potential energy of the van der Waals is converted to kinetic energy (the rest to the energy of elastic deformation), which is kind of stimulant to support and "falling domino" – following the collapse of the rings, which form the nanotube, with each coagulated ring reducing the total potential energy of the system.

Finally, the third segment (from 35 ps to 45 ps) corresponds to the ended "domino" process – carbon tube collapsed completely. We emphasize that the nanotube, which collapsed (as seen from the Fig. 26a), has less potential energy than it was before beginning of the "domino effect".

In other words, the spread "domino effect" waves – a process that goes with the release of energy: about 0.01 eV per atom of carbon. This is certainly not comparable in any way with the degree of energy yield in nuclear reactions, but the fact of power generation carbon nanotube is obvious.

Later scientists analyzed the kinematic characteristics of the process – what the rate of propagation of the wave of destruction or collapse of carbon nanotubes and their characteristics is it determined?

Calculations show that the wave of dominoes in a tube diameter of 4–5 nm is about 1 km/s (as seen from the Fig. 26b) and depends on its geometry – the diameter and chirality in a non-linear manner. The maximum effect should be observed in the tube with a diameter slightly less than 4.5 nm – it will be carbon rings to collapse at a speed of 1.28 km/sec. The theoretical dependence obtained by the author, shows the blue solid line. And now an example of how energy is released in such a system with a "domino effect" can be used in nanodevices. The author offers an original way to use – "nanogun" (Fig. 27a). Imagine that at our disposal there is a carbon nanotube with chirality of (55.0) and corresponding to observation of the dominoes diameter.

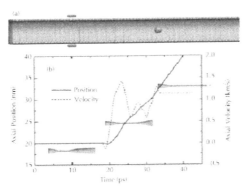

FIGURE 26 (a) Time dependences of potential energy and the length of the collapsed nanotube, (b) The velocity of propagation of the domino wave for carbon nanotubes with different diameters and chirality.

Put a C_{60} fullerene inside a nanotube. A little imagination, and a carbon nano-tube can be considered as the gun trunk, and the C_{60} molecule – as its shell.

The molecule located inside will extruded from it into the other, open end under the influence of squeezing nanotube (see Fig. 5.6).

The question is, what is the speed of the "core"? Chang estimated that, de-pending on the initial position of the fullerene molecule when leaving a nanotube, it can reach speeds close to the velocity of "domino effect" waves – about 1 km/s (Fig. 27b). Interestingly, this speed is reached by the "core" for just 2 picoseconds and at a distance of 1 nm. It is easy to calculate that the observed acceleration is of great value $0,5 \cdot 10^{15}{}_M / c^2$. For comparison, the speed of bullets in an AK-47 is 1.5 times lower than the rate of fullerene emitted from a gun.

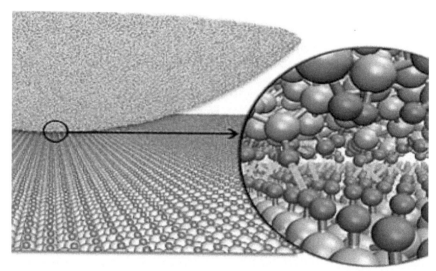

FIGURE 27 Nanocannon scheme acting on the basis of "domino effect" in the incision. (a) Inside a carbon nanotube (55.0) is the fullerene molecule C_{60}. (b) The initial position and velocity of the departure of the "core" (a fullerene molecule), depending on the time. The highest rate of emission of C_{60} (1.13 km/s) comparable to the velocity of the domino wave.

Necessary be noted that the simulation took place in nanogun at assumption of zero temperature in Kelvin. However, this example is not so abstract and may be used in the injecting device.

Thus, for the first time been demonstrated, albeit only in theory, the use of single-walled carbon nanotubes as energy sources.

6.12 FLUID FLOW IN NANOTUBES

The friction of surface against the surface in the absence of the interlayer between the liquid material (so-called dry friction) is created by irregularities in the given surfaces that rub one another, as well as the interaction forces between the particles that make up the surface.

As part of their study, the researchers built a computer model that calculates the friction force between nanosurfaces (Fig. 28). In the model, these surfaces were presented simply as a set of molecules for which forces of intermolecular interactions were calculated.

As a result, scientists were able to establish that the friction force is directly proportional to the number of interacting particles. The researchers propose to consider this quantity by analogue of so-called true macroscopic contact area. It is known that the friction force is directly proportional to this area (it should not be confused with common area of the contact surfaces of the bodies).

In addition, the researchers were able to show that the friction surface of the nanosurfaces can be considered within the framework of the classical theories of friction of non-smooth surfaces.

A literature review shows that nowadays molecular dynamics and mechanics of the continuum in are the main methods of research of fluid flow in nanotubes.

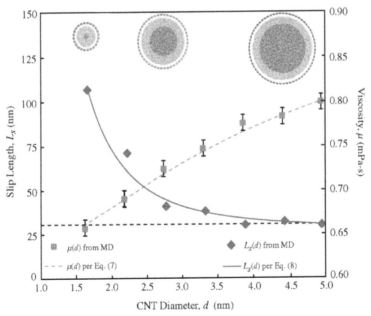

FIGURE 28 Computer model of friction at the nanoscale (the right shows the surfaces of interacting particles).

Although the method of molecular dynamics simulations is effective, it at the same time requires enormous computing time especially for large systems. Therefore, simulation of large systems is more reasonable to carry out nowadays by the method of continuum mechanics.

The fluid flow in the channel can also be considered in the framework of the continuum hypothesis. The Navier–Stokes equation can be used and the velocity profile can be determined for Poiseuille flow.

However the water flow by means of pressure differential through the carbon nanotubes with diameters ranging from 1.66 to 4.99 nm is studies using molecular dynamics simulation study. For each nanotube the value enhancement predicted by the theory of liquid flow in the carbon nanotubes is calculated. This formula is defined as a ratio of the observed flow in the experiments to the theoretical values without considering slippage on the model of Hagen–Poiseuille. The calculations showed that the enhancement decreases with increasing diameter of the nanotube.

Important conclusion is that by constructing a functional dependence of the viscosity of the water and length of the slippage on the diameter of carbon nanotubes, the experimental results in the context of continuum fluid mechanics can easily be described. The aforementioned is true even for carbon nanotubes with diameters of less than 1.66 nm.

The theoretical calculations use the following formula for the steady velocity profile of the viscosity η of the fluid particles in the CNT under pressure gradient $\partial p / \partial z$:

$$v(r) = \frac{R^2}{4\eta}\left[1 - \frac{r^2}{R^2} + \frac{2L_S}{R}\right]\frac{\partial p}{\partial z} \tag{23}$$

The length of the slip, which expresses the speed heterogeneity at the boundary of the solid wall and fluid is defined as:

$$L_S = \frac{v(r)}{dv/dr}\bigg|_{r=R} \tag{24}$$

Then the volumetric flow rate, taking into account the slip Q_s is defined as:

$$Q_S = \int_0^R 2\pi r \cdot v(r)dr = \frac{\pi\left[(d/2)^4 + 4(d/2)^3 \cdot L_S\right]}{8\eta}\cdot\frac{\partial p}{\partial z} \tag{25}$$

Equation (25) is a modified Hagen–Poiseuille equation, taking into account slippage. In the absence of slip $L_S = 0$ (Eq. (25)) coincides with the Hagen–

Poiseuille flow (Eq. (20)) for the volumetric flow rate without slip Q_P. In some works the parameter enhancement flow ε is also introduced. It is defined as the ratio of the calculated volumetric flow rate of slippage to Q_p (calculated using the effective viscosity and the diameter of the CNT). If the measured flux is modeled using Eq. (25), the degree of enhancement takes the form:

$$\varepsilon = \frac{Q_S}{Q_P} = \left[1 + 8\frac{L_S(d)}{d}\right]\frac{\eta_\infty}{\eta(d)} \tag{26}$$

where $d = 2R$ – diameter of CNT, η_∞ – viscosity of water, $L_S(d)$ – CNT slip length depending on the diameter, $\eta(d)$ – the viscosity of water inside CNTs depending on the diameter.

If $\eta(d)$ finds to be equal to η_∞, then the influence of the effect of slip on ε is significant, if $L_S(d) \geq d$. If $L_S(d) < d$ and $\eta(d) = \eta_\infty$, then there will be no significant difference compared to the Hagen–Poiseuille flow with no slip.

Table 2 shows the experimentally measured values of the enhancement water flow. Enhancement flow factor and the length of the slip were calculated using the equations given above.

TABLE 2 Experimentally measured values of the enhancement water flow.

Nanosystems	Diameter (nm)	Enhancement, ε	slip length, L_S, (nm)
carbon nanotubes	300–500	1	0
	44	22–34	113–177
carbon nanotubes	7	10^4–10^5	3900–6800
	1,6	560–9600	140–1400

Figure 29 depicts the change in viscosity of the water and the length of the slip in diameter. As can be seen from the figure, the dependence of slip length to the diameter of the nanotube is well described by the empirical relation

$$L_S(d) = L_{S,\infty} + \frac{C}{d^3} \tag{27}$$

where $L_{S,\infty} = 30$ nm – slip length on a plane sheet of graphene, a C – const.

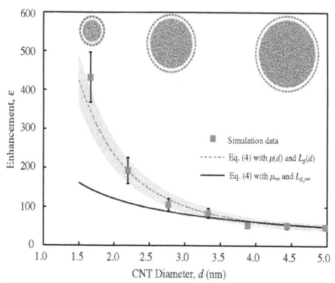

FIGURE 29   Variation of viscosity and slip length with CNT diameter.

Figure 30 shows dependence of the enhancement of the flow rate ε on the diameter for all seven CNTs.

There are three important features in the results. First, the enhancement of the flow decreases with increasing diameter of the CNT. Second, with increasing diameter, the value tends to the theoretical value of Eqs. (26) and (27) with a slip $L_{s\infty}$ = 30 nm and the effective viscosity $\mu(d) = \mu_\infty$. The dotted line shown the curve of 15% in the second error in the theoretical data of viscosity and slip length. Third, the change ε in diameter of CNTs cannot be explained only by the slip length.

To determine the dependence the volumetric flow of water from the pressure gradient along the axis of single-walled nanotube with the radii of 1.66, 2.22, 2.77, 3.33, 3.88, 4.44 and 4.99 nm in the method of molecular modeling was used. Snapshot of the water–CNT is shown in Fig. 31.

Figure 31 shows the results of calculations to determine the pressure gradient along the axis of the nanotube with the diameter of 2.77 nm and a length of 20 nm. Change of the density of the liquid in the cross sections was less than 1%.

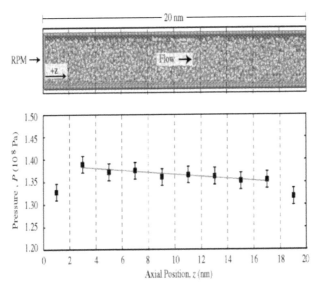

FIGURE 30 Flow enhancement as predicted from MD simulations.

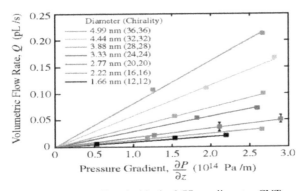

FIGURE 31 Axial pressure gradient inside the 2.77 nm diameter CNT.

Figure 32 shows the dependence of the volumetric flow rate from the pressure gradient for all seven CNTs. The flow rate ranged from 3–14 m/sec. In the range considered here the pressure gradient $(0-3)\times10^9\,atm/m$ Q ($pl/sek = 10^{-15}\,m^3/sek$) is directly proportional $\partial p/\partial z$. Coordinates of chirality for each CNT are indicated in the figure legend. The linearity of the relations between flow and pressure gradient confirms the validity of calculations of the Eq. (25).

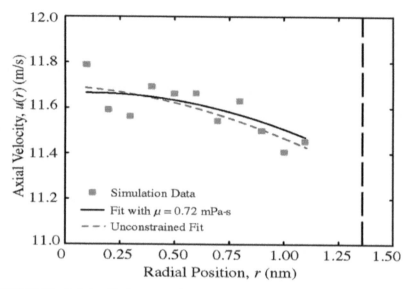

FIGURE 32   Relationship between volumetric liquid flow rate in carbon nanotubes with different diameters and applied pressure gradient.

Figure 33 shows the profile of the radial velocity of water particles in the CNT with diameter 2.77 nm. The vertical dotted line at 1.38 nm marked surface of the CNT. It is seen that the velocity profile is close to a parabolic shape.

Researchers consider the flow of water under a pressure gradient in the single-walled nanotubes of "chair" type of smaller radii: 0.83, 0.97, 1.10, 1.25, 1.39 and 1.66 nm.

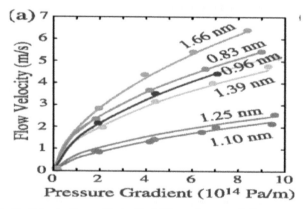

FIGURE 33   Radial velocity profile inside 2.77 nm diameter CNT.

Figure 34 shows the dependence of the mean flow velocity $\bar{v}$ on the applied pressure gradient $\Delta P/L$ in the long nanotubes – 75 nm at 298 K. A similar picture pattern occurs in the tube with the length of 150 nm.

As we can see, there is conformance with the Darcy law, the average flow rate for each CNT increases with increasing pressure gradient. For a fixed value of $\Delta p/L$, however, the average flow rate does not increase monotonically with increasing diameter of the CNTs, as follows from Poiseuille equation. Instead, when at the same pressure gradient, decrease of the average speed in a CNT with the radius of 0.83 nm to a CNT with the radius of 1.10 nm, similar to the CNTs 1.10 and 1.25 nm, then increases from a CNT with the radius of 1.25 nm to a CNT of 1.66 nm.

The nonlinearity of the relationship between $\bar{v}$ and $\Delta P/L$ are the result of inertia losses (i.e., insignificant losses) in the two boundaries of the CNT. Inertial losses depend on the speed and are caused by a sudden expansion, abbreviations, and other obstructions in the flow.

We note the important conclusion of the work of scientists in which the method of molecular modeling shows that the Eq. (23) (Poiseuille parabola) correctly describes the velocity profile of liquid in a nanotube when the diameter of a flow is 5–10 times more than the diameter of the molecule (≈ 0.17 nm for water).

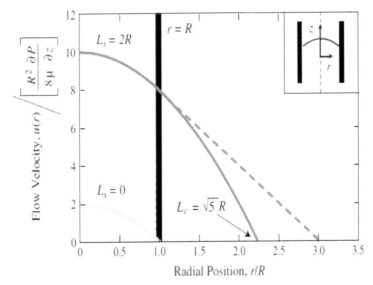

FIGURE 34   Relationship between average flow velocity and applied pressure gradient for the 75 nm long CNTs.

In Fig. 35, we can see the effect of slip on the velocity profile at the boundary of radius R of the pipe and fluid. When $L_S = 0$ the fluid velocity at the wall vanishes, and the maximum speed (on the tube axis) exceeds flow speed twice.

The figure shows the velocity profiles for Poiseuille flow without slip ($L_S = 0$) and with slippage $L_S = 2R$. The flow rate is normalized to the speed corresponding to the flow without slip. Thick vertical lines indicate the location of the pipe wall. The thick vertical lines indicate the location of the tube wall. As the length of the slip, the flow rate increases, decreases the difference between maximum and minimum values of the velocity and the velocity profile becomes more like a plug.

Velocity of the liquid on a solid surface can also be quantified by the coefficient of slip L_c. The coefficient of slippage – there is a difference between the radial position in which the velocity profile would be zero and the radial position of the solid surface. Slip coefficient is equal to $L_C = \sqrt{R^2 + 2RL_S} = \sqrt{5}R$.

For linear velocity profiles (e.g., Couette flow), the length of the slip and slip rate are equal. These values are different for the Poiseuille flow.

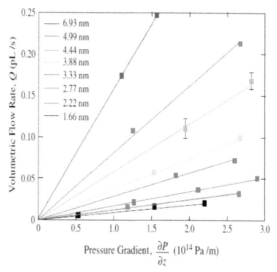

FIGURE 35   No-slip Poiseuille flow and slip Poiseuille flow through a tube.

Figure 36 shows dependence of the volumetric flow rate Q from the pressure gradient $\partial p / \partial z$ in long nanotubes with diameters between 1.66 nm and 6.93 nm. As can be seen in the studied range of the pressure gradient Q is proportional to $\partial p / \partial z$. As in the Poiseuille flow, volumetric flow rate increases monotonically with the diameter of CNT at a fixed pressure gradient. Magnitude of calculations error for all the dependencies are similar to the error for the CNT diameter 4.44 nm, marked in the figure.

Many researchers considered steady flow of incompressible fluids in a channel width $2h$ under action of the force of gravity ρg or pressure gradient $\partial p / \partial y$, which is described by the Navier–Stokes equations. The velocity profile has a parabolic form:

$$U_y(z) = \frac{\rho g}{2\eta} \cdot \left[(\delta + h)^2 - z^2 \right]$$

where δ – length of the slip, which is equal to the distance from the wall to the point at which the velocity extrapolates to zero.

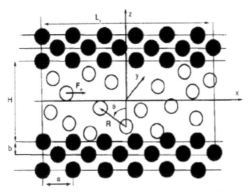

FIGURE 36   Volumetric flow rate in CNTs versus pressure gradient.

6.13   SOME OF THE IDEAS AND APPROACHES FOR MODELING IN NANOHYDROMECANICS

Let's consider the fluid flow through the nanotube. Molecules of a substance in a liquid state are very close to each other (Fig. 37).

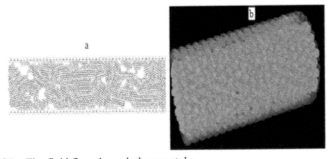

FIGURE 37   The fluid flow through the nanotube.

Most liquid's molecules have a diameter of about 0,1 nm. Each molecule of the fluid is "squeezed" on all sides by neighboring molecules and for a period of time $\left(10^{-10} - 10^{-13} s\right)$ fluctuates around certain equilibrium position, which itself from time to time is shifted in distance commensurating with the size of molecules or the average distance between molecules l_{cp}:

$$l_{cp} \approx \sqrt[3]{\frac{1}{n_0}} = \sqrt[3]{\frac{\mu}{N_A \rho}},$$

where: n_0 – number of molecules per unit volume of fluid, N_A – Avogadro's number, ρ – fluid density, μ – molar mass.

Estimates show that one cubic of nanowater contains about 50 molecules. This gives a basis to describe the mass transfer of liquid in a nanotube-based continuum model. However, the specifics of the complexes, consisting of a finite number of molecules, should be kept in mind. These complexes, called clusters in literatures, are intermediately located between the bulk matter and individual particles – atoms or molecules. The fact of heterogeneity of water is now experimentally established.

There are groups of molecules in liquid – "microcrystals" containing tens or hundreds of molecules. Each microcrystal maintains solid form. These groups of molecules, or "clusters" exist for a short period of time, then break up and are recreated again. Besides, they are constantly moving so that each molecule does not belong at all times to the same group of molecules, or "cluster".

Modeling predicts that gas molecules bounce off the perfectly smooth inner walls of the nanotubes as billiard balls, and water molecules slide over them without stopping. Possible cause of unusually rapid flow of water is maybe due to the small-diameter nanotube molecules move on them orderly, rarely colliding with each other. This "organized" move is much faster than usual chaotic flow. However, while the mechanism of flow of water and gas through the nanotubes is not very clear, and only further experiments and calculations can help understand it.

The model of mass transfer of liquid in a nanotube proposed in this paper is based on the availability of nanoscale crystalline clusters in it.

A similar concept was developed in which the model of structured flow of fluid through the nanotube is considered. It is shown that the flow character in the nanotube depends on the relation between the equilibrium crystallite size and the diameter of the nanotube.

Figure 38 shows the results of calculations by the molecular dynamics of fluid flow in the nanotube in a plane (a) and three-dimensional (b) statement. The figure shows the ordered regions of the liquid.

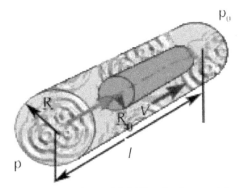

FIGURE 38 The results of calculations of fluid flow in the nanotube.

The typical size of crystallite is 1–2 nm, i.e. compared, for example, with a diameter of silica nanotubes of different composition and structure. The flow model proposed in the present work is based on the presence of "quasi-solid" phase in the central part of the nanotube and liquid layer, non-autonomous phases.

Consideration of such a structure that is formed when fluid flows through the nanotube, is also justified by the aforementioned results of the experimental studies and molecular modeling.

When considering the fluid flow with such structure through the nanotube, we will take into account the aspect ratio of "quasi-solid" phase and the diameter of the nanotube so that a character of the flow is stable and the liquid phase can be regarded as a continuous medium with viscosity η.

Let's establish relationship between the volumetric flow rate of liquid Q flowing from a liquid layer of the nanotube length l, the radius R and the pressure drop $\Delta p / l$, $\Delta p = p - p_0$, where p_0 is the initial pressure in the tube (Fig. 39).

Let R_0 be a radius of the tube from the "quasi-solid" phase, v – velocity of fluid flow through the nanotube.

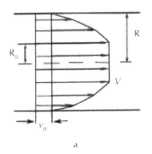
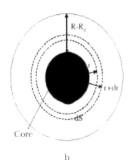

FIGURE 39 Flow through liquid layer of the nanotube.

Structural regime of fluid flow (Fig. 40) implies the existence of the continuous laminar layer of liquid (the liquid layer in the nanotube) along the walls of a pipe. In the central part of a pipe a core of the flow is observed, where the fluid moves, keeping his former structure, i.e. as a solid ("quasi-solid" phase in the nanotube). The velocity slip is indicated in Fig. 40a through v_0.

Let's find the velocity profile $v(r)$ in a liquid interlayer $R_0 \le r \le R$ of the nanotube. We select a cylinder with radius r and length l in the interlayer, located symmetrically to the center line of the pipe (see Fig. 40b).

At the steady flow, the sum of all forces acting on all the volumes of fluid with effective viscosity η, is zero.

The following forces are applied on the chosen cylinder: the pressure force and viscous friction force affects the side of the cylinder with radius r, calculated by the Newton formula.

Thus,

$$\left(p - p_0\right)\pi r^2 = -\eta \frac{dv}{dr} 2\pi r l \qquad (28)$$

Integrating expression (28) between r to R with the boundary conditions $r = R : V = V_0$, we obtained a formula to calculate the velocity of the liquid layers located at a distance r from the axis of the tube:

$$v(r) = \left(p - p_0\right)\frac{R^2 - r^2}{4\eta l} + v_0 \qquad (29)$$

Maximum speed V_s has the core of the nanotube $0 \le r \le R_0$ and is equal to:

$$V_s = \left(p - p_0\right)\frac{R^2 - R_0^2}{4\eta l} + v_0 \qquad (30)$$

Such structure of the liquid flow through nanotubes considering the slip is similar to a behavior of viscoplastic liquids in the tubes. Indeed, as we know, for viscoplastic fluids a characteristic feature is that they are to achieve a certain critical internal shear stresses τ_0 behave like solids, and only when internal stress exceeds a critical value above begin to move as normal fluid. Scientists shown that liquid behaves in the nanotube the similar way. A critical pressure drop is also needed to start the flow of liquid in a nanotube.

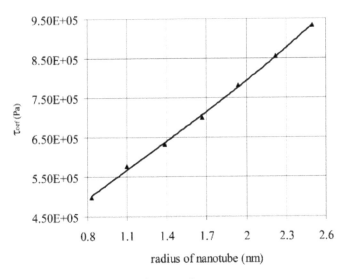

FIGURE 40 Structure of the flow in the nanotube.

Structural regime of fluid flow requires existence of continuous laminar layer of liquid along the walls of pipe. In the central part of the pipe is observed flow with core radius R_{00}, where the fluid moves, keeping his former structure, i.e. as a solid.

The velocity distribution over the pipe section with radius R of laminar layer of viscoplastic fluid is expressed as follows:

$$v(r) = \frac{\Delta p}{4\eta l}\left(R^2 - r^2\right) - \frac{\tau_0}{\eta}(R - r) \tag{31}$$

The speed of flow core in $0 \le r \le R_{00}$ is equal

$$v_{st} = \frac{\Delta p}{4\eta l}\left(R^2 - R_{00}^2\right) - \frac{\tau_0}{\eta}(R - R_{00}) \tag{32}$$

Let's calculate the flow or quantity of fluid flowing through the nanotube cross-section S at a time unit. The liquid flow dQ for the inhomogeneous velocity field flowing from the cylindrical layer of thickness dr, which is located at a distance r from the tube axis is determined from the relation

$$dQ = v(r)dS = v(r)2\pi r dr \tag{33}$$

where ds – the area of the cross-section of cylindrical layer (between the dotted lines in Fig. 40).

Let's place Eq. (29) in Eq. (33), integrate over the radius of all sections from R_0 to R and take into account that the fluid flow through the core flow is determined from the relationship $Q_s = \pi R_0^2 v_s$. Then we get the formula for the flow of liquid from the nanotube:

$$Q = Q_P \left[v_* \frac{8l\eta R_*^2}{\Delta p R^4} + \left(\frac{R_*}{R} \right)^4 \right] \qquad (34)$$

If $(R_0 / R)^4 \ll 1$ (no nucleus) and $v_0 \Delta p R^2 / 8l\eta \ll 1$ (no slip), then Eq. (34) coincides with Poiseuille formula (Eq. 20). When $R_0 \approx R$ (no of a viscous liquid interlayer in the nanotube), the flow rate Q is equal to volumetric flow $Q \approx \pi R^2 v_0$ of fluid for a uniform field of velocity (full slip).

Accordingly, flow rate of the viscoplastic fluid flowing with a velocity (4.7), is equal to:

$$Q = -\frac{\pi R^3 \tau_0}{3\eta} \left[1 - \left(\frac{R_{00}}{R} \right)^3 \right] + Q_P \left[1 - \left(\frac{R_{00}}{R} \right)^4 \right] \qquad (35)$$

Comparing Eqs. (29)–(32) and (34), (35), we can see that the structure of the flow of the liquid through the nanotubes considering the slippage, is similar to that of the flow of viscoplastic fluid in a pipe of the same radius R.

Given that the size of the central core flow of viscoplastic fluid (radius R_{00}) is defined by

$$R_{00} = \frac{2\tau_0 l}{\Delta p} \qquad (35)$$

for viscoplastic fluid flow we obtain Buckingham formula:

$$Q = Q_P \left[1 + \frac{1}{3} \left(\frac{2l\tau_0}{R\Delta p} \right)^4 - \frac{4}{3} \left(\frac{2l\tau_0}{R\Delta p} \right) \right] \qquad (36)$$

We'll establish a conformity of the pipe that implements the flow of a viscoplastic fluid with a fluid-filled nanotube, the same size and with the same pressure drop. We say that an effective internal critical shear stress τ_{0ef} of viscoplastic fluid flow, which ensures the coincidence rate with the flow of fluid in the nano-tube. Then from Eq. (36) we obtain equation of fourth order to determine τ_{0ef} :

$$\left(\frac{2l\tau_{0ef}}{R\Delta p}\right)^4 - 4\left(\frac{2l\tau_{0ef}}{R\Delta p}\right) = A, \ A = 3(\varepsilon-1), \varepsilon = Q/Q_P \tag{37}$$

The solution of Eq. (37) can be found, for example, the iteration method of Newton:

$$\overline{\tau}_{0ef\,n} = \overline{\tau}_{0ef\,n-1} - \frac{\overline{\tau}_{0ef\,n-1}^{-4} - 4\overline{\tau}_{0ef\,n-1} - A}{4\overline{\tau}_{0ef\,n-1}^{-3} - 4}, \ \overline{\tau}_{0ef} = \frac{2l\tau_{0ef}}{R\Delta p} \tag{38}$$

The first component in Eq. (34) represents the contribution to the fluid flow due to the slippage, and it becomes clear that the slippage significantly enhances the flow rate in the nanotube, when $l\eta v_0 \gg \Delta p R^2$.

This result is consistent with experimental and theoretical results (which show that water flow in nanochannels can be much higher than under the same conditions, but for the liquid continuum.

In the absence of slippage $\varepsilon = 1$ the equation (4.16) has a trivial solution $\overline{\tau}_{0ef} = 0$.

THE RESULTS OF THE CALCULATION:

Let's determine the dependence of the effective critical inner shear stress τ_{0ef} on the radius of the nanotubes, by taking necessary values for calculations $\varepsilon = Q/Q_P$. The results of calculations at $\Delta p/l = 2,1 \cdot 10^{14}$ Pa/m are in the table below:

R , M	τ_{0ef} (Па)	$\varepsilon = Q/Q_P$
$0,83 \cdot 10^{-9}$	498498	350
$0,83 \cdot 10^{-9}$	577500	200
$0,83 \cdot 10^{-9}$	632599	114
$1,665 \cdot 10^{-9}$	699300	84
$1,94 \cdot 10^{-9}$	782208	68
$2,22 \cdot 10^{-9}$	855477	57
$2,495 \cdot 10^{-9}$	932631	50

Calculations show that the value of effective internal shear stress depends on the size of the nanotube.

Figure 41 shows the dependence τ_{0ef} on the nanotube radius.

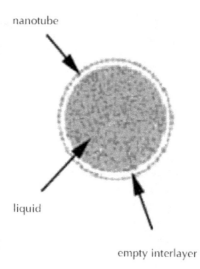

FIGURE 41   Dependence of the effective inner shear stress from the radius of the nanotube.

As you can see, this dependence $\tau_{0ef}(R)$ is almost linear. Within the range of the considered nanotube sizes τ_{0ef} has relatively low values, which indicates the smoothness of the surface of carbon nanotubes.

THE FLOW OF FLUID WITH AN EMPTY INTERLAYER:

The works of EM Kotsalis et al. (2004) and Xi Chen et al. (2008) were analyzed in the aforementioned analysis of the structure of liquid flow in carbon nanotubes. The results of the calculations of the cited works (Figs. 42 and 43) showed that during the flow of the liquid particles, an empty layer between the fluid and the nanotube is formed. The area near the walls of the carbon nanotube $R_* \leq r \leq R$ becomes inaccessible for the molecules of the liquid due to van der Waals repulsion forces of the heterogeneous particles of the carbon and water (Fig. 2). Moreover, according to the results of EM Kotsalis et al. (2004) and Xi Chen et al. (2008) thicknesses of the layers $R_* \leq r \leq R$ regardless of radiuses of the nanotubes are practically identical: $R_* / R \approx 0.8$.

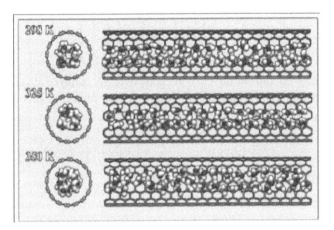

FIGURE 42   The structure of the flow.

A similar result was obtained in Hongfei Ye et al. (2011), which is an image (Fig. 43) of the configuration of water molecules inside (8, 8) single-walled carbon nanotubes at different temperatures: 298, 325, and 350 ^0K.

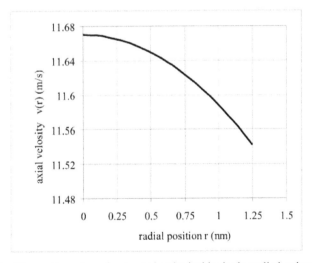

FIGURE 43   The configuration of water molecules inside single-walled carbon nanotubes.

Integrating expression (28) between r to $R_.$ at the boundary conditions $r = R_*$: $v = v_*$, we obtain a formula to calculate the velocity of the liquid layers located at a distance r from the axis of the tube:

$$v(r) = v_* + \frac{2Q_p}{\pi R^2}\left[\left(\frac{R_*}{R}\right)^2 - \left(\frac{r}{R}\right)^2\right] \tag{39}$$

Let's insert Eq. (39) in Eq. (33), integrate over the radius of all sections from 0 to R_*. Then we get a formula for the flow of liquid from the nanotube:

$$Q = Q_P\left[v_* \frac{8l\eta R_*^2}{\Delta p R^4} + \left(\frac{R_*}{R}\right)^4\right] \tag{40}$$

or

$$\varepsilon = \frac{8l\eta v_*}{\Delta p R^2}\left(\frac{R_*}{R}\right)^2 + \left(\frac{R_*}{R}\right)^4$$

from which we can determine the unknown v_*:

$$v_* = \frac{Q_P}{\pi R^2}\left[\varepsilon - \left(\frac{R_*}{R}\right)^4\right]\left(\frac{R}{R_*}\right)^2 \tag{41}$$

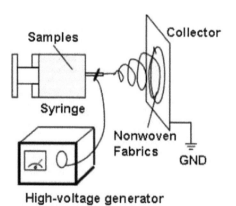

FIGURE 44   Profile of the radial velocity of water in carbon nanotubes.

Figure 44 shows the profile of the radial velocity of water particles in a carbon nanotube with a diameter of 2.77 nm, calculated using the formula (39) at $Q_P = 4,75 \cdot 10^{-19} n/m^2$, $\varepsilon = 114$. The velocity at the border v_* is equal to 11.55 m/s.

It is seen that the velocity profile is similar to a parabolic shape, and at the same time agrees with the numerical values.

The calculations suggest the following conclusions. Flow of liquid in a nanotube was investigated using synthesis of the methods of the continuum theory and molecular dynamics. Two models are considered. The first is based on the fact that fluid in the nanotube behaves like a viscoplastic. A method of calculating the value of limiting shear stress is proposed, which was dependent on the nanotube radius. A simplified model agrees quite well with the results of the molecular simulations of fluid flow in carbon nanotubes. The second model assumes the existence of an empty interlayer between the liquid molecules and wall of the nanotube. This formulation of the task is based on the results of experimental works known from the literature. The velocity profile of fluid flowing in the nanotube is practically identical to the profile determined by molecular modeling.

As seen from the results of the calculations, the velocity value varies slightly along the radius of the nanotube. Such a velocity distribution of the fluid particles can be explained by the lack of friction between the molecules of the liquid and the wall due to the presence of an empty layer. This leads to an easy slippage of the liquid and, consequently, anomalous increase in flow compared to the Poiseuille flow.

6.14 MODELING AND SIMULATION TECHNIQUES FOR NANOFIBERS

Symbols	Definition	Units
R	Radius of jet	m
n	Jet velocity	m/s
Q	Flow rate	m^3/s
I	Jet current	A
J	Current density	A/m^2
$A(s)$, S	Cross-sectional area	m^2
K	The conductivity of the liquid	S/m
E	Electric field	V/m
L	Spinning distance	M
s	Electric density	C/m^2
P	Linear momentum	kg m/s, (N.s)
F	Force	N
t	Time	S
m	Mass	Kg

r	Density	kg/m^3
z	Jet axial position	m
p	Pressure	N/m^2
t_{zz}	Axial viscous normal stress	N/m^2
g	Surface tension	N/m
R'	Slope of the jet surface	—
t_t^e	Tangential electric force	N/m^2
t_n^e	Normal electric force	N/m^2
ε	Dielectric constant of the jet	—
$\bar{\varepsilon}$	Dielectric constant of the ambient air	—
W_e	Electric work	J
l	Distance	M
U_e	Electric potential energy	J
t_{rr}	Radial normal stress	N/m^2
$r_{p'}$	Polarized charge density	C/m^2
P'	Polarization	C/m^2
q, q_0 & Q_b	Charge	C
r	Distance between two charges	m
V	Electric potential	kV
t	Shear stress	N/m^2
K'	Flow consistency index	—
m	Flow behavior index	—
m	Constant	—
$\dot{\hat{\gamma}}$	Rate of strain tensor	s^{-1}
S_k	Excess stress	N/m^2
C_k	Configurational tensor	N/m^2
h	Viscosity	P
E_k	Young's modulus	N/m^2
m_k	Shear modulus	N/m^2

b_k	Constitutive mobility	$m^2/(V \cdot s)$
N	Number of beads per unit volume	m^{-3}
T	Temperature	K
G	Elastic modulus	N/m^2
l	Filament	m
ε'	Lagrangian axial strain	—
b	Power exponent	—

6.14.1 INTRODUCTION

Electrospinning is a procedure in which an electrical charge to draw very fine (typically on the micro or nanoscale) fibers from polymer solution or molten. Electrospinning shares characteristics of both electrospraying and conventional solution dry spinning of fibers. The process does not require the use of coagulation chemistry or high temperatures to produce solid threads from solution. This makes the process more efficient to produce the fibers using large and complex molecules. Recently, various polymers have been successfully electrospun into ultrafine fibers mostly in solvent solution and some in melt form [1–2]. Optimization of the alignment and morphology of the fibers is produced by fitting the composition of the solution and the configuration of the electrospinning apparatus such as voltage, flow rate, and etc. As a result, the efficiency of this method can be improved [3]. Mathematical and theoretical modeling and simulating procedure will assist to offer an in-depth insight into the physical understanding of complex phenomena during electrospinningand might be very useful to manage contributing factors toward increasing production rate [4–5].

Despite the simplicity of the electrospinning technology, industrial applications of it are still relatively rare, mainly due to the notable problems of very low fiber production rate and difficulties in controlling the process [6].

Modeling and simulation (M&S) give information about how something will act without actually testing it in real. The model is a representation of a real object or system of objects for purposes of visualizing its appearance or analyzing its behavior. Simulation is transition from a mathematical or computational model to description of the system behavior based on sets of input parameters [7–8]. Simulation is often the only means for accurately predicting performance of the modeled system [9]. Using simulation is generally cheaper and safer than conducting experiments with a prototype of the final product. Also simulation can often be even more realistic than traditional experiments, as they allow the free configuration of environmental and operational parameters and can often be run faster than in real time. In a situation with different alternatives analysis, simulation can improve the efficiency, in particular when the necessary data to initialize can easily be obtained from operational data. Applying simulation adds decision support systems to the toolbox of traditional decision support systems [10].

Simulation permits set up a coherent synthetic environment that allows for integration of systems in the early analysis phase for a virtual test environment in the final system. If managed correctly, the environment can be migrated from the development and test domain to the training and education domain in real system under realistic constraints [11].

A collection of experimental data and their confrontation with simple physical models appears as an effective approach towards the development of practical tools for controlling and optimizing the electrospinning process. On the other hand, it is necessary to develop theoretical and numerical models of electrospinning because of demanding a different optimization procedure for each material [12]. Utilizing a model to express the effect of electrospinning parameters will assist researchers to make an easy and systematic way of presenting the influence of variables and by means of that, the process can be controlled. Additionally, it causes to predict the results under a new combination of parameters. Therefore, without conducting any experiments, one can easily estimate features of the product under unknown conditions [13].

6.14.2   ELECTROSPINNING

Spinning is the processes applied for drawing a fiber from polymer into filaments by passing through a spinneret, which is classified into solution spinning (wet or dry) and melt spinning. Conventional fiber-forming techniques have limitations of controlling the fiber diameter due to their dependency on the devices such as spinneret diameter. The major technique that can be used to make fibers thinner than 100 μm is electrospinning which is capable of giving very long continuous fibers by electrostatically drawing a polymer jet through a virtual spatio-temporal orifice [14]. An overview on invention history of this technique have reported for obtaining more clear vision and then the principle of electrospinning methodology is discussed.

6.14.2.1 THE HISTORY OF THE SCIENCE AND TECHNOLOGY OF ELECTROSPINNING

William Gilbert discovered the first record of the electrostatic attraction of a liquid in 1600 [15]. The first electrospinning patent was submitted by John Francis Cooley in 1900 [16]. After that in 1914 John Zeleny studied on the behavior of fluid droplets at the end of metal capillaries, which caused the beginning of the mathematically model the behavior of fluids under electrostatic forces [17]. Between 1931 and 1944 Anton Formhals took out at least 22 patents on electrospinning [16]. In 1938, N.D. Rozenblum and I.V. Petryanov–Sokolov generated electrospun fibers, which they developed into filter materials [18]. Between 1964 and 1969 Sir Geoffrey Ingram Taylor produced the beginnings of a theoretical foundation of electrospinning by mathematically modeling the shape of the (Taylor) cone formed by the fluid droplet under the effect of an electric field [19–20]. In the early 1990s several research groups (such as Reneker) demonstrated electrospun nanofibers. Since 1995, the number of publications about electrospinning has been increasing exponentially every year [16].

6.14.2.2 THE BASIC PRINCIPLES OF ELECTROSPINNING

As mentioned before, electrospinning gives the impression of being a very simple and easily controlled technique for the production of fibers with dimensions down to the nanometer range. Electrospinning fiber precursors are classified in two polymers as fiber forming substantial and materials such as metals, ceramics, and glasses. In a typical electrospinning experiment in a laboratory, a polymer solution or melt is pumped through a thin nozzle with an inner diameter on the order of 100 mm. In most of laboratory systems, the nozzle simultaneously serves as an electrode, to which a high electric field of 100–500 kVm-1 is applied, and the distance to the counter electrode is 10–25 cm [21]. The currents that flow during electrospinning range from a few hundred nanoamperes to microamperes. A high voltage is applied to the solution such that at a critical voltage, typically more than 5 kV, the repulsive force within the charged solution is larger than its surface tension and a jet would erupt from the tip of the spinneret. Although the jet is stable near to the tip of the spinneret, it soon enters a bending instability stage with further stretching of the solution jet under the electrostatic forces in the solution as the solvent evaporates [22–23]. The substrate on which the electrospun fibers are collected is typically brought into contact with the counter electrodes that are rotating and flat types [22]. The vertical alignment of the electrodes "from top to bottom" is not insignificant with respect to the process, but in principal, electrospinning can also be carried out "from bottom to top" or horizontally (Fig. 45) [24].

The formation of nanofibers is determined by many operating parameters, which are included in Table 1.

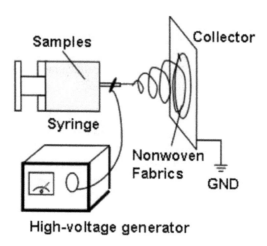

FIGURE 45   Electrospinning set up.

TABLE 1   Classification of affecting parameters and nozzle configuration types on electrospinning.

Affecting parameters		
Process parameters	System parameters	Ambient parameters
✓ electric potential	✓ molecular weight	✓ temperature
✓ flow rate	✓ molecular weight distribution	✓ humidity
✓ concentration	✓ architecture of the polymer	✓ air velocity
✓ spinning distance		

Type of electrospinning nozzle configuration

✓ single
✓ side by side
✓ co axial

6.14.3 *MODELING AND SIMULATION CONCEPTS*

Numerous processing operations of complex fluids involve free surface deformations: examples include spraying and atomization of fertilizers and pesticides, fiber-spinning operations, paint application, roll coating of adhesives and food processing operations. Systematically understanding such flows in various processes such as electrospinning can be extremely difficult because of the large number of different forces that may be involved: including capillarity, viscosity, inertia, gravity as well as the additional stresses resulting from the extensional deformation of the microstructure within the fluid [25].

Theoretical and numerical understanding of flows in these processes can help to dominate their drawbacks. For example, in our study, inspite of individual applications of electrospinning process, its mass production is still presented a challenge [6]. For achieving higher mass production and orientation of nanofibers for special application like tissue engineering and microelectronics, it is necessary to comprehend and control dynamic and mechanic behavior of electrospinning jet [26–27]. Modeling and simulations will give a better understanding of electrospinning jet mechanics. For example, the effect of secondary external field can be surveyed using simulation studies. As well, poor deposition control may be in part owing to the lack of understanding of mechanisms of dynamic interactions of the fast moving jets with the electric field and collectors [27].

Electrospinning modeling and simulation are discussed in detail in following sections.

6.14.3.1 ELECTROSPINNING MODELING

The electrospinning process is a fluid dynamics related problem. Controlling the property, geometry, and mass production of the nanofibers, is essential to comprehend quantitatively how the electrospinning process transforms the fluid solution through a millimeter diameter capillary tube into solid fibers which are four to five orders smaller in diameter [28]. Although information on the effect of various processing parameters and constituent material properties can be obtained experimentally, theoretical models offer in-depth scientific understanding which can be useful to clarify the affecting factors that cannot be exactly measured experimentally. Results from modeling also explained how processing parameters and fluid behavior lead to the nanofiber of appropriate properties. The term "properties" refers to basic properties (such as fiber diameter, surface roughness, fiber connectivity, etc.), physical properties (such as stiffness, toughness, thermal conductivity, electrical resistivity, thermal expansion coefficient, density, etc.) and specialized properties (such as biocompatibility, degradation curve, etc. for biomedical applications) [23, 29].

For example, the developed models can be used for the analysis of mechanisms of jet deposition and alignment on various collecting devices in arbitrary electric fields [27].

The various method formulated by researchers are prompted by several applications of nanofibers. It would be sufficient to briefly describe some of these methods to observed similarities and disadvantages of these approaches. An abbreviated literature review of these models will be discussed in Section 6.14.5.

6.14.3.2 ELECTROSPINNING SIMULATION

Electrospun polymer nanofibers demonstrate outstanding mechanical and thermo-dynamic properties as compared to macroscopic-scale structures. These features are attributed to nanofiber microstructure [30–31]. Theoretical modeling predicts the nanostructure formations during electrospinning. This prediction could be verified by various experimental condition and analysis methods, which called simulation. Numerical simulations can be compared with experimental observations as the last evidence [27, 32].

Parametric analysis and accounting complex geometries in simulation of electrospinning are extremely difficult due to the non-linearity nature in the problem. Therefore, a lot of researches have done to develop an existing electrospinning simulation for viscoelastic liquids [33].

6.14.4   THE BASIC OF ELECTROSPINNING MODELING

Balance of the producing accumulation is, particularly, a basic source of quantitative models of phenomena or processes. Differential balance equations are formulated for momentum, mass and energy through the contribution of local rates of transport expressed by principle of Newton's, Fick's and Fourier laws. For description of more complex systems like electrospinning that involved strong turbulence of the fluid flow, characterization of product property is necessary and various balances are required [34].

The basic principle used in modeling of chemical engineering process is a concept of balance of momentum, mass and energy, which can be expressed in a general form as:

$$A = I + G - O - C \qquad (42)$$

where: A = accumulation built up within system, I = input entering through system surface, G = generation produced in system volume, O = output leaving through system boundary, and C = consumption used in system volume.

The form of expression depends on the level of the process phenomenon description [34–35].

According to the electrospining models, the jet dynamics is governed by a set of three equations representing mass, energy and momentum conservation for the electrically charge jet [36].

6.14.4.1   MASS CONSERVATION

The concept of mass conservation is widely used in many fields such as chemistry, mechanics, and fluid dynamics. Historically, mass conservation was discovered in chemical reactions by Antoine Lavoisier in the late 18th century, and was of decisive importance in the progress from alchemy to the modern natural science of chemistry. The concept of matter conservation is useful and sufficiently accurate for most chemical calculations, even in modern practice [37].

The equations for the jet follow from Newton's Law and the conservation laws obey, namely, conservation of mass and conservation of charge [38].

According to the conservation of mass equation

$$\pi R^2 v = Q \tag{43}$$

For incompressible jets, by increasing the velocity the radius of the jet decreases. At the maximum level of the velocity, the radius of the jet reduces. The macromolecules of the polymers are compacted together closer while the jet becomes thinner as it shown in Fig. 46. When the radius of the jet reaches the minimum value and its speed becomes maximum to keep the conservation of mass equation, the jet dilates by decreasing its density, which called electrospinning dilation [39–40].

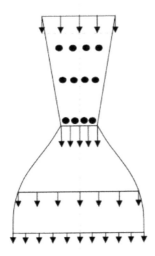

FIGURE 46   Macromolecular chains are compacted during the electrospinning [40].

6.14.4.2   ELECTRIC CHARGE CONSERVATION

An electric current is a flow of electric charge. Electric charge flows when there is voltage present across a conductor. In physics, charge conservation is the principle that electric charge can neither be created nor destroyed. The net quantity of electric charge, the amount of positive charge minus the amount of negative charge in the universe, is always conserved. The first written statement of the principle was by American scientist and statesman Benjamin Franklin in 1747 [41]. Charge conservation is a physical law, which states that the change in the amount of electric charge in any volume of space is exactly equal to the amount of charge in a region and the flow of charge into and out of that region [42].

During the electrospinning process, the electrostatic repulsion between excess charges in the solution stretches the jet. This stretching also decreases the jet diameter that this leads to the law of charge conservation as the second governing equation [43].

In electrospinning process, the electric current, which induced by electric field included two parts, conduction and convection.

The conventional symbol for current is I:

$$I = I_{conduction} + I_{convection} \tag{44}$$

Electrical conduction is the movement of electrically charged particles through a transmission medium. The movement can form an electric current in response to an electric field. The underlying mechanism for this movement depends on the material.

$$I_{conduction} = J_{cond} \times S = KE \times \pi R^2 \tag{45}$$

$$J = \frac{I}{A(s)} \tag{46}$$

$$I = J \times S \tag{47}$$

Convection current is the flow of current with the absence of an electric field.

$$I_{convection} = J_{conv} \times S = 2\pi R(L) \times \sigma v \tag{48}$$

$$J_{conv} = \sigma v \tag{49}$$

So, the total current can be calculated as:

$$\pi R^2 KE + 2\pi R v \sigma = I \tag{50}$$

6.14.4.3 MOMENTUM BALANCE

In classical mechanics, linear momentum or translational momentum is the product of the mass and velocity of an object. Like velocity, linear momentum is a vector quantity, possessing a direction as well as a magnitude:

$$P = mv \tag{51}$$

Linear momentum is also a conserved quantity, meaning that if a closed system (one that does not exchange any matter with the outside and is not acted on by outside forces) is not affected by external forces, its total linear momentum cannot change. In classical mechanics, conservation of linear momentum is implied by Newton's laws of motion; but it also holds in special relativity (with a modified formula) and, with appropriate definitions, a (generalized) linear momentum conservation law holds in electrodynamics, quantum mechanics, quantum field theory, and general relativity [35, 44]. For example, according to the third law, the forces between two particles are equal and opposite. If the particles are numbered 1 and 2, the second law states:

$$F_1 = \frac{dP_1}{dt} \tag{52}$$

$$F_2 = \frac{dP_2}{dt} \tag{53}$$

Therefore:

$$\frac{dP_1}{dt} = -\frac{dP_2}{dt} \tag{54}$$

$$\frac{d}{dt}(P_1 + P_2) = 0 \tag{55}$$

If the velocities of the particles are v_{11} and v_{12} before the interaction, and afterwards they are v_{21} and v_{22}, then

$$m_1 v_{11} + m_2 v_{12} = m_1 v_{21} + m_2 v_{22} \tag{56}$$

This law holds no matter how complicated the force is between particles. Similarly, if there are several particles, the momentum exchanged between each pair of particles adds up to zero, so the total change in momentum is zero. This conservation law applies to all interactions, including collisions and separations caused by explosive forces. It can also be generalized to situations where Newton's laws do not hold, for example in the theory of relativity and in electrodynamics [44–45].

The momentum equation for electrospinning modeling is formulated by considering the forces on a short segment of the jet [46–47].

$$\frac{d}{dz}(\pi R^2 \rho \upsilon^2) = \pi R^2 \rho g + \frac{d}{dz}\left[\pi R^2 (-p + \tau_{zz})\right] + \frac{\gamma}{R}.2\pi RR' + 2\pi R(t_t^e - t_n^e R') \quad (57)$$

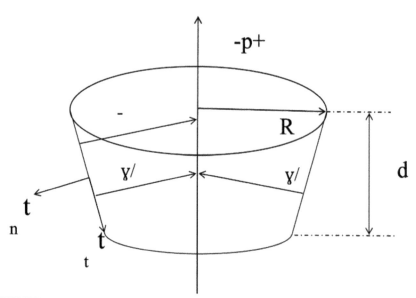

FIGURE 47   Momentum balance on a short section of the jet.

As it is shown in the Fig. 47, the element's angels could be defined as α and β. According to the mathematical relationships, it is obvious that:

$$\alpha + \beta = \pi \quad (58)$$

$$\sin \alpha = \tan \alpha \tag{59}$$
$$\cos \alpha = 1$$

The relationships (Fig. 47) between these electric forces are given below:

$$t_n^e \sin \alpha \cong t_n^e \tan \alpha \cong -t_n^e \tan \beta \cong -\frac{dR}{dz} t_n^e = -R' t_n^e \tag{60}$$

$$t_t^e \cos \alpha \cong t_t^e \tag{61}$$

So the effect of the electric forces in the momentum balance equation can be presented as:

$$2\pi RL(t_t^e - R't_n^e)dz \tag{62}$$

(Notation: In the main momentum equation, final formula is obtained by dividing into dz)

Generally, the normal electric force is defined as:

$$t_n^e \cong \frac{1}{2}\overline{\varepsilon}E_n^2 = \frac{1}{2}\overline{\varepsilon}(\frac{\sigma}{\overline{\varepsilon}})^2 = \frac{\sigma^2}{2\overline{\varepsilon}} \tag{63}$$

A few amount of electric forces is perished in vicinity of the air.

$$E_n = \frac{\sigma}{\overline{\varepsilon}} \tag{64}$$

The electric force can be presented by:

$$F = \frac{\Delta We}{\Delta l} = \frac{1}{2}(\varepsilon - \overline{\varepsilon})E^2 \times \Delta S \tag{65}$$

The force per surface unit is:

$$\frac{F}{\Delta S} = \frac{1}{2}(\varepsilon - \overline{\varepsilon})E^2 \tag{66}$$

Generally the electric potential energy is obtained by:

$$Ue = -We = -\int F.ds \tag{67}$$

$$\Delta We = \frac{1}{2}(\varepsilon - \bar{\varepsilon})E^2 \times \Delta V = \frac{1}{2}(\varepsilon - \bar{\varepsilon})E^2 \times \Delta S.\Delta l \tag{68}$$

So, finally it could be resulted:

$$t_n^e = \frac{\sigma^2}{2\bar{\varepsilon}} - \frac{1}{2}(\varepsilon - \bar{\varepsilon})E^2 \tag{69}$$

$$t_t^e = \sigma E \tag{70}$$

6.14.4.4   COULOMB'S LAW

Coulomb's law is a mathematical description of the electric force between charged objects which is formulated by the 18th-century French physicist Charles–Augustin de Coulomb. It is analogous to Isaac Newton's law of gravity. Both gravitational and electric forces decrease with the square of the distance between the objects, and both forces act along a line between them [48]. In Coulomb's law, the magnitude and sign of the electric force are determined by the electric charge, more than the mass of an object. Thus, charge which is a basic property matter determines how electromagnetism affects the motion of charged targets [49].

Coulomb force is thought to be the main cause for the instability of the jet in the electrospinning process [50]. This statement is based on the Earnshaw's theorem, named after Samuel Earnshaw [51] which claims that "A charged body placed in an electric field of force cannot rest in stable equilibrium under the influence of the electric forces alone". This theorem can be notably adapted to the electrospinning process [50]. The instability of charged jet influences on jet deposition and as a consequence on nanofiber formation. Therefore, some researchers applied developed models to the analysis of mechanisms of jet deposition and alignment on various collecting devices in arbitrary electric fields [52].

The equation for the potential along the centerline of the jet can be derived from Coulomb's law. Polarized charge density is obtained:

$$\rho_{p'} = -\vec{\nabla}.\vec{P}' \tag{71}$$

Where P' is polarization:

$$\vec{P}' = (\varepsilon - \bar{\varepsilon})\vec{E} \tag{72}$$

By substituting P' in equation:

$$\rho_{P'} = -(\bar{\varepsilon} - \varepsilon)\frac{dE}{dz'} \tag{73}$$

Beneficial charge per surface unit can be calculated as below:

$$\rho_{P'} = \frac{Q_b}{\pi R^2} \tag{74}$$

$$Q_b = \rho_b.\pi R^2 = -(\bar{\varepsilon} - \varepsilon)\pi R^2 \frac{dE}{dz'} \tag{75}$$

$$Q_b = -(\bar{\varepsilon} - \varepsilon)\pi \frac{d(ER^2)}{dz'} \tag{76}$$

$$\rho_{sb} = Q_b.dz' = -(\bar{\varepsilon} - \varepsilon)\pi \frac{d}{dz'}(ER^2)dz' \tag{77}$$

The main equation of Coulomb's law:

$$F = \frac{1}{4\pi\varepsilon_0}\frac{qq_0}{r^2} \tag{78}$$

The electric field is:

$$E = \frac{1}{4\pi\varepsilon_0}\frac{q}{r^2} \tag{79}$$

The electric potential can be measured:

$$\Delta V = -\int E.dL \tag{80}$$

$$V = \frac{1}{4\pi\varepsilon_0}\frac{Q_b}{r} \tag{81}$$

According to the beneficial charge equation, the electric potential could be rewritten as:

$$\Delta V = Q(z) - Q_\infty(z) = \frac{1}{4\pi\bar{\varepsilon}} \int \frac{(q - Q_b)}{r} dz' \tag{82}$$

$$Q(z) = Q_\infty(z) + \frac{1}{4\pi\bar{\varepsilon}} \int \frac{q}{r} dz' - \frac{1}{4\pi\bar{\varepsilon}} \int \frac{Q_b}{r} dz' \tag{83}$$

$$Q_b = -(\bar{\varepsilon} - \varepsilon)\pi \frac{d(ER^2)}{dz'} \tag{84}$$

The surface charge density's equation is:

$$q = \sigma.2\pi RL \tag{85}$$

$$r^2 = R^2 + (z - z')^2 \tag{86}$$

$$r = \sqrt{R^2 + (z - z')^2} \tag{87}$$

The final equation, which obtained by substituting the mentioned equations is:

$$Q(z) = Q_\infty(z) + \frac{1}{4\pi\bar{\varepsilon}} \int \frac{\sigma.2\pi R}{\sqrt{(z - z')^2 + R^2}} dz' - \frac{1}{4\pi\bar{\varepsilon}} \int \frac{(\bar{\varepsilon} - \varepsilon)\pi}{\sqrt{(z - z')^2 + R^2}} \frac{d(ER^2)}{dz'} \tag{88}$$

It is assumed that β is defined:

$$\beta = \frac{\varepsilon}{\bar{\varepsilon}} - 1 = -\frac{(\bar{\varepsilon} - \varepsilon)}{\bar{\varepsilon}} \tag{89}$$

So, the potential equation becomes:

$$Q(z) = Q_\infty(z) + \frac{1}{2\bar{\varepsilon}} \int \frac{\sigma.R}{\sqrt{(z - z')^2 + R^2}} dz' - \frac{\beta}{4} \int \frac{1}{\sqrt{(z - z')^2 + R^2}} \frac{d(ER^2)}{dz'} \tag{90}$$

The asymptotic approximation of χ is used to evaluate the integrals mentioned above:

$$\chi = \left(-z + \xi + \sqrt{z^2 - 2z\xi + \xi^2 + R^2} \right) \tag{91}$$

Where χ is "aspect ratio" of the jet (L = length, R_0= Initial radius).

This leads to the final relation for the axial electric field:

$$E(z) = E_\infty(z) - \ln \chi \left(\frac{1}{\varepsilon} \frac{d(\sigma R)}{dz} - \frac{\beta}{2} \frac{d^2(ER^2)}{dz^2} \right) \tag{92}$$

6.14.4.5 CONSTITUTIVE EQUATIONS

In modern condensed matter physics, the constitutive equation plays a major role. In physics and engineering, a constitutive equation or relation is a relation between two physical quantities that is specific to a material or substance, and approximates the response of that material to external stimulus, usually as applied fields or forces [53]. There are a sort of mechanical equation of state, and describe how the material is constituted mechanically. With these constitutive relations, the vital role of the material is reasserted [54]. There are two groups of constitutive equations: Linear and nonlinear constitutive equations [55]. These equations are combined with other governing physical laws to solve problems; for example in fluid mechanics the flow of a fluid in a pipe, in solid state physics the response of a crystal to an electric field, or in structural analysis, the connection between applied stresses or forces to strains or deformations [53].

The first constitutive equation (constitutive law) was developed by Robert Hooke and is known as Hooke's law. It deals with the case of linear elastic materials. Following this discovery, this type of equation, often called a "stress-strain relation" in this example, but also called a "constitutive assumption" or an "equation of state" was commonly used [56]. Walter Noll advanced the use of constitutive equations, clarifying their classification and the role of invariance requirements, constraints, and definitions of terms like "material", "isotropic", "aeolotropic", etc. The class of "constitutive relations" of the form stress rate = f (velocity gradient, stress, density) was the subject of Walter Noll's dissertation in 1954 under Clifford Truesdell [53]. There are several kinds of constitutive equations, which are applied commonly in electrospinning. Some of these applicable equations are discussed as following:

6.14.4.5.1 OSTWALD-DE WAELE POWER LAW

Rheological behavior of many polymer fluids can be described by power law constitutive equations [55]. The equations that describe the dynamics in electrospinning constitute, at a minimum, those describing the conservation of mass, momentum and charge, and the electric field equation. Additionally a constitutive equation for the fluid behavior is also required [57]. A Power-law fluid, or the

Ostwald-de Waele relationship, is a type of generalized Newtonian fluid for which the shear stress, τ, is given by:

$$\tau = K\left(\frac{\partial v}{\partial y}\right)^{m} \tag{93}$$

Which $\partial v/\partial y$ is the shear rate or the velocity gradient perpendicular to the plane of shear. The power law is only a good description of fluid behavior across the range of shear rates to which the coefficients are fitted. There are a number of other models that better describe the entire flow behavior of shear-dependent fluids, but they do so at the expense of simplicity, so the power law is still used to describe fluid behavior, permit mathematical predictions, and correlate experimental data [47, 58].

Nonlinear rheological constitutive equations applicable for polymer fluids (Ostwald-de Waele power law) were applied to the electrospinning process by Spivak and Dzenis [59–61].

$$\hat{\tau}^c = \mu\left[tr\left(\dot{\hat{\gamma}}^2\right)\right]^{(m-1)/2} \dot{\hat{\gamma}} \tag{94}$$

$$\mu = K\left(\frac{\partial v}{\partial y}\right)^{m-1} \tag{95}$$

Viscous Newtonian fluids are described by a special case of equation above with the flow index $m=1$. Pseudoplastic (shear thinning) fluids are described by flow indices $0 \leq m \leq 1$. Dilatant (shear thickening) fluids are described by the flow indices $m>1$ [60].

6.14.4.5.2   GIESEKUS EQUATION

In 1966, Giesekus established the concept of anisotropic forces and motions into polymer kinetic theory. With particular choices for the tensors describing the anisotropy, one can obtained Giesekus constitutive equation from elastic dumbbell kinetic theory [62–63]. The Giesekus equation is known to predict, both qualitatively and quantitatively, material functions for steady and non-steady shear and elongational flows. However, the equation sustains two drawbacks: it predicts that the viscosity is inversely proportional to the shear rate in the limit of infinite shear rate and it is unable to predict any decrease in the elongational viscosity with increasing elongation rate in uniaxial elongational flow. The first one is not serious because of retardation time, which is included in the constitutive equation but the second one is more critical because the elongational viscosity of some polymers decrease with increasing of elongation rate [64–65].

In the main Giesekus equation, the tensor of excess stresses depending on the motion of polymer units relative to their surroundings was connected to a sequence of tensors characterizing the configurational state of the different kinds of network structures present in the concentrated solution or melt. The respective set of constitutive equations indicates [66–67]:

$$S_k + \eta \frac{\partial C_k}{\partial t} = 0 \tag{96}$$

The equation below indicates the upper convected time derivative (Oldroyd derivative):

$$\frac{\partial C_k}{\partial t} = \frac{DC_k}{Dt} - \left[C_k \nabla \upsilon + (\nabla \upsilon)^T C_k \right] \tag{97}$$

(Note: The upper convective derivative is the rate of change of some tensor property of a small parcel of fluid that is written in the coordinate system rotating and stretching with the fluid.)

C_k also can be measured as following:

$$C_k = 1 + 2E_k \tag{98}$$

According to the concept of "recoverable strain" S_k may be understood as a function of E_k and vice versa. If linear relations corresponding to Hooke's law are adopted.

$$S_k = 2\mu_k E_k \tag{99}$$

So:

$$S_k = \mu_k (C_k - 1) \tag{100}$$

The Eq. (96) becomes:

$$S_k + \lambda_k \frac{\partial S_k}{\partial t} = 2\eta D \tag{101}$$

$$\lambda_k = \frac{\eta}{\mu_k} \tag{102}$$

As a second step in order to rid the model of the shortcomings is the scalar mobility constants B_k, which are contained in the constants η. This mobility constant can be represented as:

$$\tfrac{1}{2}(\beta_k S_k + S_k \beta_k) + \bar{\eta}\frac{\partial C_k}{\partial t} = 0 \tag{103}$$

The two parts of Eq. (103) reduces to the single constitutive equation:

$$\beta_k + \bar{\eta}\frac{\partial C_k}{\partial t} = 0 \tag{104}$$

The excess tension tensor in the deformed network structure where the well-known constitutive equation of a so-called Neo-Hookean material is proposed [66, 68]:

Neo-Hookean equation: $S_k = 2\mu_k E_k = \mu_k(C_k - 1)$ $\qquad$ (105)

$$\mu_k = NKT$$

$$\beta_k = 1 + \alpha(C_k - 1) = (1 - \alpha) + \alpha C_k \tag{106}$$

where K is Boltzmann's constant.

By substitution Eqs. (105) and (106) in the equation (105), it can obtained where the condition 0≤α≤1 must be fulfilled, the limiting case α=0 corresponds to an isotropic mobility [69].

$$0 \le \alpha \le 1 \quad [1 + \alpha(C_k - 1)](C_k - 1) + \lambda_k \frac{\partial C_k}{\partial t} = 0 \tag{107}$$

$$\alpha = 1 \quad C_k(C_k - 1) + \lambda_k \frac{\partial C_k}{\partial t} = 0 \tag{108}$$

$$0 \le \alpha \le 1 \quad C_k = \frac{S_k}{\mu_k} + 1 \tag{109}$$

By substituting equations above in Eq. (102), it becomes:

$$\left[1+\frac{\alpha S_k}{\mu_k}\right]\frac{S_k}{\mu_k}+\lambda_k\frac{\partial C_k}{\partial t}=0 \tag{110}$$

$$\frac{S_k}{\mu_k}+\frac{\alpha S_k^2}{\mu_k^2}+\lambda_k\frac{\partial(S_k/\mu_k+1)}{\partial t}=0 \tag{111}$$

$$\frac{S_k}{\mu_k}+\frac{\alpha S_k^2}{\mu_k^2}+\frac{\lambda_k}{\mu_k}\frac{\partial S_k}{\partial t}=0 \tag{112}$$

$$S_k+\frac{\alpha S_k^2}{\mu_k}+\lambda_k\frac{\partial S_k}{\partial t}=0 \tag{113}$$

D means the rate of strain tensor of the material continuum [66].

$$D=\frac{1}{2}\left[\nabla v+(\nabla v)^T\right] \tag{114}$$

The equation of the upper convected time derivative for all fluid properties can be calculated as:

$$\frac{\partial\otimes}{\partial t}=\frac{D\otimes}{Dt}-\left[\otimes.\nabla v+(\nabla v)^T.\otimes\right] \tag{115}$$

$$\frac{D\otimes}{Dt}=\frac{\partial\otimes}{\partial t}+\left[(v.\nabla).\otimes\right] \tag{116}$$

By replacing S_k instead of the symbol:

$$\lambda_k\frac{\partial S_k}{\partial t}=\lambda_k\frac{DS_k}{Dt}-\lambda_k\left[S_k\nabla v+(\nabla v)^T S_k\right]=\lambda_k\frac{DS_k}{Dt}-\lambda_k(v.\nabla)S_k \tag{117}$$

By simplification the equation above:

$$S_k+\frac{\alpha S_k^2}{\mu_k}+\lambda_k\frac{DS_k}{Dt}=\lambda_k(v.\nabla)S_k \tag{118}$$

$$S_k=2\mu_k E_k \tag{119}$$

The assumption of $E_k=1$ would lead to the next equation:

$$S_k + \frac{\alpha \lambda_k S_k^2}{\eta} + \lambda_k \frac{DS_k}{Dt} = \frac{\eta}{\mu_k}(2\mu_k)D = 2\eta D = \eta\left[\nabla v + (\nabla v)^T\right] \qquad (120)$$

In electrospinning modeling articles τ is used commonly instead of S_k [36, 39, 70].

$$S_k \leftrightarrow \tau$$

$$\tau + \frac{\alpha \lambda_k \tau^2}{\eta} + \lambda_k \tau_{(1)} = \eta\left[\nabla v + (\nabla v)^T\right] \qquad (121)$$

6.14.4.5.3   MAXWELL EQUATION

Maxwell's equations are a set of partial differential equations that, together with the Lorentz force law, form the foundation of classical electrodynamics, classical optics, and electric circuits. These fields are the bases of modern electrical and communications technologies. Maxwell's equations describe how electric and magnetic fields are generated and altered by each other and by charges and currents. They are named after the Scottish physicist and mathematician James Clerk Maxwell who published an early form of those equations between 1861 and 1862 [71–72].

The simplest model of flexible macromolecules in a dilute solution is the elastic dumbbell (or bead-spring) model. This has been widely used for purely mechanical theories of the stress in electrospinning modeling [73].

Maxwell constitutive equation was first applied by Reneker et al. (2000). Consider an electrified liquid jet in an electric field parallel to its axis. They modeled a segment of the jet by a viscoelastic dumbbell. They used Gaussian electrostatic system of units. According to this model each particle in the electric field exerts repulsive force on another particle. Therefore the stress between these particles can be measured by [52]:

$$\dot{\tau} = G\left(\varepsilon' - \frac{\tau}{\eta}\right) \qquad (122)$$

The stress can be calculated by a Maxwell viscoelastic constitutive equation [74]:

$$\dot{\tau} = G\left(\varepsilon' - \frac{\tau}{\eta}\right) \qquad (123)$$

Where ε' is the Lagrangian axial strain:

$$\varepsilon' \equiv \frac{\partial \dot{x}}{\partial \xi} . \hat{i}.$$

(124)

6.14.4.6 SCALING

Physical aspect of a phenomenon can use the language of differential equation, which represents the structure of the system by selecting the variables that characterize the state of it, and certain mathematical constraint on the values of those variables can take on. These equations can predict the behavior of the system over a quantity like time. For an instance, a set of continuous functions of time that describe the way the variables of the system developed over time starting from a given initial state [75]. In general, the renormalization group theory, scaling and fractal geometry, are applied to the understanding of the complex phenomena in physics, economics and medicine [76].

In more recent times, in statistical mechanics, the expression "scaling laws" has referred to the homogeneity of form of the thermodynamic and correlation functions near critical points, and to the resulting relations among the exponents that occur in those functions. From the viewpoint of scaling, electrospinning modeling can be studied in two ways, allometric and dimensionless analysis. Scaling and dimensional analysis actually started with Newton, and allometry exists everywhere in our day life and scientific activity [76–77].

6.14.4.6.1 ALLOMETRIC SCALING

Electrospinning applies electrically generated motion to spin fibers. So, it is difficult to predict the size of the produced fibers, which depends on the applied voltage in principal. Therefore, the relationship between radius of jet and the axial distance from nozzle is always the subject of investigation [78–79]. It can be described as an allometric equation by using the values of the scaling exponent for the initial steady, instability and terminal stages [80].

The relationship between r and z can be expressed as an allometric equation of the form:

$$r \approx z^{b}$$

(125)

When the power exponent, $b = 1$ the relationship is isometric and when $b \neq 1$ the relationship is allometric [78, 81]. In another view, $b = -1/2$ is considered for the straight jet, $b = -1/4$ for instability jet and $b = 0$ for finally stage [55, 79].

Due to high electrical force acting on the jet, it can be illustrated [78]:

$$\frac{d}{dz}\left(\frac{v^2}{2}\right) = \frac{2\sigma E}{\rho r} \tag{126}$$

Equations of mass and charge conservations applied here as mentioned before [78, 81–82]

From the above equations it can be seen that [39, 78]

$$v \approx r^{-2}, \; \sigma \approx r, \; E \approx r^{-2}, \; \frac{d v^2}{dz} \approx r^{-2}, \; r \approx z^{-\frac{1}{2}} \tag{127}$$

The charged jet can be considered as a one-dimensional flow as mentioned. If the conservation equations modified, they would change as [78]:

$$2\pi r \sigma^{\alpha} v + K \pi r^2 E = I \tag{128}$$

$$r \approx z^{-\alpha/(\alpha+1)} \tag{129}$$

Where α is a surface charge parameter, the value of α depends on the surface charge in the jet. When $\alpha = 0$ no charge in jet surface, and in $\alpha = 1$ use for full surface charge.

Allometric scaling equations are more widely investigated by different researchers. Some of the most important allometric relationships for electrospinning are presented in Table 2.

TABLE 2   Investigated scaling laws applied in electrospinning model.

Parameters	Equation	Ref.
The conductance and polymer concentration	$g \approx c^{\beta}$	[39]
The fiber diameters and the solution viscosity	$d \approx \eta^{\alpha}$	[79]
The mechanical strength and threshold voltage	$\sigma \approx E_{threshold}^{-\alpha}$	[83]
The threshold voltage and the solution viscosity	$E_{threshold} \approx \eta^{1/4}$	[83]
The viscosity and the oscillating frequency	$\eta \approx \omega^{-0.4}$	[83]
the volume flow rate and the current	$I \approx Q^b$	[82]
The current and the fiber radius	$I \approx r^2$	[84]

TABLE 2 *(Continued)*

Parameters	Equation	Ref.
The surface charge density and the fiber radius	$\sigma \approx r^3$	[84]
The induction surface current and the fiber radius	$\phi \approx r^2$	[84]
The fiber radius and AC frequency	$r \approx \Omega^{1/4}$	[55]

β, α and b= scaling exponent

6.14.4.6.2 DIMENSIONLESS ANALYSIS

One of the simplest, yet most powerful, tools in the physics is dimensional analysis in which there are two kinds of quantities: dimensionless and dimensional. Dimensionless quantities, which are without associated physical dimensions, are widely used in mathematics, physics, engineering, economics, and in everyday life (such as in counting). Numerous well-known quantities, such as π, e, and φ, are dimensionless. They are "pure" numbers, and as such always have a dimension of 1 [85–86].

Dimensionless quantities are often defined as products or ratios of quantities that are not dimensionless, but whose dimensions cancel out when their powers are multiplied [87].

In non-dimensional scaling, there are two key steps:
(a) Identify a set of physically-relevant dimensionless groups, and
(b) Determine the scaling exponent for each one.

Dimensional analysis will help you with step (a), but it cannot be applicable possibly for step (b).

A good approach to systematically getting to grips with such problems is through the tools of dimensional analysis (Bridgman, 1963). The dominant balance of forces controlling the dynamics of any process depends on the relative magnitudes of each underlying physical effect entering the set of governing equations [88]. Now, the most general characteristics parameters, which used in dimensionless analysis in electrospinning are introduced in Table 3.

TABLE 3 Characteristics parameters employed and their definitions.

Parameter	Definition
Length	R_0
Velocity	$v_0 = \dfrac{Q}{\pi R_0^2 K}$

TABLE 3 *(Continued)*

Parameter	Definition
Electric field	$E_0 = \dfrac{I}{\pi R_0^2 K}$
Surface charge density	$\sigma_0 = \bar{\varepsilon} E_0$
Viscose stress	$\tau_0 = \dfrac{\eta_0 \upsilon_0}{R_0}$

For achievement of simplified form of equations and reduction a number of unknown variables, the parameters should be subdivided into characteristics scales in order to become dimensionless. Electrospinning dimensionless groups are shown in Table 4 [89].

TABLE 4   Dimensionless groups employed and their definitions.

Name	Definition	Field of application
Froude number	$Fr = \dfrac{\upsilon_0^2}{g R_0}$	The ratio of inertial to gravitational forces
Reynolds number	$Re = \dfrac{\rho \upsilon_0 R_0}{\eta_0}$	The ratio of the inertia forces to the viscos forces
Weber number	$We = \dfrac{\rho \upsilon_0^2 R_0}{\gamma}$	The ratio of the surface tension forces to the inertia forces
Deborah number	$De = \dfrac{\lambda \upsilon_0}{R_0}$	The ratio of the fluid relaxation time to the instability growth time
Electric Peclet number	$Pe = \dfrac{2 \bar{\varepsilon} \upsilon_0}{K R_0}$	The ratio of the characteristic time for flow to that for electrical conduction
Euler number	$Eu = \dfrac{\varepsilon_0 E^2}{\rho \upsilon_0^2}$	The ratio of electrostatic forces to inertia forces

TABLE 4   *(Continued)*

Name	Definition	Field of application
Capillary number	$Ca = \dfrac{\eta v_0}{\gamma}$	The ratio of inertia forces to viscose forces
Ohnesorge number	$oh = \dfrac{\eta}{\left(\rho \gamma R_0\right)^{1/2}}$	The ratio of viscose force to surface force
Viscosity ratio	$r_\eta = \dfrac{\eta_p}{\eta_0}$	The ratio of the polymer viscosity to total viscosity
Aspect ratio	$\chi = \dfrac{L}{R_0}$	The ratio of the length to the primary radius of jet
Electrostatic force parameter	$\varepsilon = \dfrac{\overline{\varepsilon} E_0^2}{\rho v_0^2}$	The relative importance of the electrostatic and hydrodynamic forces
Dielectric constant ratio	$\beta = \dfrac{\varepsilon}{\overline{\varepsilon}} - 1$	The ratio of the field without the dielectric to the net field with the dielectric

The governing and constitutive equations can be transformed into dimensionless form using the dimensionless parameters and groups.

6.14.5   SOME OF ELECTROSPINNING MODELS

The most important mathematical models for electrospinning process are classified in the Table 5. According to the year, advantages and disadvantages of the models:

TABLE 5   The most important mathematical models for electrospinning.

Researchers	Model	Year	Ref.
Taylor, G. I.	Leaky dielectric model	1969	[90]
Melcher, J. R.	• Dielectric fluid • Bulk charge in the fluid jet considered to be zero • Only axial motion • Steady state part of jet		

TABLE 5   *(Continued)*

Researchers	Model	Year	Ref.
Ramos	Slender body	1996	[91]
	• Incompressible and axi-symertric and viscose jet under gravity force • No electrical force • Jet radius decreases near zero • Velocity and pressure of jet only change during axial direction • Mass and volume control equations and Taylor expansion were applied to predict jet radius		
Saville, D. A.	Electrohydrodynamic model	1997	[90]
	• The hydrodynamic equations of dielectric model was modified • Using dielectric assumption • This model can predict drop formation • Considering jet as a cylinder (ignoring diameter reduction) • Only for steady state part of the jet		
Spivak, A. Dzenis, Y.	Spivak and Dzenis model	1998	[92]
	• The motion of viscose fluid jet with the low conductivity were surveyed in a external electric field • Single Newtonian Fluid jet • the electric field assumed to be uniform and constant, unaffected by the charges carried by the jet • Use asymptotic approximation were applied in a long distance from nozzle • Tangential electric force assumed to be zero • Using non-linear rheological constitutive equation (Ostwald_dewaele law), non-linear behavior of fluid jet were investigated		

TABLE 5   *(Continued)*

Researchers	Model	Year	Ref.
Jong Wook	Droplet formation model • Droplet formation of charged fluid jet was studied in this model • The ratio of mass, energy and electric charge transition are the most important parameters on droplet formation • Deformation and break-up of droplets were investigated too • Newtonian and Non–Newtonian fluids • Only for high conductive and viscose fluids	2000	[93]
Reneker, D. H. Yarin, A. L.	Reneker model • For description of instabilities in viscoelastic jets • Using molecular chains theory, behavior of polymer chain by spring-bead model in electric field was studied • Electric force based on electric field cause instability of fluid jet while repulsion force between surface charges make perturbation and bending instability • The motion path of these two cases were studied • Governing equations: momentum balance, motion equations for each bead, Maxwell tension and columbic eqs.	2000	[52]

TABLE 5 *(Continued)*

Researchers	Model	Year	Ref.
Hohman, M. Shin, M.	Stability theory • This model is based on dielectric model with some modification for Newtonian fluids. • This model can describe whipping, bending and Rayleigh instabilities and introduced new ballooning instability. • 4 motion regions were introduced: dipping mode, spindle mode, oscillating mode, precession mode. • Surface charge density introduced as the most effective parameter on instability formation. • Effect of fluid conductivity and viscosity on nanofibers diameter were discussed. • Steady solutions may be obtained only if the surface charge density at the nozzle is set to zero or a very low value	2001	[94]
Feng, J. J	Modifying Hohman model • For both Newtonian and non–Newtonian fluids • Unlike Hohman model, the initial surface charge density was not zero, so the "ballooning instability" did not accrue. • Only for steady state part of the jet • Simplifying the electric field equation which Hohman used in order to eliminating Ballooning instability.	2002	[46]
Wan-Guo-Pan	Wan–Guo–Pan model • They introduced thermo-electro-hydro dynamics model in electrospinning process • This model is modification on Spivak model which mentioned before • The governing equations in this model: Modified Maxwell equation, Navier–Stocks equations, and several rheological constitutive equation.	2004	[61]

TABLE 5   *(Continued)*

Researchers	Model	Year	Ref.
Ji-Haun	AC-electrospinning model	2005	[55]
	• Whipping instability in this model was distinguished as the most effective parameter on uncontrollable deposition of nanofibers • Applying AC current can reduce this instability so make oriented nanofibers • This model found a relationship between axial distance from nozzle and jet diameter • This model also connected AC frequency and jet diameter		
Roozemond (Eindhoven University and Technology)	Combination of slender body and dielectric model • In this model, a new model for viscoelastic jets in electrospinning were presented by combining these two models • All variables were assumed uniform in cross section of the jet but they changed in during z direction • Nanofiber diameter can be predicted	2007	[95]
Wan	Electromagnetic model • Results indicated that the electromagnetic field which made because of electrical field in charged polymeric jet is the most important reason of helix motion of jet during the process	2012	[26]
Dasri	Dasri model • This model was presented for description of unstable behavior of fluid jet during electrospinning • This instability causes random deposition of nanofiber on surface of the collector • This model described dynamic behavior of fluid by combining assumption of Reneker and Spivak models	2012	[96]

The most frequent numeric mathematical methods, which were used in different models are listed in Table 6.

TABLE 6 Applied numerical methods for electrospinning.

Method	Ref.
Relaxation method	[33, 36, 46]
Boundary integral method (boundary element method)	[74, 93]
Semi-inverse method	[36, 55]
(Integral) control-volume formulation	[91]
Finite element method	[90]
Kutta-Merson method	[97]
Lattice Boltzmann method with finite difference method	[43]

6.14.6 ELECTROSPINNING SIMULATION EXAMPLE

In order to survey of electrospinning modeling application, its main equations were applied for simulating the process according to the constants, which summarized in Table 7.

Mass and charge conservations allow v and σ to be expressed in terms of R and E, and the momentum and E-field equations can be recast into two second-order ordinary differential equations for R and E. The slender-body theory (the straight part of the jet) was assumed to investigate jet behavior during the spinning distance. The slope of the jet surface (R') is maximum at the origin of the nozzle. The same assumption has been used in most previous models concerning jets or drops. The initial and boundary conditions, which govern the process are introduced as:

Initial values (z=0):

$$R(0) = 1$$
$$E(0) = E_0$$
$$\tau_{prr} = 2r_\eta \frac{R_0'}{R_0^3}$$
$$\tau_{pzz} = -2\tau_{prr}$$

(130)

Feng [46] indicated that E(0) effect is limited to a tiny layer below the nozzle which its thickness is a few percent of R_0. It was assumed that the shear inside the nozzle is effective in stretching of polymer molecules as compered with the following elongation.

Boundary values ($z=\chi$):

$$R(\chi) + 4\chi R'(\chi) = 0$$
$$E(\chi) = E_\chi$$
$$\tau_{prr} = 2r_\eta \frac{R_\chi'}{R_\chi^3} \tag{90}$$
$$\tau_{pzz} = -2\tau_{prr}$$

The asymptotic scaling can be stated as [46]:

$$R(z) \propto z^{-1/4} \tag{91}$$

Just above the deposit point ($z=\chi$), asymptotic thinning conditions applied. R drops towards zero and E approaches E. The electric field is not equal to E, so we assumed a slightly larger value, E_χ.

TABLE 7 Constants used in electrospinning simulation.

Constant	Quantity
Re	$2.5. 10^{-3}$
We	0.1
Fr	0.1
Pe	0.1
De	10
ε	1
β	40
χ	20
E_0	0.7
E_χ	0.5
r_η	0.9

The momentum, electric field and stress equations could be rewritten into a set of four coupled first order ordinary differential equations (ODE's) with the above-mentioned boundary conditions. Numerical relaxation method has been chosen to solve the generated boundary value problem.

The results of these systems of equations are presented in Figs. 48 and 49 that matched quite well with the other studies that have been published [36, 38, 46, 70, 98].

The variation prediction of R, R', ER², ER²' and E versus axial position (z) are shown in Fig. 48. Physically, the amount of counductable charges reduces with decreasing jet radius. Therefore, to maintain the same jet current, more surface charges should be carried by the convection. Moreover, in the considered simulation region, the density of surface charge gradually increased. As the jet gets thinner and faster, electric conduction gradually transfers to convection. The electric field is mainly induced by the axial gradient of surface charge, thus, it is insensitive to the thinning of the electrospun jet:

$$\frac{d(\sigma R)}{dz} \approx -\left(2R\frac{dR}{dz}\right)/Pe \qquad (131)$$

Therefore, the variation of E versus z can be written as:

$$\frac{d(E)}{dz} \approx \ln \chi \left(\frac{d^2 R^2}{dz^2}\right)/Pe \qquad (132)$$

Downstream of the origin, E shoots up to a peak and then relax due to the decrease of electrostatic pulling force in consequence of the reduction of surface charge density, if the current was held at a constant value. However, in reality, the increase of the strength of the electric field also increases the jet current, which is relatively linearly [46, 98]. As the jet becomes thinner downstream, the increase of jet speed reduces the surface charge density and thus E, so the electric force exerted on the jet and thus R' become smaller. The rates of R and R' are maximum at z=0, and then relaxes smoothly downstream toward zero [36, 46]. According to the relation between R, E and z, ER² and ER²' vary in accord with parts (c) and (d) in Fig. 48.

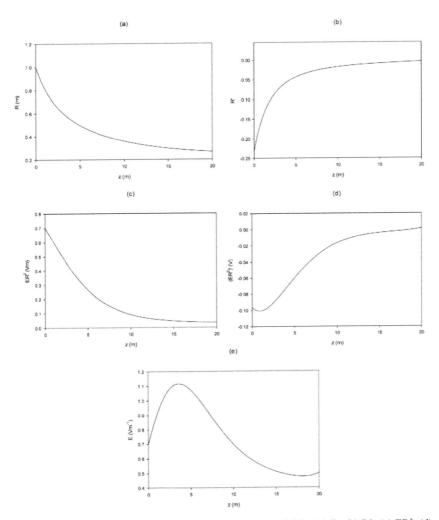

FIGURE 48 Solutions given by the electrospinning model for (a) R; (b) R'; (c) ER^2; (d) $ER^{2'}$ and (e) E.

Figure 49 shows the changes of axial, radial shear stress and the difference between them, the tensile force (T) versus z. The polymer tensile force is much larger in viscoelastic polymers because of the strain hardening. T also has an initial rise, because the effect of strain hardening is so strong that it overcomes the shrinking radius of the jet. After the maximum value of T, it reduces during the jet thinning. As expected, the axial polymer stress rises, because the fiber is stretched in axial direction, and the radial polymer stress declines. The variation of T along the jet can be nonmonotonic, however, meaning the viscous normal stress may promote or resist stretching in a different part of the jet and under different conditions [36, 46].

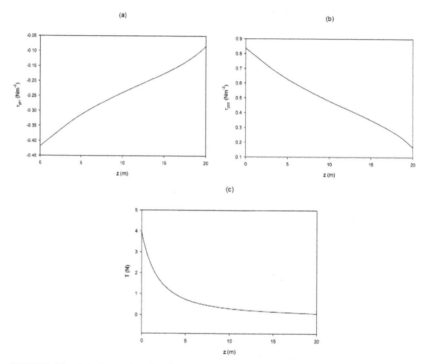

FIGURE 49   Solutions given by the electrospinning model for (a) t_{prr}; (b) t_{pzz} and (c) T.

6.15   MECHANISM OF NANOSTRUCTURE FORMATION

The notion "cluster" assumes energy-wise compensated nucleus with a shell, the surface energy of which is rather small, as a result under given conditions the cluster represents a stable formation. Nanocomposites of various shapes can be obtained either by the dispersion of a substance in a definite medium, or condensation or synthesis from low-molecular chemical particles (Fig. 50). Thus, the techniques for obtaining nanostructures can be classified by the mechanism of their formation. Other features, by which the methods for nanostructure production can be classified, comprise the variants of nanoproduct formation process by the change of energy consumption. A temperature or an energy factor is usually evident here. Besides, a so-called apparatus factor plays an important role together with the aforesaid features. At present, nanostructural "formations" of various shapes and compositions are obtained in a rather wide region of actions upon the chemical particles and substances.

In chemical literature a cluster is equated with a complex compound containing a nucleus and a shell. Usually a nucleus consists of metal atoms combined with metallic bond, and a shell of ligands. Manganese carbonyls $[(Co)_5MnMn(Co)_5]$ and cobalt carbonyls $[Co_6(Co)_{18}]$, nickel pentadienyls $[Ni_6(C_5H_6)_6]$ belong to elementary clusters.

In recent years, the notion "cluster" has got an extended meaning. At the same time the nucleus can contain not only metals or not even contain metals. In some clusters, for instance, carbon ones there is no nucleus at all. In this case their shape can be characterized as a sphere (icosahedron, to be precise)—fullerenes, or as a cylinder—fullerene tubules. Surely a certain force field is formed by atoms on internal walls inside such particles. It can be assumed that electrostatic, electromagnetic, and gravitation fields conditioned by corresponding properties of atoms contained in particle shells can be formed inside tubules and fullerenes. If analyze papers recently published, it should be noted that a considerable exceeding of surface size over the volume and, consequently, a relative growth of the surface energy in comparison with the growth of volume and potential energy is the main feature of clusters. If particle dimensions (diameters of "tubes" and "spheres") change from 1 up to several hundred nanometers, they would be called nanoparticles. In some papers, the area of nanoclusters existence is within 1–10 nm.

Based on classical definitions, in given paper, metal nanoparticles and nanocrystals are referred to as nanoclusters. Apparently, the difference of nanoparticles from other particles (smaller or larger) is determined by their specific characteristics. The search of nanoworld distinctions from atomic-molecular, micro- and macroworld can lead to finding analogies and coincidences in colloid chemistry, chemistry of polymers, and coordination compounds. Firstly, it should be noted that nanoparticles usually represent a small collective aggregation of atoms being within the action of adjacent atoms, thus conditioning the shape of nanoparticles. A nanoparticle shape can vary depending upon the nature of adjacent atoms and character of formation medium. Obviously, the properties of separate atoms and molecules (of small size) are determined by their energy and geometry characteristics, the determinative role being played by electron properties. In particular, electron interactions determine the geometry of molecules and atomic structures of small size and mobility of these chemical particles in media, as well as their activity or reactivity.

When the number of atoms in chemical particle exceeds 30, a certain stabilization of its shape being also conditioned by collective influence of atoms constituting the particle is observed. Simultaneously, the activity of such a particle remains high but the processes with its participation have a directional character. The character of interactions with the surroundings of such structures is determined by their formation mechanism.

During polymerization or co-polymerization the influence of macromolecule growth parameters changes with the increase of the number of elementary acts of its growth. According to scientists, after 7–10 acts the shape or geometry of nanoparticles formed becomes the main determinative factor providing the further growth of macromolecule (chain development). A nanoparticle shape is usually determined not only by its structural elements but also by its interactions with surrounding chemical particles.

Due to the overlapping of different classification features, it is appropriate to present a set of diagrams by main features. For instance, the methods for nanoparticle formation by substance dispersion and chemical particle condensation can be identified with physical and chemical methods, though such decision is incorrect, since substance destruction methods can contain both chemical and physical impacts. In turn, when complex nanostructures are formed from simple ones, both purely physical and chemical factors are possible. However, in the process of substance dispersion high-energy sources, such as electric arc, laser pyrolysis, plasma sources, mechanical crushing or grinding should be applied.

From the aforesaid, it can be concluded that the possibility of self-organization of nanoparticles with the formation of corresponding nanosystems and nanomaterials is the main distinction of nanoworld from pico-, micro-, and macroworld. Recently, much attention has been paid to synergetics or the branch of science dealing with self-organization processes since these processes, in many cases, proceeds with small energy consumption and, consequently, is more ecologically clear in comparison with existing technological processes.

In turn, nanoparticle dimensions are determined by its formation conditions. When the energy consumed for macroparticle destruction or dispersion over the surface increases, the dimensions of nanomaterials are more likely to decrease. The notion "nanomaterial" is not strictly defined. Several researchers consider nanomaterials to be aggregations of nanocrystals, nanotubes, or fullerenes. Simultaneously, there is a lot of information available that nanomaterials can represent materials containing various nanostructures. The most attention researchers pay to metallic nanocrystals. Special attention is paid to metallic nanowires and nanofibers with different compositions.

Here are some names of nanostructures:

1) fullerenes, 2) gigantic fullerenes, 3) fullerenes filled with metal ions, 4) fullerenes containing metallic nucleus and carbon (or mineral) shell, 5) one-layer nanotubes, 6) multi-layer nanotubes, 7) fullerene tubules, 8) "scrolls", 9) conic nanotubes, 10) metal-containing tubules, 11) "onions", 12) "Russian dolls", 13) bamboo-like tubules, 14) "beads", 15) welded nanotubes, 16) bunches of nanotubes, 17) nanowires, 18) nanofibers, 19) nanoropes, 20) nanosemi-spheres (nanocups), 21) nanobands and similar nanostructures, as well as various derivatives from enlisted structures. It is quite possible that a set of such structures and notions will be enriched.

In most cases nanoparticles obtained are bodies of rotation or contain parts of bodies of rotation. In natural environment there are minerals containing fullerenes or representing thread-like formations comprising nanometer pores or structures. In the first case, it is talked about schungite that is available in quartz rock in unique deposit in Prionezhje. Similar mineral can also be found in the river Lena basin, but it consists of micro- and macro-dimensional cones, spheroids, and complex fibers. In the second case, it is talked about kerite from pegmatite on Volyn (Ukraine) that consists of polycrystalline fibers, spheres, and spirals mostly of micron dimensions, or fibrous vetcillite from the state of Utah (USA); globular anthraxolite and asphaltite.

Diameters of some internal channels are up to 20–50 nm. Such channels can be of interest as nanoreactors for the synthesis of organic, carbon, and polymeric substances with relatively low energy consumption. In case of directed location of internal channels in such matrixes and their inner-combinations the spatial structures of certain purpose can be created. Terminology in the field of nanosystems existence is still being developed, but it is already clear that nanoscience obtains qualitatively new knowledge that can find wide application in various areas of human practice thus, significantly decreasing the danger of people's activities for themselves and environment.

The system classification by dimensional factor is known, based on which we consider the following:

- microobjects and microparticles $10^{-6}–10^{-3}$ m in size;
- nanoobjects and nanoparticles $10^{-9}–10^{-6}$ m in size;
- picoobjects and picoparticles $10^{-12}–10^{-9}$ m in size.

Assuming that nanoparticle vibration energies correlate with their dimensions and comparing this energy with the corresponding region of electromagnetic waves, we can assert that energy action of nanostructures is within the energy region of chemical reactions. System self-organization refers to synergetics. Quite often, especially recently, the papers are published, for example, it is considered that nanotechnology is based on self-organization of metastable systems. As assumed, self-organization can proceed by dissipative (synergetic) and continual (conservative) mechanisms. Simultaneously, the system can be arranged due to the formation of new stable ("strengthening") phases or due to the growth provision of the existing basic phase. This phenomenon underlies the arising nanochemistry. Below is one of the possible definitions of nanochemistry.

Nanochemistry is a science investigating nanostructures and nanosystems in metastable ("transition") states and processes flowing with them in near-"transition" state or in "transition" state with low activation energies.

To carry out the processes based on the notions of nanochemistry, the directed energy action on the system is required, with the help of chemical particle field as well, for the transition from the prepared near-"transition" state into the process product state (in our case-into nanostructures or nanocomposites). The perspective area of nanochemistry is the chemistry in nanoreactors. Nanoreactors can be compared with specific nanostructures representing limited space regions in which chemical particles orientate creating "transition state" prior to the formation of the desired nanoproduct. Nanoreactors have a definite activity, which predetermines the creation of the corresponding product. When nanosized particles are formed in nanoreactors, their shape and dimensions can be the reflection of shape and dimensions of the nanoreactor.

In the last years a lot of scientific information in the field of nanotechnology and science of nanomaterials appeared. Scientists defined the interval from 1 to 1000 nm as the area of nanostructure existence, the main feature of which is to regulate the system self-organization processes. However, later some scientists limited the upper threshold at 100 nm. At the same time, it was not well substantiated. Now many nanostructures varying in shapes and sizes are known. These nanostructures have sizes that fit into the interval, determined by Smally, and are active in the processes of self-organization, and also demonstrate specific properties.

Problems of nanostructure activity and the influence of nanostructure super small quantities on the active media structural changes are explained.

The molecular nanotechnology ideology is analyzed. In accordance with the development tendencies in self-organizing systems under the influence of nanosized excitations the reasons for the generation of self-organization in the range $10^{-6}-10^{-9}$m should be determined.

Based on the law of energy conservation the energy of nanoparticle field and electromagnetic waves in the range 1–1,000 nm can transfer, thus corresponding to the range of energy change from soft X-ray to near IR radiation. This is the range of energies of chemical reactions and self-organization (structuring) of systems connected with them.

Apparently the wavelengths of nanoparticle oscillations near the equilibrium state are close or correspond to their sizes. Then based on the concepts of ideologists of nanotechnology in material science the definition of nanotechnology can be as follows:

Nanotechnology is a combination of knowledge in the ways and means of conducting processes based on the phenomenon of nanosized system self-organization and utilization of internal capabilities of the systems that results in decreasing the energy consumption required for obtaining the targeted product while preserving the ecological cleanness of the process.

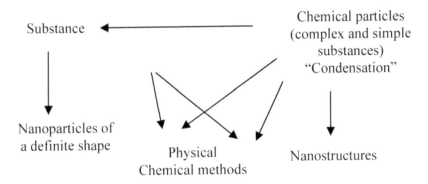

FIGURE 50   Classification diagram of nanostructure formation techniques by the features "dispersion" and "condensation."

At the same time, the conceptions "dispersion" and "condensation" are conditional and can be explained in various ways (Figs. 51 and 52). Among the physical methods of "dispersion" high-temperature methods and methods with relatively low temperatures or high-energy and low-energy ones are distinguished.

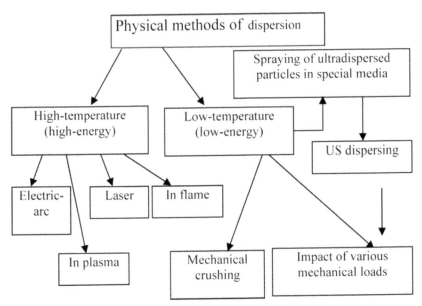

FIGURE 51   Classification diagram of physical methods of dispersion.

Considerably fewer chemical methods of dispersion are known, though, in the physical methods listed chemical processes are surely present, since it is difficult to imagine spraying and grinding of a substance without chemical reactions of destruction. Therefore, it is more appropriate to speak about physical methods of impact upon the substance that lead to their dispersion (decomposition, destruction at high temperatures, radiations or mechanical loads). Then, chemical methods are identified with methods of impact upon the substances of chemical particles and media.

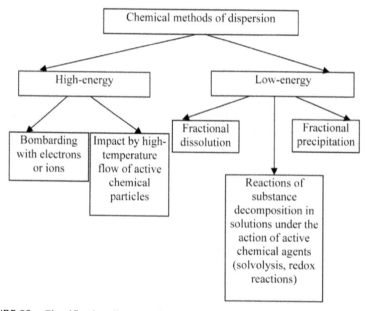

FIGURE 52   Classification diagram of dispersion chemical methods.

Basically nanostructures are obtained from pico-sized chemical particles by means of chemical or physical-chemical techniques, in which the activity of a chemical particle but not its activation during "condensation" is determinant (Fig. 52). Therefore, the classification diagram of chemical techniques for nanostructure formation under the action of chemical media can be given as follows:

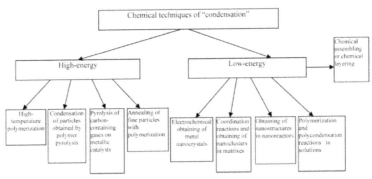

FIGURE 53 Classification diagram of chemical techniques of "condensation."

Actually, such diagrams are approximate in the same way as conditional are separate points of separation and difference between them. Several techniques can be referred to combined or physical methods of impact upon active chemical particles. For instance, CVD method comprises high-energy technique that leads to the formation of gaseous phase that can be attributed to chemical methods of dispersion, and then active chemical particles formed during the pyrolysis "transform" into nanostructures. The polymerization processes of gaseous phase particles are carried out by means of probe technological stations, and this can be referred to the physical techniques of "condensation." So, under the physical methods of formation of various nanostructures, including nanocrystals, nanoclusters, fullerenes, nanotubes, and so on, it means the techniques in which physical impact results in the formation of nanostructures from active pico-sized chemical particles (Fig. 54).

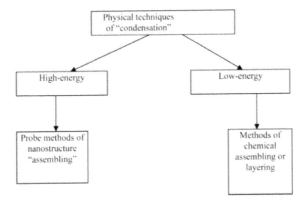

FIGURE 54 Physical impact results in the formation of nanostructures from active pico-sized chemical particles.

The proposed classification does not reflect multi-vicissitude of methods for nanostructure formation. Usually nanostructure production comprises the following:

1)   preparation of "embryos" or precursors of nanostructures;
2)   production of nanostructures; and
3)   isolation and refining of nanostructures of a definite shape.

The production of nanostructures by various methods, including mechanical ones, for instance, extrusion, grinding and similar operations can proceed in several stages.

Mechanical crushing, combined methods for grinding in media and influence of action power and medium, where the substance is crushed and sprayed, upon the size and shape of nanostructures formed will be discussed in this chapter. At the same time, multi-stepped combined methods for obtaining nanostructures with different shapes and sizes have been applied in recent years. Ways of substance dispersion by mechanical methods till nanoproducts are obtained (fine powders or ultradispersed particles) can be given in the following diagram (Fig. 55):

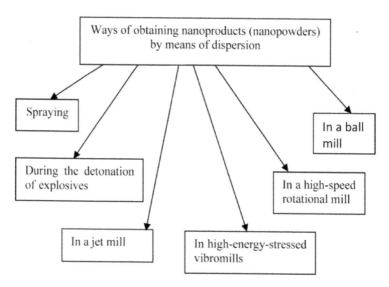

FIGURE 55   Diagram of classification of ways for obtaining powder-like nanoproducts by means of dispersion.

One of the most widely applied methods used for grinding different substances is the crushing and grinding in mills: ball, rod, colloid combined with the action of shock force, abrasion force, centrifugal force and vibration upon the materials. Since, initially, carbon nanostructures, fullerenes and nanotubes were extracted from the carbon dust obtained in electrical charge, the possibility of the formation of corresponding nanostructures when using conventional mechanical methods of action upon substances seems doubtful. However, the investigation of the products obtained after graphite crushing in ball mills allows making the conclusion that the method proposed can be quite competitive high-temperature way for forming nanostructures when the mechanical power provides practically the same conditions as the heat. In this case, it is difficult to imagine a directed action of the combined forces upon the substance. However, at the set speed of the mill drum the directed action of the "stream" of steel balls upon the material can be predicted. As an example, let us give the description of one of such processes during graphite grinding. Carbon nanostructures resembling "onions" by their shape are obtained when grinding graphite in ball mills. The product is ground in planetary ball mill in the atmosphere of pure argon. Mass ratio of steel balls to the powder of pure graphite equals 40. The rotation speed of the drum is 270 rpm. The grinding time changed from 150 to 250 hrs. It was observed that after 150 hrs of grinding the nanoproduct obtained resembles by characteristics and appearance the nanoparticles obtained in electric-arc method.

Iron-containing nanostructures were also obtained in planetary ball mills after milling the iron powders in heptane adding oleic acid.

The distinctive feature of the considered technique for producing metal/carbon nanocomposites is a wide application of independent, modern, experimental and theoretical analysis methods to substantiate the proposed technique and investigation of the composites obtained (quantum-chemical calculations, methods of transmission electron microscopy and electron diffraction, method of X-ray photoelectron spectroscopy, X-ray phase analysis, etc.). The technique developed allows synthesizing a wide range of metal/carbon nanocomposites by composition, size and morphology depending on the process conditions. In its application it is possible to use secondary metallurgical and polymer raw materials. Thus, the nanocomposite structure can be adjusted to extend the function of its application without pre-functionalization. Controlling the sizes and shapes of nanostructures by changing the metal-containing phase, to some extent, apply completely new, practicable properties to the materials which sufficiently differ from conventional materials.

The essence of the method consists in coordination interaction of functional groups of polymer and compounds of 3d-metals as a result of grinding of metal-containing and polymer phases Further, the composition obtained undergoes thermolysis following the temperature mode set with the help of thermogravimetric and differential thermal analyses. At the same time, one observes the polymer carbonization, partial or complete reduction of metal compounds and structuring of carbon material in the form of nanostructures with different shapes and sizes.

Metal/carbon nanocomposite (Me/C) represents metal nanoparticles stabilized in carbon nanofilm structures. In turn, nanofilm structures are formed with carbon amorphous nanofibers associated with metal containing phase. As a result of stabilization and association of metal nanoparticles with carbon phase, the metal chemically active particles are stable in the air and during heating as the strong complex of metal nanoparticles with carbon material matrix is formed.

6.16   WET SPINNING TECHNIQUE (A CASE STUDY)

Low voltage actuating materials ("artificial muscles") are required for many applications in robotics, medical devices and machines. One of the biggest limitations to date for the application of conducting polymers, such as polyaniline, is their low breaking strength- typically less than 10 MPa during an electrochemical actuation cycle under external load. We report here the significant improvement in actuator strength, stress generation and work-per-cycle through the incorporation of small amounts (up to 0.6% w/w) of carbon nanotubes as reinforcement in the polyaniline matrix. A wet spinning and drawing process has been developed to produce continuous lengths of these carbon nanotube reinforced fibers. As actuators, these composites fibers continue to operate at applied stresses in excess of 100 MPa producing a maximum work per cycle of over 300 kJ/m^3. This performance is 3 times higher than previously produced conducting polymer actuators and exceeds skeletal muscle in terms of stress generation by 300 times.

Actuating materials capable of producing useful movement and forces are recognized as the "missing link" in the development of a wide range of frontier technologies including haptic devices, microelectromechanical systems (MEMS) and even molecular machines. Immediate uses for these materials include an electronic Braille screen, a rehabilitation glove, tremor suppression and a variable camber propeller. Most of these applications could be realized with actuators that have equivalent performance to natural skeletal muscle. Although many actuator materials are available, none have the same mix of speed, movement and force as skeletal muscle. Indeed the actuator community was challenged to produce a material capable of beating a human in an arm wrestle. This challenge remains to be met.

One class of materials that has received considerable attention as actuators is the low voltage electrochemical systems utilizing conducting polymers and carbon nanotubes. Low voltage sources are convenient and safe and power inputs are potentially low. One deficiency of conducting polymers and nanotubes compared with skeletal muscle is their low actuation strains: less than 15% for conducting polymers and less than 1% for nanotubes. It has been argued that the low strains can be mechanically amplified (levers, bellows, hinges, etc.) to produce useful movements, but higher forces are needed to operate these amplifiers.

In recent studies of the forces and displacements generated from conducting polymer actuators, it has become obvious that force generation is limited by the breaking strength of the actuator material. Researchers predicted that the maximum stress generated by an actuator can be estimated as 50% of the breaking stress so that for highly drawn polyaniline (PANi) fibers stresses of the order of 190 MPa should be achievable. However, in practice the breaking stresses of conducting polymer fibers when immersed in electrolyte and operated electromechanically are significantly lower than their dry-state strengths. The reasons for the loss of strength are not well known, but the limitations on actuator performance are severe. The highest reported stress that can be sustained by conducting polymers during actuator work cycles is in the range 20–34 MPa for polypyrrole (PPy) films. However, the maximum stress that can be sustained by PPy during actuation appears very sensitive to the dopant ion used and the preparation conditions, with many studies showing maximum stress values of less than 10 MPa.

The low stress generation from conducting polymers, limited by the low breaking strengths, mean that the application of mechanical amplifiers is also very limited. To improve the mechanical performance, we have investigated the use of carbon nanotubes as reinforcing fibers in a polyaniline (PANi) matrix. Previous work has shown that the addition of single wall nanotubes (SWNTs) and multi-wall nanotubes (MWNTs) to various polymer matrices have produced significant improvements in strength and stiffness. It has been shown that the modulus of PAni can be increased by up to 4 times with the addition of small (<2%) amounts of nanotubes. Similar improvements in the modulus of actuating polymers may lead to significant increases in the stress generated and work per cycle. Other previous studies have shown that PANi can be wet spun into continuous fibers and that these may be used as actuators. Isotonic strains of 0.3% and isometric stresses of 2 MPa were obtained from these fibers when operated in ionic liquid electrolytes. The aim of the present study was to develop methods for incorporating carbon nanotubes into PAni fibers and to determine the effects on actuator performance.

A wet spinning technique was used to prepare the composite fibers. Firstly, the NTs (HIPCO single wall carbon nanotubes from Carbon Nanotechnology Inc.) were dispersed by sonication for 30 minutes in a mixture of 2-acrylamido-2-methyl-1-propane sulphonic acid (AMPSA, Aldrich 99%) and dichloroacetic acid (DCAA). Polyaniline (PAni, Santa Fe Science and Technology Inc.) and additional AMPSA were then dissolved in the dispersion by high speed mixing. After degassing, the spinning solution was injected using N_2 pressure into an acetone coagulation bath. The spun fibers were hand drawn to approximately 5 times their original length across a soldering iron wrapped in Teflon® tape at 100°C.

The isotonic actuation strains were measured using a Dual Mode lever system (Aurora Scientific) and were determined at different (constant) stresses as shown in Fig. 56 for the neat PANi and NT/PANi composite fibers. During isotonic actuation testing, each sample was tested at increasing loads until rupture occurred. It is shown that the addition of nanotubes greatly increases the breaking strength of the composite fibers under actuation conditions. The NT-reinforced fibers could sustain stresses up to 120 MPa without failure during electrochemical cycling. Even the neat PANi could sustain stresses to 90 MPa during work cycles.

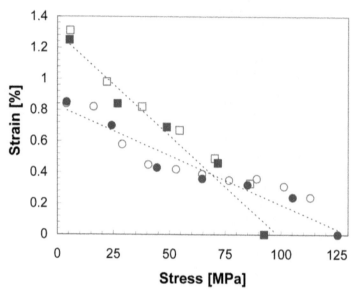

FIGURE 56   Isotonic strains measured during cyclic voltammetry in M HCl at 50 mV/s between 0.0 and 0.6V (vs. Ag/AgCl reference). Open and filled symbols show the results from two separate samples. Square symbols are for neat Pani fibers and circles are for NT-reinforced fibers.

The reinforcing effect afforded by small additions of NTs to the PANi was demonstrated by dry-state tensile testing, as shown in Fig. 57 and summarized in Table 8. A very large increase in the breaking strength from 170 MPa for neat PAni to 255 MPa was produced with the addition of 0.6 wt% NTs. Similarly, the modulus of the composite fibers was doubled from 3.4 GPa to 7.3 GPa. The change in modulus can be analyzed by the rule-of-mixtures approach used in fiber composites. As a starting point, the NTs are treated as fully dispersed and aligned in the fiber direction. Taking densities of 1.4 g/cm³ and 1.5 g/cm³ for the PAni and NTs, respectively [28, 29], the volume fraction of NTs in the composite fibers was 0.56%. The simple rule of mixtures approach predicts that the moduli of the composites fibers should be 7 GPa using a modulus of 640 GPa for the isolated NTs. The measured modulus is in good agreement with the calculated value, suggesting a high degree of separation, alignment and bonding of the NTs to the PAni matrix.

The results in Fig. 56 also show that the addition of nanotubes to the PAni matrix affects the actuation strain at low stress levels. The "free stroke" (strain at zero external load) for the NT/PAni sample was approximately 60% of the neat PAni free stroke. In this case the higher modulus nanotubes restrict the volume changes occurring within the PANi matrix. The actuator strains of the composite fibers may be calculated using the same iso-strain assumption inherent in the rule-of-mixtures approach. The expected actuator strain in the composite materials is estimated from:

$$\varepsilon_c^* = \frac{V_m \varepsilon_m^* E_m + V_f \varepsilon_f^* E_f}{E_c} \tag{133}$$

where subscripts c, m and f refer to the composite, PAni matrix and NT fibers and E and V are the moduli and volume fraction for each phase. Actuation in both the matrix and the NT fibers are considered in the analysis, since it is known that the NTs also produce appreciable strains when electrochemically charged. In previous work on NT/PAni composites, where unaligned mats of NTs were dip coated with PANi to give composites of 75% (w/w) NTs, it has been shown that both the nanotubes and the polymer contribute to the actuation strain.

The calculated composite actuation strain (at zero applied stress) was found to be 0.57% where the matrix actuation strain was taken as 1.3% and a NT actuation strain was assumed to be –0.06% for an anodic current pulse. The measured actuation strain (0.85%) was appreciably larger than the calculated strain. Clearly, the stiffening effect caused by the nanotubes cannot on its own account for the observed actuation behavior.

The larger than expected actuation in the composite fibers suggest that the NTs increase the efficiency of the actuation mechanism, perhaps through improved charge transfer. The magnitude of the increase in efficiency is demonstrated by calculating the matrix strain needed to account for the measured composite actuation strain. Thus, a matrix strain of 1.8% would be required to generate a composite strain of 0.85%. The higher matrix strains that occur in the NT composites compared with the neat PAni fibers, suggest a higher charge transfer efficiency possibly due to the higher conductivity of the NT–PAni composites (Table 8). For comparison, neat PAni fibers were also sputter coated with platinum to improve their conductivity and these samples showed actuation strains of 2% compared with 1.3% for uncoated PAni fibers.

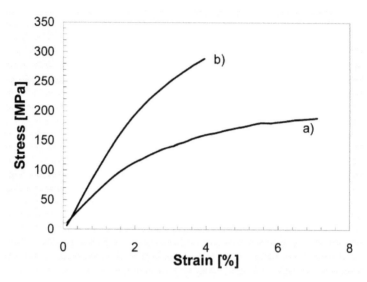

FIGURE 57   Stress-strain curves (dry state) for (a) neat Pani and (b) NT-reinforced Pani fibers.

TABLE 8   Measured and calculated properties of NT/PANi fiber composites.

	0% NT	0.62% NT
Elastic Modulus (GPa)	3.4 ± 0.4	7.3 ± 0.4
Tensile Strength (MPa)	170 ± 22	255 ± 32
Elongation at Break (%)	9 ± 3	4 ± 0.6
Max. Work Per Cycle (kJ/m3)	365	320
Adjusted PANi matrix actuation (%)	NA	1.8%
4-probe conductivity (S/cm)	497 ± 55	716 ± 36

Improved electrical conductivity has previously been shown to increase actuation performance by reducing iR losses along the fiber length and so increasing the active portion of the fiber.

Furthermore, strong interactions between the PAni and NTs has been previously suggested as enhancing the electroactivity of PAni/NT composites. As shown in Fig. 58, the electroactivity of the composites is improved by the addition of only 0.62% NTs. The more pronounced redox processes occurring in the composite fibers indicate an improved electrochemical efficiency that leads to an increase in the PAni matrix actuation.

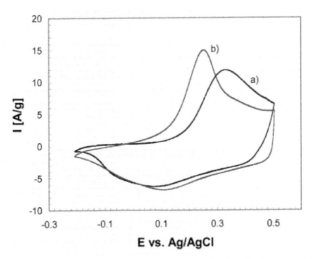

FIGURE 58 Cyclic voltammograms showing current density (I) for potential (E) scans between –0.2V and +0.5V (vs. Ag/AgCl) in M HCl at 50 MV/s: (a) neat Pani and (b) NT-reinforced Pani fibers.

Enhancement of electroactivity of PAni/NT composite films has been attributed to a possible increase in the degree of protonation of the PAni due to strong-stacking interactions between the NTs and the PAni. Clearly the improved electroactivity of the composite fibers will induce larger actuation strains in the PAni matrix, as described above.

The higher load capacity of the PAni/NT fibers also translates to higher work-per-cycles (Table 8). The maximum work-per-cycle of the neat PAni was 365 kJ/m^3 compared with other reported values of up to 83 kJ/m^3 for polypyrrole. With the addition of 0.62% NTs the work-per-cycle decreases slightly to 325 kJ/m^3, which is still more than 10 times higher than the work capacity of skeletal muscle. The dramatic increase in work per cycle compared with previous results from conducting polymers is a consequence of the higher strains at much higher stresses that are produced by the PAni and composite fibers.

In summary, it has been shown that the addition of small amounts of carbon nanotubes to PAni fibers produces significant improvements in their electroactivity, which translates to enhanced actuation performance. While the neat PAni produced an actuation strain of 1.3% at near zero loads, the presence of the NTs decreased this actuation to 0.85%. The resultant actuation strain of the composite

fibers was determined by a balance between the increased electroactivity leading to higher strains in the PAni matrix (up to 1.8%) and the increased modulus,
restricting matrix deformation and producing lower strains. The most significant
effect of NT additions was, however, the much improved breaking strength and
much higher operating stress levels. A 5-fold increase in work-per-cycle of compared with other conducting polymer actuators was achieved with the composite
fibers. Useful actuation strains could be obtained at up to 100 MPa applied stress,
which is 3 times higher than other conducting polymer actuators and 300 times
higher than skeletal muscle. The improved strength and stiffness of the composite fibers can be utilized in various applications where high force operation is
required, such as in strain amplification systems or biomimetic musculoskeletal
systems.

6.17   ELECTROSPINNING OF BLEND NANOFIBERS

Polyaniline was synthesized by the oxidative polymerization of aniline in acidic
media. 3 ml of distilled aniline was dissolved in 150 mL of 1N HCl and kept at
0–5°C. 7.325g of $(NH_4)_2S_2O_8$ was dissolved in 35 mL of 1N HCl and added drop
wise under constant stirring to the aniline/HCl solution over a period of 20 minutes. The resulting dark green solution was maintained under constant stirring for
4 hours. The prepared suspension was dialyzed in a cellulose tubular membrane
(Dialysis Tubing D9527, molecular cutoff = 12,400, Sigma) against distilled water for 48 hours. Then it was filtered and washed with water and methanol. The
synthesized Polyaniline was added to 150 mL of 1N (NH4) OH solution. After
an additional 4 hours the solution was filtered and a deep blue emeraldine base
form of Polyaniline was obtained (PANIEB). The synthesized Polyaniline was
dried and crushed into fine powder and then passed trough a 100 mesh. Intrinsic
viscosity of the synthesized Polyaniline dissolved at Sulfuric acid (98%) was 1.18
dl/g at 25°C.

The PANI solution with concentration of 5% (W/W) was prepared by dissolving exact amount of PANI in NMP. The PANI was slowly added to the NMP with
constant stirring at room temperature. This solution was then allowed to stir for
1 hour in a sealed container. 20% (W/W) solution of PAN in NMP was prepared
separately and was added drop wise to the well-stirred PANI solution. The blend
solution was allowed to stir with a mechanical stirrer for an additional 1 hour.

Various polymer blends with PANI content ranging from 10 wt% to 30 wt%
were prepared by mixing different amount of 5% PANI solution and 20% PAN
solution. Total concentrations of the blend solutions were kept as 12.5%.

Polymeric nanofibers can be made using the electrospinning process, which
has been described in the literature and patent. Electrospinning uses a high electric field to draw a polymer solution from tip of a capillary toward a collector. A
voltage is applied to the polymer solution, which causes a jet of the solution to

be drawn toward a grounded collector. The fine jets dry to form polymeric fibers, which can be collected as a web.

Our electrospinning equipment used a variable high voltage power supply from Gamma High Voltage Research (USA). The applied voltage can be varied from 1–30 kV. A 5-mL syringe was used and positive potential was applied to the polymer blend solution by attaching the electrode directly to the outside of the hypodermic needle with internal diameter of 0.3 mm. The collector screen was a 20×20 cm aluminum foil, which was placed 10 cm horizontally from the tip of the needle. The electrode of opposite polarity was attached to the collector. A metering syringe pump from New Era pump systems Inc. (USA) was used. It was responsible for supplying polymer solution with a constant rate of 20 µL/min.

Electrospinning was done in a temperature-controlled chamber and temperature of electrospinning environment was adjusted on 25, 50 and 75°C. Schematic diagram of the electrospinning apparatus was shown in Fig. 59. Factorial experiment was designed to investigate and identify the effects of parameters on fiber diameter and morphology (Table 9).

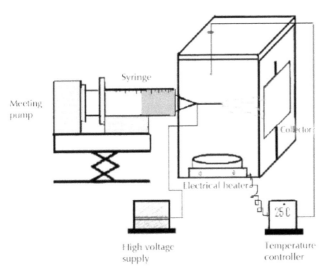

FIGURE 59   Schematic diagram of electrospinning apparatus.

6.17.1   CHARACTERIZATION
Shear viscosities of the fluids were measured at shear rate of 500 sec^{-1} and 22°C using a Brookfield viscometer (DVII+, USA). Fiber formation and morphology of the electrospun PANI/PAN fibers were determined using a scanning electron microscope (SEM) Philips XL-30A (Holland). Small section of the prepared samples was placed on SEM sample holder and then coated with gold by a BAL–TEC

SCD 005 sputter coater. The diameter of electrospun fibers was measured with image analyzer software (manual microstructure distance measurement). For each experiment, average fiber diameter and distribution were determined from about 100 measurements of the random fibers. Electrical conductivity of the electrospun mats was measured by the standard four- probe method after doping with HCl vapor.

TABLE 9   Factorial design of experiment.

Factor	Factor Level
PANI Content (wt%)	10,20,30
Electrospinning temperature (°C)	25,50,75
Applied voltage (kV)	20,25,30

6.18   RESULTS AND DISCUSSION

Published literature have shown that in the electrospinning process, the system configuration and operation conditions differ vastly from one material to another, depending on the material and the choice of solvent. Physical and chemical parameters of polymer solution such as viscosity, electrical conductivity, surface tension and air temperature can determinedly affect the formability and morphology of electrospun fibers. In the following sections effects of some electrospinning parameters on the fiber formation and morphology of PANI/PAN blend solutions were discussed and the best condition for obtaining PANI/PAN fibers was examined.

6.19   EFFECT OF PANI CONTENT

We were not able to obtain the fibers from the pure PANI solution because a stable drop at the end of the needle was not maintained. Figure 60 shows SEM micrographs of PANI nanoparticles electrospun from pure PANI solution. As seen in Fig. 60, most of PANI particles have a round shape, while the fibrous structure is not observed. The major complication in electrospinning of PANI is the poor solubility of PANI. At low polymer concentration, the solution does not contain sufficient material to produce stable solid fibers. With increasing polymer concentration, insoluble PANI particles in the solution increase rapidly, result the unspinnable solution. Therefore, we prepared PANI/PAN blend solutions with different PANI content using NMP as solvent. At PANI content above 30% regardless of electrospinning conditions drops were formed instead of fibers. A series of experiments were carried out when the PANI weight percent was varied from 10% to 30%. The applied voltage was 20 to 30 kV and the chamber temperature was held at 25, 50 and 75°C. Figure 60 shows the SEM micrographs and the surface morphology of obtained fibers at 25°C and 25 kV. At a solution containing 30% PANI, the fibrous structure was not completely stabilized and a bead-on-string structure with non-uniform morphology was obtained. The fibers between the beads had a circular cross section, with a diameter typically between 60 nm and 460 nm and mean fiber diameter of 164 nm. As the PANI content decreases to lower than 20%, a fibrous structure was stabilized. At 20% PANI content, fibers mean diameter increased to 425 nm with some beads on the fibers. At 10% PANI content, continuous fibers without beads were resulted regardless of electric field with the mean fiber diameter of 602 nm at 25 kV. Smooth and uniform fibers with average diameter of 652 nm were electrospun from PAN solution at the same electrospinning condition. These results reveal that as the PANI contents in the blends increase up to 30% the average diameter of blend nanofiber gradually decreases from 602 to 164 nm and its distribution becomes significantly broader with higher standard deviation as shown in Fig 60. It is also observed that fibers with not uniform morphology are electrospun at 25°C. Figure 61 shows SEM photomicrographs of electrospun PANI/PAN blend fibers at 50°C at various blend ratios. This figure shows that fibers with uniform morphology without remarkable beads are formed regardless of PANI content. It is also observed that at 50°C average diameter of electrospun fibers decreases from 194 nm at 10% PANI content to 124 nm at 30% PANI content at 50°C. Similar to the results obtained at 25°C fiber formation from pure PANI solution and blends containing more than 30% PANI was not possible. In electrospinning, the coiled polymer chains in the solution are transformed by the elongational flow of the jet into oriented entangled networks. Experimental observations in electrospinning confirm that for fiber formation to occur, a minimum chain entanglement is required. Below this critical chain entanglement, application of voltage results beads and droplets due to jet instability. The gradual increase in fiber diameter with content of PAN in the blends may be explained by the increase of solution viscosity due to higher viscosity of PAN solution. The ranges of shear viscosity of the PANI/PAN blends are shown at Table 10. It is obvious that shear viscosity of the solutions decrease with PANI content in the blends. Therefore, as the concentration of PAN in the blend is increased; the solution viscosity and resulted polymer chain entanglements increase significantly. During electrospinning, the stable jet ejected from Taylor's cone is subjected to tensile stresses and may undergo significant elengational flow. The nature of this elongational flow may determine the degree of stretching of the jet. The characteristics of this elengational flow can be determined by elasticity and viscosity of the solution. The results show that viscosity of the PAN solution is higher than PANI solution. Hence viscosity of the blend solution decreases with an increase in PANI content. Therefore, jet stretching during the electrospinning is more effective at higher PANI content. As a result, the fibers diameters decrease with increasing PANI content in the blends. On the other hand, at the high PANI content, an insufficiently deformable entangled network of polymer chain exists and the ejected jet reaches the collector before the solvent fully evaporates. Therefore, at low solution viscosity ejected jet breaks into droplets and a mixture of beads and fibers is obtained. This explains the formation of droplets and beads at high PANI content. Effect of electrospinning temperature is discussed in the following section. Researchers showed that the diameters of electrospun nanofibers are greatly affected by solution viscosity, and solution viscosity has an allometric relationship with its concentration. Our results shows that the electrospun nanofibers diameters (d) of PANI/PAN blends has a relationship with PANI content in the form of

$$d \propto (PANI\%)^2 \qquad\qquad (134)$$

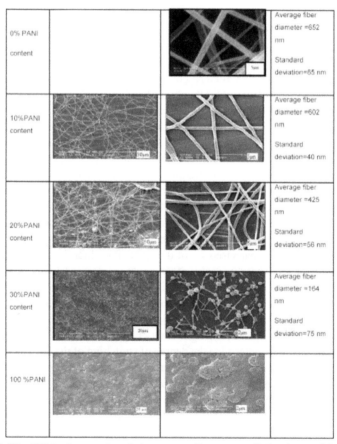

FIGURE 60 SEM micrographs of electrospun fibers at applied voltage of 25 kV and temperature of 25°C with a constant spinning distance of 10 cm.

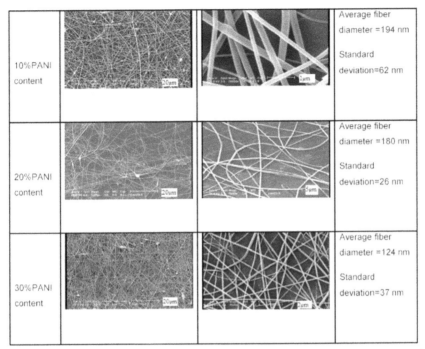

FIGURE 61   SEM micrographs of electrospun nanofibers at applied voltage of 25 kV and temperature of 50°C with a constant spinning distance of 10 cm.

6.20   EFFECT OF ELECTROSPINNING TEMPERATURE

Studies on the electrospinning show that many parameters may influence the transformation of polymer solution into nanofibers. Some of these parameters include (1) the solution related properties such as viscosity and surface tension, (2) process variables such as electric potential at the capillary tip and (3) ambient parameters such as air temperature in the electrospinning chamber. In order to study the effect of electrospinning temperature on the morphology and texture of electrospun PANI/PAN nanofibers, solution containing 20% PANI was electrospun at temperatures 25, 50 and 75°C. SEM micrographs of electrospun fibers at 20 kV are shown in Fig. 62. Interestingly, the electrospinning of the solution shows bead free fiber morphology at 50°C and 75°C, whereas fibers with large beads are observed at 25°C especially at high PANI contents (Fig. 62). The electrospun sample at 25°C shows fibers with several beads and not uniform surface morphology. With an increase in electrospinning temperature fibers morphology changes gradually from mixture of beads and fibers through uniform fibers. As shown in Fig. 62 at 50°C continuous fibers with uniform morphology were obtained while increasing the electrospinning temperature to 75°C caused bead free but fragile and cracked fibers. Diameter measurement of electrospun fibers at 25°C showed a size range of approximately 400 to 700 nm with 480 nm being the most frequently occurring. They were within the same range of reported size for electrospun PANI/PEO nanofibers [18]. With increasing the electrospinning temperature to 50°C, fiber diameter was decreased to a range of approximately 110 to 290 nm with 170 nm the most occurring frequency. At 75°C, fibers dimensions were 70 to 170 nm with 110 nm the most occurring frequency. It was obvious that diameter of electrospun fibers were decreased with increasing of electrospinning temperature. The distributions of fibers diameters electrospun at 25, 50 and 75°C are shown in Fig. 63. At 25°C broad distribution of fibers diameters was obtained, while a narrow distribution in fibers diameters was observed at 50 and 75°C.

TABLE 10   Shear viscosity of the PANI/PAN blend solutions at 22°C and shear rate of 500 sec^{-1} and average diameter of electrospun nanofibers.

PANI/PAN blend ratio % (w/w)	Shear viscosity (Pa.s)	Average nanofiber diameter (nm)
100/0 (5% solution)	0.159	No fiber
30/70	0.413	164
20/80	0.569	425
10/90	0.782	602
0/100	1.416	652

Several factors with PANI/PAN blends may explain the effects of electrospinning temperature and PANI content on morphology of the electrospun fibers. Since nanofibers are resulted from evaporation of solvent from polymer solution jets, the fiber diameters will depend on the jet sizes, elongation of the jet and evaporation rate of the solvent [24]. At a constant PANI content, as the electrospinning temperature is increased, the rate of solvent evaporation from the ejected jet increases significantly. In the case of electrospinning at 25°C due to the high boiling point of NMP (approximately 202°C), the fibers with relatively high solvent content travels during electrospinning process and reach the collector. Therefore, the collected fibers have irregular morphology due to contraction of the fibers during the electrospinning and on the collector. At higher electrospinning temperature rate of solvent evaporation from the ejected jet increases significantly and a skin is formed on the surface of the jet, which results collection of dry fiber with smooth surface. Presence of a thin, mechanically distinct polymer skin on the liquid jet during electrospinning has been discussed by researchers. On the other hand higher electrospinning temperature results higher degree of stretching and more uniform elongation of the ejected jet due to higher mobility and lower Viscosity of the solution. Therefore fibers with smaller diameters and narrower diameters distribution will be electrospun at higher electrospinning temperature.

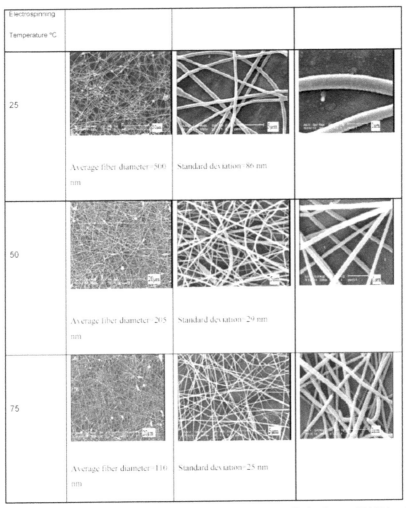

FIGURE 62   SEM micrographs of electrospun nanofibers at applied voltage of 20 kV and PANI content of 20% with a constant spinning distance of 10 cm.

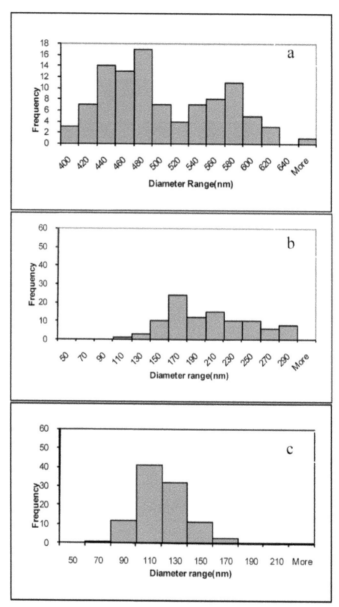

FIGURE 63   Distribution of fiber diameter electrospun at PANI content of 20%, applied voltage of 20 kV, spinning distance of 10 cm and electrospinning temperature of (a) 25°C, (b) 50°C and (c) 75°C.

6.21   EFFECT OF APPLIED VOLTAGE

In order to study the effects of applied voltage, the blend solutions were electrospun at various applied voltages and temperatures. From the results shown in Fig 64, it is obvious that the diameter of electrospun PANI/PAN fibers at 50°C decreased as the applied voltage increased. Similar results were observed for electrospun fibers at 25 and 70°C (results were not shown).

6.22   ELECTRICAL CONDUCTIVITY

Figure 64 shows electrical conductivity of the electrospun mats at various PANI/PAN blend ratios. As expected, electrical conductivity of the mats was found to increase with an increase in PANI content in the blends. Figure 65 shows that the electrical conductivity of the mats increases sharply when the PANI content in the blends is less than 5%, after which it will gradually reach to 10^{-1} S/cm at higher PANI content. This result is in agreement with the observations of Yang and co workers [35], which reported the electrical conductivity of PANI/PAN blend composites. Yang et al. [35] proposed the classical law of percolation theory, $\sigma(f)=c(f-f_p)^t$, where c is a constant, t is critical exponent of the equation, f is the volume fraction of the filler particle and f_p is the volume fraction at percolation threshold. The results of Fig. 65 indicate that the conductivity of the mats follows the scaling law of percolation theory mentioned above as shown in Eq. (135), which results a value of 0.5 wt% of PANI for f_p. This value for the percolation threshold is much lower than that reported by some scientists which may be due to the difference in the studied sample form. Their measurements were performed on the prepared films whereas our measurements were performed on the nanofiber mats. It is worth noting that the classical percolation theory predicts a percolation threshold of f_p =0.16 for conducting particles dispersed in an insulating matrix in three dimensions which is in agreement of our finding.

$$\sigma = 9 \times 10^{-7} (f - 0.5)^{3.91} \quad R^2 = 0.99 \qquad (135)$$

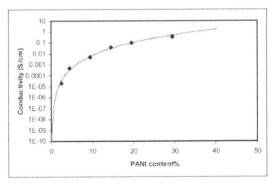

FIGURE 64   Electrical conductivity of electrospun mats at various PANI contents.

The electrospinning of PANI/PAN blend in NMP was processed and fibers with diameter ranging from 60 to 600 nm were obtained based on electrospinning conditions. Morphology of fibers was investigated at various blends ratios and electrospinning temperature. At 30% PANI content and 25°C fibers with average diameter of 164 nm were formed with beads (droplets of polymer over the woven mat) and not uniform morphology. At this condition solution viscosity and chain entanglements may not be enough, resulting in spraying of large droplets connected with very thin fibers. Averages of fibers diameters were decreased with PANI content in the solutions but PANI/PAN solution containing more than 30% PANI did not form a stable jet regardless of applied voltage and electrospinning temperature. For pure PANI solution, since the viscosity is too low to get stable drops and jets, we could not get the fibers. It was found that at 25°C fiber morphology was changed to beaded fibers when PANI content was higher than 20%. With increasing the electrospinning temperature, the morphology was changed from beaded fibers to uniform fibrous structure and the fiber diameter was also decreased from 500 nm to 100 nm when the electrospinning temperature changes from 25°C to 75°C. The mean of fiber diameter is the smallest and the fiber diameter distribution is the narrowest for the electrospun fibers at 75°C. However some cracks are observed on the surface of the electrospun fibers. There was a slightly decrease in average fiber diameter with increasing applied voltage. It is concluded that the optimum condition for nanoscale and uniform PANI/PAN fiber formation is 20% PANI content and 50°C electrospinning temperature regardless of the applied voltage. The conductivity of the mats follows the scaling law of percolation theory, which predicts a value of 0.5 wt% of PANI as percolation threshold for the blend of PANI/PAN.

KEYWORDS

- **electrospinning**
- **mathematical modeling**
- **nanoelements**
- **nanofibers**
- **nanotubes**
- **polymers**

REFRENCES

1. Reneker, D.H. and A.L. Yarin, *Electrospinning Jets and Polymer Nanofibers.* Polymer, 2008, **49**(10): p. 2387–2425.
2. Reneker, D.H., et al., *Electrospinning of Nanofibers from Polymer Solutions and Melts.* Advances in Applied Mechanics, 2007, **41**: p. 343–346.

3. Haghi, A.K. and G. Zaikov, *Advances in Nanofiber Research*. 2012: Smithers Rapra Technology. 194.
4. Haghi, A.K., *Electrospinning of nanofibers in textiles*. 2011, North Calorina: Apple Academic PressInc. 132.
5. Maghsoodloo, S., et al., *A Detailed Review on Mathematical Modeling of Electrospun Nanofibers*. Polymers Research Journal **6**: p. 361–379.
6. Frenot, A. and I.S. Chronakis, *Polymer Nanofibers Assembled by Electrospinning*. Current Opinion in Colloid and Interface Science, 2003, **8**(1): p. 64–75.
7. Fritzson, P., *Principles of object-oriented modeling and simulation with Modelica 2.1*. 2010: Wiley–IEEE Press.
8. Collins, A.J., et al., *The Value of Modeling and Simulation Standards*. 2011, Virginia Modeling, Analysis and Simulation Center, Old Dominion University: Virginia. p. 1–8.
9. Robinson, S., *Simulation: the practice of model development and use*. 2004: Wiley. 722.
10. Carson, I.I. and S. John, *Introduction to modeling and simulation*, in *Proceedings of the 36th conference on Winter simulation*. 2004, Winter Simulation Conference: Washington, DC. p. 9–16.
11. Banks, J., *Handbook of simulation*. 1998: Wiley Online Library. 342.
12. Pritsker, A.B. and B. Alan, *Principles of Simulation Modeling*. 1998, New York: Wiley. 426.
13. Carroll, C.P., *The development of a comprehensive simulation model for electrospinning*. Vol. 70. 2009, Cornell University 300.
14. Menon, A. and K. Somasekharan, *Velocity, acceleration & jerk in electrospinning*, in *The Internet Journal of Bioengineering*. 2009.
15. Gilbert, W., *De Magnete* Transl. PF Mottelay, Dover, UK. 1958, New York: Dover Publications, Inc. 366.
16. Tucker, N., et al., *The History of the Science and Technology of Electrospinning from 1600 to 1995*. Journal of Engineered Fibers and Fabrics, 2012, 7: p. 63–73.
17. Zeleny, J., *The electrical discharge from liquid points, and a hydrostatic method of measuring the electric intensity at their surfaces*. Physical Review, 1914, **3**(2): p. 69–91.
18. Hassounah, I., *Melt electrospinning of thermoplastic polymers*. 2012: Aachen: Hochschulbibliothek Rheinisch–Westfälische Technischen Hochschule Aachen. 650.
19. Taylor, G.I., *The Scientific Papers of Sir Geoffrey Ingram Taylor*. Mechanics of Fluids, 1971, **4**.
20. Yeo, L.Y. and J.R. Friend, *Electrospinning Carbon Nanotube Polymer Composite Nanofibers*. Journal of Experimental Nanoscience, 2006, **1**(2): p. 177–209.
21. Miao, J., et al., *Electrospinning of nanomaterials and applications in electronic components and devices*. Journal of Nanoscience and Nanotechnology, 2010, **10**(9): p. 5507–5519.
22. Teo, W.E. and S. Ramakrishna, *A review on electrospinning design and nanofiber assemblies*. Nanotechnology, 2006, **17**(14): p. R89–R106.
23. Bhardwaj, N. and S.C. Kundu, *Electrospinning: a fascinating fiber fabrication technique*. Biotechnology Advances, 2010, **28**(3): p. 325–347.

24. Greiner, A. and J.H. Wendorff, *Electrospinning: a fascinating method for the preparation of ultrathin fibers.* Angewandte Chemie International Edition, 2007, **46**(30): p. 5670–5703.

25. Thoppey, N.M., et al., *Effect of Solution Parameters on Spontaneous Jet Formation and Throughput in Edge Electrospinning from a Fluid–Filled Bowl.* Macromolecules, 2012, **45**: p. 6527–6537.

26. Wan, Y., et al., *Modeling and Simulation of the Electrospinning Jet with Archimedean Spiral.* Advanced Science Letters, 2012, **10**(1): p. 590–592.

27. Liu, L. and Y.A. Dzenis, *Simulation of Electrospun Nanofiber Deposition on Stationary and Moving Substrates.* Micro & nano–Letters, 2011, **6**(6): p. 408–411.

28. Huang, Z.M., et al., *A review on polymer nanofibers by electrospinning and their applications in nanocomposites.* Composites Science and Technology, 2003, **63**(15): p. 2223–2253.

29. Ramakrishna, S., *An Introduction to Electrospinning and Nanofibers.* 2005: World Scientific Publishing Company. 396.

30. Arinstein, A., et al., *Effect of supramolecular structure on polymer nanofiber elasticity.* Nature Nanotechnology, 2007, **2**(1): p. 59–62.

31. Lu, C., et al., *Computer Simulation of Electrospinning. Part I. Effect of Solvent in Electrospinning.* Polymer, 2006, **47**(3): p. 915–921.

32. Greenfeld, I., et al., *Polymer dynamics in semidilute solution during electrospinning: A simple model and experimental observations.* Physical Review 2011, **84**(4): p. 41806–41815.

33. Solberg, R.H.M., *Position-controlled deposition for electrospinning.* 2007, Eindhoven University of Technology: Eindhoven. p. 75.

34. Gradoń, L., *Principles of Momentum, Mass and Energy Balances.* Chemical Engineering and Chemical Process Technology. **1**: p. 1–6.

35. Bird, R.B., W.E. Stewart, and E.N. Lightfoot, *Transport Phenomena.* Vol. 2. 1960, New York: Wiley & Sons, Incorporated, John 808.

36. Peters, G.W.M., M.A. Hulsen, and R.H.M. Solberg, *A Model for Electrospinning Viscoelastic Fluids,* in *Department of Mechanical Engineering.* 2007, Eindhoven University of Technology: Eindhoven p. 26.

37. Whitaker, R.D., *An historical note on the conservation of mass.* Journal of Chemical Education, 1975, **52**(10): p. 658.

38. Hohman, M.M., et al., *Electrospinning and electrically forced jets. I. Stability theory.* Physics of Fluids, 2001, **13**: p. 2201–2221.

39. He, J.H., et al., *Mathematical models for continuous electrospun nanofibers and electrospun nanoporous microspheres.* Polymer International, 2007, **56**(11): p. 1323–1329.

40. Xu, L., F. Liu, and N. Faraz, *Theoretical model for the electrospinning nanoporous materials process.* Computers and Mathematics with Applications, 2012, **64**(5): p. 1017–1021.

41. Heilbron, J.L., *Electricity in the 17th and 18th Century: A Study of Early Modern Physics.* 1979: Univ of California Press.

42. Orito, S. and M. Yoshimura, *Can the universe be charged?* Physical review letters, 1985, **54**(22): p. 2457–2460.

43. Karra, S., *Modeling electrospinning process and a numerical scheme using Lattice Boltzmann method to simulate viscoelastic fluid flows.* 2012.

44. Feynman, R.P., et al., *The Feynman Lectures on Physics; Vol. I.* American Journal of Physics, 1965, **33**: p. 750.
45. Bennett, C.O. and J.E. Myers, *Momentum, Heat, and Mass Transfer.* Vol. 370. 1982, New York: McGraw–Hill 848.
46. Feng, J.J., *The stretching of an electrified non–Newtonian jet: A model for electrospinning.* Physics of Fluids, 2002, **14**(11): p. 3912–3927.
47. Hou, S.H. and C.K. Chan, *Momentum Equation for Straight Electrically Charged Jet.* Applied Mathematics and Mechanics, 2011, **32**(12): p. 1515–1524.
48. Maxwell, J.C., *Electrical Research of the Honorable Henry Cavendish, 426,* in *Cambridge University Press,* Cambridge, Editor. 1878, Cambridge University Press, Cambridge, UK: UK.
49. Heilbron, J.L., *Electricity in the 17th and 18th Century: A Study of Early Modern Physics.* 1979: University of California Press. 437.
50. Vught, R.V., *Simulating the dynamical behavior of electrospinning processes,* in *Department of Mechanical Engineering.* 2010, Eindhoven University of Technology: Eindhoven. p. 68.
51. Jeans, J.H., *The Mathematical Theory of Electricity and Magnetism.* 1927, London: Cambridge University Press. 536.
52. Reneker, D.H., et al., *Bending Instability of Electrically Charged Liquid Jets of Polymer Solutions in Electrospinning.* Journal of Applied physics, 2000, **87**: p. 4531.
53. Truesdell, C. and W. Noll, *The non-linear field theories of mechanics.* 2004: Springer. 579.
54. Roylance, D., *Constitutive equations,* in *Lecture Notes. Department of Materials Science and Engineering.* 2000, Massachusetts Institute of Technology: Cambridge. p. 10.
55. He, J.H., Y. Wu, and N. Pang, *A mathematical model for preparation by AC-electrospinning process.* International Journal of Nonlinear Sciences and Numerical Simulation, 2005, **6**(3): p. 243–248.
56. Little, R.W., *Elasticity.* 1999: Courier Dover Publications. 431.
57. Bhattacharjee, P., V. Clayton, and A.G. Rutledge, *Electrospinning and Polymer Nanofibers: Process Fundamentals,* in *Comprehensive Biomaterials.* 2011, Elsevier. p. 497–512.
58. Clauset, A., C.R. Shalizi, and M.E.J. Newman, *Power-law Distributions in Empirical Data.* SIAM Review, 2009, **51**(4): p. 661–703.
59. Garg, K. and G.L. Bowlin, *Electrospinning jets and nanofibrous structures.* Biomicrofluidics, 2011, **5**: p. 13403–13421.
60. Spivak, A.F. and Y.A. Dzenis, *Asymptotic decay of radius of a weakly conductive viscous jet in an external electric field.* Applied Physics Letters, 1998, **73**(21): p. 3067–3069.
61. Wan, Y., Q. Guo, and N. Pan, *Thermo-electro-hydrodynamic model for electrospinning process.* International Journal of Nonlinear Sciences and Numerical Simulation, 2004, **5**(1): p. 5–8.
62. Giesekus, H., *Die elastizität von flüssigkeiten.* Rheologica Acta, 1966, **5**(1): p. 29–35.
63. Giesekus, H., *The physical meaning of Weissenberg's hypothesis with regard to the second normal-stress difference,* in *The Karl Weissenberg 80th Birthday Celebration*

Essays, J. Harris and K. Weissenberg, Editors. 1973, East African Literature Bureau p. 103–112.

64. Wiest, J.M., *A differential constitutive equation for polymer melts*. Rheologica Acta, 1989, **28**(1): p. 4–12.

65. Bird, R.B. and J.M. Wiest, *Constitutive Equations for Polymeric Liquids*. Annual Review of Fluid Mechanics, 1995, **27**(1): p. 169–193.

66. Giesekus, H., *A simple constitutive equation for polymer fluids based on the concept of deformation-dependent tensorial mobility*. Journal of Non–Newtonian Fluid Mechanics, 1982, **11**(1): p. 69–109.

67. Oliveira, P.J., *On the Numerical Implementation of Nonlinear Viscoelastic Models in a Finite–Volume Method*. Numerical Heat Transfer: Part B: Fundamentals, 2001, **40**(4): p. 283–301.

68. Simhambhatla, M. and A.I. Leonov, *On the Rheological Modeling of Viscoelastic Polymer Liquids with Stable Constitutive Equations*. Rheologica Acta, 1995, **34**(3): p. 259–273.

69. Giesekus, H., *A unified approach to a variety of constitutive models for polymer fluids based on the concept of configuration-dependent molecular mobility*. Rheologica Acta, 1982, **21**(4–5): p. 366–375.

70. Feng, J.J., *Stretching of a straight electrically charged viscoelastic jet*. Journal of Non–Newtonian Fluid Mechanics, 2003, **116**(1): p. 55–70.

71. Eringen, A.C. and G.A. Maugin, *Electrohydrodynamics*, in *Electrodynamics of Continua II*. 1990, Springer. p. 551–573.

72. Hutter, K., *Electrodynamics of Continua (A. Cemal Eringen and Gerard A. Maugin)*. SIAM Review, 1991, **33**(2): p. 315–320.

73. Marrucci, G., *The free energy constitutive equation for polymer solutions from the dumbbell model*. Journal of Rheology, 1972, **16**: p. 321–331.

74. Kowalewski, T.A., S. Barral, and T. Kowalczyk, *Modeling Electrospinning of Nanofibers*, in *IUTAM Symposium on Modeling Nanomaterials and Nanosystems*. 2009, Springer: Aalborg, Denmark. p. 279–292.

75. Kuipers, B., *Qualitative reasoning: modeling and simulation with incomplete knowledge*. 1994: the MIT press. 554.

76. West, B.J., *Comments on the renormalization group, scaling and measures of complexity*. Chaos, Solitons and Fractals, 2004, **20**(1): p. 33–44.

77. De Gennes, P.G. and T.A. Witten, *Scaling Concepts in Polymer Physics*. Vol. Cornell University Press. 1980, 324.

78. He, J.H. and H.M. Liu, *Variational approach to nonlinear problems and a review on mathematical model of electrospinning*. Nonlinear Analysis, 2005, **63**: p. e919-e929.

79. He, J.H., Y.Q. Wan, and J.Y. Yu, *Allometric scaling and instability in electrospinning*. International Journal of Nonlinear Sciences and Numerical Simulation 2004, **5**(3): p. 243–252.

80. He, J.H., Y.Q. Wan, and J.Y. Yu, *Allometric Scaling and Instability in Electrospinning*. International Journal of Nonlinear Sciences and Numerical Simulation, 2004, **5**: p. 243–252.

81. He, J.H. and Y.Q. Wan, *Allometric scaling for voltage and current in electrospinning*. Polymer, 2004, **45**: p. 6731–6734.

82. He, J.H., Y.Q. Wan, and J.Y. Yu, *Scaling law in electrospinning: relationship between electric current and solution flow rate.* Polymer, 2005, **46**: p. 2799–2801.
83. He, J.H., Y.Q. Wanc, and J.Y. Yuc, *Application of vibration technology to polymer electrospinning.* International Journal of Nonlinear Sciences and Numerical Simulation, 2004, **5**(3): p. 253–262.
84. Kessick, R., J. Fenn, and G. Tepper, *The use of AC potentials in electrospraying and electrospinning processes.* Polymer, 2004, **45**(9): p. 2981–2984.
85. Boucher, D.F. and G.E. Alves, *Dimensionless numbers, part 1 and 2.* 1959.
86. Ipsen, D.C., *Units Dimensions And Dimensionless Numbers.* 1960, New York: McGraw Hill Book Company Inc. 466.
87. Langhaar, H.L., *Dimensional analysis and theory of models.* Vol. 2. 1951, New York: Wiley. 166
88. McKinley, G.H., *Dimensionless groups for understanding free surface flows of complex fluids.* Bulletin of the Society of Rheology, 2005, **2005**: p. 6–9.
89. Carroll, C.P., et al., *Nanofibers from Electrically Driven Viscoelastic Jets: Modeling and Experiments.* Korea–Australia Rheology Journal, 2008, **20**(3): p. 153–164.
90. Saville, D.A., *Electrohydrodynamics: the Taylor–Melcher leaky dielectric model.* Annual Review of Fluid Mechanics, 1997, **29**(1): p. 27–64.
91. Ramos, J.I., *Force Fields on Inviscid, Slender, Annular Liquid.* International Journal for Numerical Methods in Fluids, 1996, **23**: p. 221–239.
92. Spivak, A. and Y. Dzenis, *Asymptotic decay of radius of a weakly conductive viscous jet in an external electric field.* Applied Physics Letters, 1998, **73**(21): p. 3067–3069.
93. Ha, J.W. and S.M. Yang, *Deformation and breakup of Newtonian and non–Newtonian conducting drops in an electric field.* Journal of Fluid Mechanics, 2000, **405**: p. 131–156.
94. Hohman, M.M., et al., *Electrospinning and electrically forced jets. I. Stability theory.* Physics of Fluids, 2001, **13**: p. 2201.
95. Peters, G., M. Hulsen, and R. Solberg, *A Model for Electrospinning Viscoelastic Fluids.*
96. Dasri, T., *Mathematical Models of Bead–Spring Jets during Electrospinning for Fabrication of Nanofibers.* Walailak Journal of Science and Technology, 2012, **9**.
97. Holzmeister, A., A.L. Yarin, and J.H. Wendorff, *Barb formation in electrospinning: Experimental and theoretical investigations.* Polymer, 2010, **51**(12): p. 2769–2778.
98. Angammana, C.J. and S.H. Jayaram, *A Theoretical Understanding of the Physical Mechanisms of Electrospinning,* in *Proc. ESA Annual Meeting on Electrostatics.* 2011: Case Western Reserve University, Cleveland OH, p. 1–9.

CHAPTER 7

POLYACETYLENE

V. A. BABKIN, G. E. ZAIKOV, M. HASANZADEH, and A. K. HAGHI

CONTENTS

7.1 SYNTHESIS, STRUCTURE, PHYSICOCHEMICAL PROPERTIES AND APPLICATION OF POLYACETYLENE

7.1.1 INTRODUCTION

Features of synthesis, structure, properties and use of polyacetylenes are considered in this monography. The catalytic polymerization of acetylene using different catalysts is shown. Plasmachemical synthesis of carbines is considered. The results of studying the structure of polyacetylenes by electron spectroscopy are presented. The results of the research of the surface morphology of polyacetylene are presented. Agency of receiving methods of polyacetylene on its properties is shown. It should be noted that the first chapter of present monography "Synthesis, structure, physicochemical properties and application of polyacetylene" prepared for publication by Professor Rakhimov A.I. and Associate Professor Titova E.S. (Volgograd State Technical University). The remaining chapters prepared for publication by Professor Ponomarev O.A. (Bashkirskii State University), Professor Babkin V.A. (Volgograd State Architect-build University, Sebrykov Departament), Associate Professor Titova E.S. and Professor Zaikov G.E. (Moscow, Institute of Biochemical Physics, Russian Academy of Sciences).

7.1.2 SYNTHESIS AND STRUCTURE OF POLYACETYLENE

The chemical element in periodic system of D.I. Mendeleev – carbon possesses a variety of unique properties. It is the reason of that, as carbon and its compounds, and materials on its basis serve as objects of basic researches and are applied in the most various areas.

Scientists' thought that two forms exists of crystal carbon only – diamond and graphite (opinion beginning of 60-th years of 20 century). These forms are widespread in the nature and they are known to mankind with the most ancient times.

The question on an opportunity of existence of forms of carbon with sp-hybridization of atoms was repeatedly considered theoretically. In 1885 German chemist A. Bayer tried to synthesize chained carbon from derivatives of acetylene by a step method. However Bayer's attempt to receive polyin has appeared unsuccessful. He received the hydrocarbon consisting from four molecules of acetylene, associated in a chain, and appeared extremely unstable.

A. M. Sladkov, V. V. Korshak, V. I. Kasatochkin and Yu. P. Kudryavcev [1] observed loss of a black sediment of polyin compound of carbon having the linear form at transmission acetylene in a water-ammonia solution of salt Cu (II) (oxidizing dehydropolycondensation of acetylene led obviously to polyacetylenides of copper). This powder blew up at heating in a dry condition, and in damp – at a detonation. Process of oxidative dehydropolycondensation of acetylene can be written down in a following kind schematically [1] at $x + y + z = n$:

$$n \ H-C\equiv C-H$$

$$\downarrow Cu^{2+}$$

$$H(-C\equiv C-)_x Cu \ + \ H(-C\equiv C-)_y H \ + \ Cu(-C\equiv C-)_z H$$

$$\downarrow FeCl_3$$

$$H(-C\equiv C-)\text{-}H$$
$$n$$

At surplus of ions Cu^{2+} the mix of various polyins and polyacetylenides of copper various molecular weights are formed. Additional oxidation of products received at this stage (with help $FeCl_3$ or $K_3[Fe\ (CN)_6]$) leads to formation polyins with the double molecular weight. The last do not blow up any more at heating and impact, but contain a plenty of copper. Possibly, trailer atoms of copper stabilize polyins by dint of to complexation.

The content by carbon was 90% of clean polyin (cleaning cleared from copper and impurity of other components of the reactionary medium). Only multi-hours heating of samples of polyin at 1000 °C in vacuum has allowed to receive analytically pure samples of α-carbyne. Similar processing results not only in purification, but also to partial crystallization of polyacetylene.

Under A.M. Sladkov's offer such polyacetylene have named "carbyne" (from Latin *carboneum* (carbon) with the termination "in", accepted in organic chemistry for a designation of acetylene bond).

By acknowledgment of polyin structures in chains is formation of oxalic acid after ozonation hydrolysis of carbine [2, 3]:

$$(-C\equiv C-)_n \ \xrightarrow{O_3} \ \left(\begin{array}{c} -C\equiv C- \\ | \ \ | \\ O \ \ O \\ \diagdown / \\ O \end{array} \right)_n \ \xrightarrow{H_2O} \ nHOOC-COOH$$

New linear polymer with cumulene bonds was received [2, 3]. It has named polycumulene. The proof of such structure became that fact that at ozonation of polycumulene is received only carbon dioxide:

$$(=C=C=)_n \ \xrightarrow{O_3} \ 2nCO_2$$

Cumulene modification of carbyne (β-carbyne) has been received on specially developed by Sladkov two-stage method [3]. At the first stage spent polycondensation of suboxide of carbon (C_3O_2) with dimagnesium dibromine acetylene as Grignard reaction with formation polymeric glycol:

$$n O{=}C{=}C{=}C{=}O + n BrMgC{\equiv}CMgBr \rightarrow \left(-C{\equiv}C-\underset{OH}{C}{=}C{=}\underset{OH}{C}-\right)_n .$$

At the second stage this polymeric glycol reduced by stannous chloride hydrochloric acid:

$$\left(-C{\equiv}C-\underset{OH}{C}{=}C{=}\underset{OH}{C}-\right)_n \xrightarrow[-\,(HCl + SnO_2)]{+\,SnCl_2} (=C{=}C{=}C{=}C{=}C{=})_n.$$

High-molecular cumulene represents an insoluble dark-brown powder with the developed specific surface (200–300 m²/g) and density 2.25 g/cm³. At multihours heating at 1000°C and the depressed pressure polycumulene partially crystallizes. Two types of monocrystals have been found out in received after such annealing a product by means of transmission electronic microscopy. Crystals corresponded to α- and β-modifications of carbyne.

One of the most convenient and accessible methods of reception carbyne or its fragments – reaction of dehydrohalogenation of the some polymers content of halogens (GP). Feature of this method is formation of the carbon chain at polymerization corresponding monomers. The problem at synthesis carbyne consists only in that at full eliminating of halogen hydride with formation of linear carbon chain. Exhaustive dehydrohalogenation is possible, if the next atoms of carbon have equal quantities of atoms of halogen and hydrogen. Therefore convenient GP for reception of carbyne were various polyvinyliden halogenides (bromides, chlorides and fluorides), poly(1,2-dibromoethylene), poly (1,1,2 or 1,2,3-trichlorobutadiene), for example:

$$(-CH2{-}CHal_2-)_n \xrightarrow[-nHHal]{+B} (-CH{-}CHal-)_n \xrightarrow[-nHHal]{+B} (=CH{=}CH=)_n$$

The reaction of dehydrohalogenation typically carry out at presence of solutions of alkalis (B⁻) in ethanol with addition of polar solvents. At use of tetrahydrofuran synthesis goes at a room temperature. This method allows to avoid course of collateral reactions. The amorphous phase only cumulene modification of carbyne is received as a result. Then, crystal of β-carbyne is synthesized from amorphous carbine by solid-phase crystallization.

Next method is dehydrogenation of polyacetylene. At interaction of polyacetylene with metallic potassium at 800°C and pressure 4 GPa led to dehydrogenation and formation of potassium hydride, the carbon matrix containing potassium. After removal potassium from products (acid processing) precipitate out brown plate crystals of β-carbyne in hexagonal forms by diameter ~1 mm and thickness up to 1 micrometer.

Carbyne also can be received by various methods of chemical sedimentation from a gas phase.

Plasmochemical synthesis of Carbyne. At thermal decomposition of hydrocarbons (acetylene, propane, heptane, benzol), carbone tetrachloride, carbon bisulphide, acetone in a stream nitrogen plasma is received the disperse carbon powders containing carbyne. Monocrystals of white color and (white or brown) polycrystals remain after selective oxidation of aromatic hydrocarbon. It is positioned, that formation of carbyne does not depend by nature initial organic compound. The moderate temperature of plasma (~3200K) and small concentration of reagents promote process.

Laser sublimations of carbon. Carbyne has been received at sedimentation on a substrate of steams of negative ions of the carbon after laser evaporation of graphite in 1971. The silvery-white layer was received on a substrate. This layer, according to data X-ray and diffraction researches, consists from amorphous and crystal particles of carbyne with the average size of crystallites $> 10^{-5}$ cm.

Arc cracking of carbon. Evaporation in electric arc spectrally pure coals with enough slow polymerization and crystallization of a carbon steam on a surface of a cold substrate yields to product in which prevail carbyne forms of carbon.

Ion-stimulated precipitation of carbyne. At ion-stimulated condensation of carbon on a lining simultaneously or alternately the stream of carbon and a stream of ions of an inert gas moves. The stream of carbon is received by thermal or ionic evaporation of graphite. This method allows receiving carbyne films with a different degree of orderliness (from amorphous up to monocrystalline layers), carbynes of the set updating, and also a film of other forms of carbon. Annealing of films of amorphous carbon with various near order leads to crystallization of various allotropic forms carbon, including carbyne.

Sladkov has drawn following conclusions on the basis of results of experiments on synthesis carbyne by methods of chemical sedimentation from gas phase:

- White sediments of carbyne are received, possibly, in the softest conditions of condensation of carbon: high enough vacuum, small intensity of a stream and low energy of flying atoms or groups of atoms, small speed of sedimentation;

- Chains, apparently, grow perpendicularly to a lining, not being cross-linked among themselves;
- Probably, being an environment monovalent heteroatoms stabilize chains, do not allow them to be cross-linked.

Reception of carbyne from carbon graphite materials leads by heating of cores from pyrolytic graphite at temperature 2700–3200K in argon medium. This leads to occurrence on the ends of cores a silvery-white strike (already through 15–20 s). This strike consists of crystals carbyne that is confirmed by data of method electron diffraction.

In 1958 Natta with employees are polymerized acetylene on catalyst system $Al(C_2H_5))_3 — Ti(OC_3H_7)_4$ [4, 5].

The subsequent researches [6–8] led to reception of films stereo regular polyacetylene. The catalyst system $Al(Et)_3 – Ti(OBu)_4$ provides reception of films of polyacetylene predominantly (up to 98%).

Films of polyacetylene are formed on a surface of the catalyst or practically to any lining moistened by a solution of the catalyst (it is preferable in toluene), in an atmosphere of the cleared acetylene [8]. The temperature and pressure of acetylene control growth of films [9, 10]. Homogeneous catalyst system before use typically maintain at a room temperature. Thus reactions of maturing of the catalyst occur [11]:

$$Ti(OBu)_4 + AlEt_3 \rightarrow EtTi(OBu)_3 \rightarrow AlEt_2(OBu)$$
$$2EtTi(OBu)_3 \rightarrow 2Ti\,(OBu)_3 + CH_4 + C_2H_6$$
$$Ti(OBu)_3 + AlEt_3 \rightarrow EtTi(OBu)_2 + Al(Et)_2OBu$$
$$EtTi(OBu)_2 + AlEt_3 \rightarrow EtTi(OBu)_2 \cdot Al\,(Et)_3$$

Ageing of the catalyst in the beginning raises its activity. However eventually the yield of polyacetylene falls because of the further reduction of the titan:

$$EtTi(OBu)_3 + Al\,(Et)_3 \rightarrow Ti(OBu)_4 + Al(Et)_2(OBu) + C_2H_4 + C_2H_6$$

The jelly-like product of red color is formed if synthesis led at low concentration of the catalyst. This product consists from confused fibrils in the size up to 800 Å. Foam material with density from 0.04 to 0.4 g/cm^3 is possible to receive from the dilute gels by sublimation of solvent at temperature below temperature of its freezing [12].

Research of speed dependence for formation of films of polyacetylene from catalyst concentration and pressure of acetylene has allowed to find an optimum parity of components for catalyst Al/Ti = 4. Increase of this parity up to 10 leads to increase in the sizes of fibrils [13]. Falling of speed of reaction in the end of process speaks deterioration of diffusion a monomer through a layer of the film formed on a surface of the catalyst [14]. During synthesis the film is formed simultaneously on walls of a flask and on a surface a catalytic solution. Gel collects in a cortex. The powder of polyacetylene settles at the bottom of a reactor. The molecular weight (M_n) a powder below, than gel, also is depressed with growth of concentration of the catalyst up to 400–500 [15]. The molecular weight of jealous polyacetylene slightly decreases with growth of concentration of the catalyst and grows from 2×10^4 up to 3.6×10^4 at rise in temperature from $-78°C$ up to $-10°C$. The molecular weight of polyacetylene in a film is twice less, than in gel [16].

Greater sensitivity of the catalyst to impurity does not allow to estimate unequivocally influence of various factors on M_p of polyacetylene. Low-molecular products with $M_p \sim -1200$ are formed at carrying out of synthesis in the medium of hydrogen [17]. Concentration of acetylene renders significant influence on M_p: at increase of its pressure up to 760 mm Hg increases M_p up to 120,000 [11, 18]. Apparently, the heterogeneity of a substratum arising because of imperfection of technics for synthesis is the reason for some irreproducibility properties of the received polymers. It is supposed, that synthesis is carried out on "surface" of catalytic clusters. Research by EPR method has allowed distinguishing in the catalyst up to four types of complexes [19]. A polyacetylene cis-transoid structures is formed as a result.

Formation of trans-structure at heats speaks thermal isomerization. The alternative opportunity – trans-disclosing of triple bond in a catalytic complex for a transitive condition is forbidden spatially [19, 20]. The structure of a complex and a kinetics of polymerization are considered in works in more detail [11, 21].

Parshakov A. S. with co-authors [22] have offered a new method of synthesis organo-inorganic composites – nanoclusters transitive metals in an organic matrix – by reactions of compounds of transitive metals of the maximum degrees of oxidation with monomers which at the first stage represent itself as a reducer. Formed thus clusters metals of the lowest degrees of oxidation are used for catalysis of polymerization a monomer with formation of an organic matrix.

Thus it is positioned, that at interaction $MoCl_5$ with acetylene in not polar mediums there is allocation HCl, downturn of a degree of oxidation of molybdenum and formation metalloorganic nanoclusters. Two distances Mo–Mo are found out in these nanoclusters by method EXAFS spectroscopy. In coordination sphere Mo there are two nonequivalent atom of chlorine and atom of carbon. On the basis of results MALDI–TOF mass-spectrometry the conclusion is made, that cluster of molybdenum has 12 or the 13-nuclear metal skeleton and its structure can be expressed by formulas $[Mo_{12}Cl_{24}(C_{20}H_{21})]^-$ or $[Mo_{13}Cl_{24}(C_{13}H_8)]^-$.

Set of results of Infrared-, Raman-, MASS-, NMR-^{13}C- and RFES has led to a conclusion, that the organic part of a composite represents polyacetylene a trans-structure. Polymeric chains lace, and alongside with the interfaced double bonds, are present linear fragments of twinned double –HC=C=CH- and triple –C≡C- bonds.

Reactions $NbCl_5$ with acetylene also is applied to synthesis of the organo-inorganic composites containing in an organic matrix clusters of transitive elements not only VI, but other groups of periodic system.

Solutions $MoCl_5$ sated at a room temperature in benzene or toluene used for reception of organo-inorganic composites with enough high concentration of metal. Acetylene passed through these solutions during 4–6 hours. Acetylene preliminary refined and drained from water and possible impurity. Color of a solution is varied from dark-yellow-green up to black in process of transmission acetylene. The solution heated up, turned to gel of black color and after a while the temperature dropped up to room. Reaction was accompanied by formation HCl. Completeness of interaction pentachloride with acetylene judged on the termination of its allocation.

The sediment, similar to gel, settled upon termination of transmission acetylene. It filtered off in an atmosphere of argon, washed out dry solvent and dried up under vacuum.

The received substances are fine-dispersed powders of black color, insoluble in water and in usual organic solvents.

Solutions after branch of a deposit represented pure solvent according to NMR. Formed compounds of molybdenum and products of oligomerization acetylene precipitated completely. The structure of products differed under the maintenance of carbon depending on speed and time of transmission acetylene a little. Thus relation C:H was conserved close to unit, and Cl:Mo – close to two. The structure of products of reaction differed slightly in benzene and toluene and was close to $MoCl_{1,9\pm0,1}(C_{30\pm1}H_{30\pm1})$.

TABLE 1   Data of the element analysis of products for reaction $MoCl_5$ with acetylene in benzene and toluene.

Solution	Percentage, weight %							
	C		H		Cl		Mo	
	Findings	Calculated	Findings	Calculated	Findings	Calculated	Findings	Calculated
Benzene	65.94	64.67	5.38	5.38	11.70	12.73	16.98	17.22
Toluene	66.23		5.24		11.87		16.66	

Presence on diffraction patterns the evolved products of a wide maximum at small corners allowed to assume X-ray amorphous or nanocrystalline a structure of the received substances. By a method of scanning electronic microscopy (SEM) it was revealed, that substances have low crystallinity and nonfibrillary morphology (Fig. 1a).

 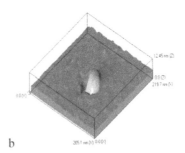

a   ⊢—9μm—⊣   b

FIGURE 1   Photomicrographes $MoCl_{1.9\pm0.1}(C_{30\pm1}H_{30\pm1})$ according to SEM (a) and ASM (b).

7.1.3   PHYSICOCHEMICAL PROPERTIES AND APPLICATION OF POLYACETYLENE

By results of atomic-power microscopy (ASM) particle size can be estimated within the limits of $10\div15$ nm (Fig. 1b). By means of translucent electronic microscopy has been positioned, that the minimal size of morphological element $MoCl_{1.9\pm0.1}(C_{30\pm1}H_{30\pm1})$ makes $1\div2$ nm.

Substances are steady and do not fly in high vacuum and an inert atmosphere up to 300°C. Formation of structures $[Mo_{12}Cl_{24}(C_{20}H_{21})]^-$ and $[Mo_{13}Cl_{24}(C_{13}H_8)]$ is supposed also on the basis of mass-spectral of researches.

Spectrum EPR of composite $MoCl_{1.9\pm0.1}(C_{30\pm1}H_{30\pm1})$ at 300 K (Fig. 1.2.a) consists of two isotropic lines. The intensive line g=1.935 is carried to unpaired electrons of atoms of molybdenum. The observable size of the g-factor is approximately equal to values for some compounds of trivalent molybdenum. For example, in $K_3[InCl_6]\cdot2H_2O$, where the ion of molybdenum Mo (+3) isomorphically substitutes In (+3), and value of the g-factor makes 1.93 ± 0.06.

The line of insignificant intensity with g=2.003, close to the g-factor free electron –2.0023, has been carried to unpaired electrons atoms of carbon of a polyacethylene matrix. Intensity of electrons signals for atoms molybdenum essentially above, than for electrons of carbon atoms of a matrix. It is possible to conclude signal strength, that the basic contribution to paramagnetic properties of a composite bring unpaired electrons of atoms of molybdenum.

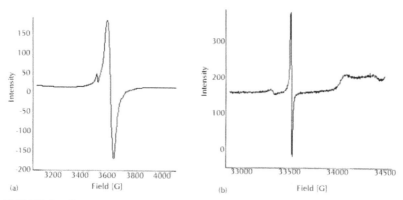

FIGURE 2   Spectrum EPR of composite $MoCl_{1,9\pm0,1}(C_{30\pm1}H_{30\pm1})$ at 300 K, removed in a continuous mode in X-(a) and W-range (b).

Spectrum EPR which has been removed in a continuous mode at 30 K, (Fig. 2b) has a little changed at transition from X to a high-frequency W-range. Observable three wide lines unpaired electrons atoms of molybdenum have been carried to three axial components with g_1=1.9528, g_2=1.9696 and g_3=2.0156, accordingly. Unpaired electrons atoms of carbon of a polyacethylene matrix the narrow signal g=2.0033 answers. In a pulse mode of shooting of spectra EPR at 30 K (Fig. 3) this line decomposes on two signals with g_1 =2.033 and g_2 =2.035. Presence of two signals EPR testifies to existence in a polyacethylene matrix of two types of the paramagnetic centers of the various natures which can be carried to distinction in their geometrical environment or to localized and delocalized unpaired electrons atoms of carbon polyacethylene chain.

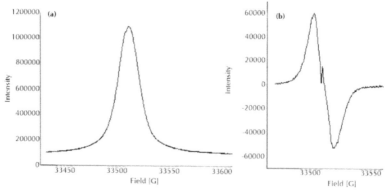

FIGURE 3   Spectrum absorption EPR of composite $MoCl_{1,9\pm0,1}(C_{30\pm1}H_{30\pm1})$, removed in a pulse mode in a W-range at 30 K (a) and its first derivative (b).

Measurement of temperature dependence of a magnetic susceptibility X_g in the field of temperatures $77 \div 300$ K has shown, that at decrease in temperature from room up to 108K the size of a magnetic susceptibility of samples is within the limits of sensitivity of the device or practically is absent. The sample started to display a magnetic susceptibility below this temperature. The susceptibility sharply increased at the further decrease in temperature.

The magnetic susceptibility a trance-polyacetylene submits to Curie law and is very small on absolute size. Comparison of a temperature course a composite and pure allows to conclude a trance-polyacetylene, that the basic contribution to a magnetic susceptibility of the investigated samples bring unpaired electrons atoms of molybdenum in cluster. Sharp increase of a magnetic susceptibility below 108K can be connected with reduction of exchange interactions between atoms of metal.

The size of electroconductivity compressed samples $MoCl_{1,9\pm0,1}(C_{30\pm1}H_{30\pm1})$, measured at a direct current at a room temperature $- (1.3 \div 3.3) \times 10^{-7} \, \Omega^{-1} \cdot cm^{-1}$ is in a range of values for a trance-polyacetylene and characterizes a composite as weak dielectric or the semiconductor. The positioned size of conductivity of samples at an alternating current $\sigma = (3.1 \div 4.7) \times 10^{-3} \, \Omega^{-1} \cdot cm^{-1}$ can answer presence of ionic (proton) conductivity that can be connected with presence of mobile atoms of hydrogen at structure of polymer.

Research of composition, structure and properties of products of interaction $NbCl_5$ with acetylene in a benzene solution also are first-hand close and differ a little with the maintenance of carbon (Table 2).

TABLE 2   Data of the element analysis of products of interaction $NbCl_5$ with acetylene in a solution and at direct interaction.

The weights content, %							
C		H		Cl		Nb	
Findings	Calculated	Findings	Calculated	Findings	Calculated	Findings	Calculated
In a solution							
45.00	45.20	3.60	3.76	20.50	22.20	26.50	28.80
At direct interaction							
41.60	45.20	4.02	3.76	22.81	22.20	27.78	28.80

To substances formula $NbCl_{2\pm0,1}(C_{12\pm1}H_{12\pm1})$ can be attributed on the basis of the received data. Interaction can be described by the equation:

$$NbCl_{5(solv/solid)} + nC_2H_2 \rightarrow NbCl_{2\pm0,1}(C_{12\pm1}H_{12\pm1})\downarrow + (n-12)C_6H_6 + 3HCl\uparrow + Q$$

The wide line was observed on diffraction pattern $NbCl_{2\pm0,1}(C_{12\pm1}H_{12\pm1})$ at $2\theta = 23-24°C$. It allowed assuming a nanocrystalline structure of the received products.

Studying of morphology of surface $NbCl_{2\pm0,1}(C_{12\pm1}H_{12\pm1})$, received by direct interaction, method SEM has shown, that particles have predominantly the spherical form, and their sizes make less than 100 nm (Fig. 4).

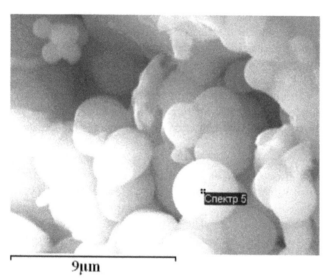

$9\mu m$

FIGURE 4   Microphoto of particles $NbCl_{2\pm0,1}(C_{12\pm1}H_{12\pm1})$, received by method SEM.

The Globular form of particles and their small size testify to the big size of their specific surface. It will be coordinated with high catalytic activity $NbCl_{2\pm0,1}(C_{12\pm1}H_{12\pm1})$.

Fibrils are formed in many cases as a result of synthesis [21]. The morphology of polyacetylene films practically does not depend on conditions of synthesis. Diameter of fibrils can change depending on these conditions and typically makes 200–800 Å [11, 21]. At cultivation of films on substrates the size fibrils decreases. The same effect is observed at reception of polyacetylene in the medium of other polymers. Time of endurance (ageing) of the catalyst especially strongly influences of the size of fibrils. The size of fibrils increases with increase in time of ageing. Detailed research of growth fibrils on thin films a method of translucent electronic microscopy has allowed to find out microfibrillar branching's on the basic fibril (the size 30–50 Å) and thickenings in places of its gearing, and also presence of rings on the ends of fibrils.

Essential changes in morphology of a film at isomerization of polyacetylene are not observed. The film consists from any way located fibrils. Fibrils sometimes are going in larger formations [20]. Formation of a film is consequence of interaction fibrils among themselves due to adhesive forces.

The big practical interest is represented catalytic system $AlR_{3-}Ti(oBu)_4$ [23]. Polyacetylene is received on it at $-60°C$, possesses fibrous structure and can be manufactured usual, accepted in technology of polymers by methods. Particle size increases from 100 up to 500 Å at use of the mixed catalyst and the increase in density of films is observed. A filtration of suspension it is possible to receive films of any sizes. On the various substrates possessing good adhesive properties, it is possible to receive films dispersion of gel. Polymer easily doping AsF_5, $FeCl_3$, I_2 and others electron acceptors. Preliminary tests have shown some advantages of such materials at their use in accumulators [23]. Gels polyacetylene with the similar properties, received on others catalytic systems, represent the big practical interest [24]. Data on technological receptions of manufacturing of polyacetylene films from gels with diameter of particles 0.01–1.00 mm are in the patent literature [25]. Films are received at presence of the mixed catalyst at an interval of temperatures from -100 up to $-48°C$. These results testify to an opportunity of transition to enough simple and cheap technology of continuous process of reception of polyacetylene films. Rather accessible catalysts are WCl_6 and $MoCl_6$. Acetylene polymerizes at $20°C$ and pressure up to 14 atm at their presence. However the received polymers contain carbonyl groups because of presence of oxygen and possess low molecular weight [26]. The complex systems including in addition to WCl_6 or $MoCl_6$ tetraphenyltin are more perspective for reception of film materials. The Film of doped predominantly (90%) a trance-structure with fibrous morphology is formed on a surface of a concentrated solution of the vanadic catalyst. Diameter of fibrils of the polymer received on catalyst $MoCl_5 - Ph_4Sn$, can vary within the limits of 300–10000 Å; in case of catalyst $WCl_6 - Ph_4Sn$ it reaches $1.2×10^5$ Å [27].

Research of a kinetics of polymerization of acetylene on catalysts $Ti(OBu)_{5-}$ $Al(Et)_3$, $WCl_{6-}PhtSn$, $MoCl_{5-}Ph_4Sn$, $Ti(CH_2CeH_5)$, has shown, that speed of process falls in the specified number [31]. Films, doped by various acceptors [CH $(SbF_5)_{0,7}]_n$, $[CH(CF_3SO_3H)_{0,8}]_m$ – had conductivity $10–20$ $\Omega^{-1}.cm^{-1}$ at $20°C$.

New original method of reception of films doped in a trance-form is polymerization of a 7,8-bis(trifluoromethyl)tricyclo[4,2,2,0]deca-3,7,9-trien (BTFM) with disclosing a cycle.

Polymerization occurs on catalytic system $WCl_6 - Sn(CH_3)_4$, precipitated on surfaces of a reactor. The film prepolymer as a result of heating in vacuum at 100–150° detaches trifluoromethylbenzene. The silvery film polyacetylene is formed. The density of polymer reaches 1.1 g/cm^3 and comes nearer to flotation density polyacetylene, received in other ways. Received this method of amorphous polyacetylene has completely a trance-configuration and does not possess fibrous structure. The rests of 1,2-bis(trifluoromethylbenzene) are present at polymer according to Infrared-spectroscopy. Crystal films of polyacetylene with monoclinic system and $\beta = 91.5°$ are received at long heating of prepolymer on networks of an electronic microscope at 100° in vacuum. Improvement of a method has allowed to receive films and completely oriented crystal polyacetylene. In the further ways of synthesis of polymers from others monomers have been developed.

Naphthalene evolves at heat treatment of prepolymer in the first case. Anthracene evolves in the second case. Purification polymer from residual impurities occurs when the temperature of heat treatment rise. According to spectral researches, the absorption caused by presence of sp-hybrid carbon is not observed in films. Absorption in the field of 1480 cm^{-1} caused by presence of C = C bonds in Spectra KP of considered polymers, is shifted compared to the absorption observed in polyacetylene, received by other methods (1460 cm^{-1}).

It is believed that it is connected with decrease of size of interface blocks. The obtained films are difficult doping in a gas phase due to its high density. Conductivity of initial films reaches 10–200 (Ω^{-1}.cm^{-1}) when doped with bromine or iodine in a solution [26].

Solutions of complex compound cyclopentadienyl-dititana in hexane and sodium cyclopentadienyl complex have high catalytic activity. Films with a metallic luster can be obtained by slowly removing the solvent from the formed gel polyacetylene in vacuum. It is assumed that the active complex has a tetrahedral structure. Polymerization mechanism is similar to the mechanism of olefin polymerization on catalyst Ziegler – Natta. Obtained at –80°C *cis*-polyacetylene films after doping had a conductivity of 240 (Ω^{-1}.cm^{-1}).

Classical methods of ionic and radical polymerization do not allow to receive high-molecular polymers with system of the conjugated bonds because of isomerization the active centers [2] connected with a polyene chain. Affinity to electron and potential of ionization considerably varies with increase in effective conjugated. One of the methods, allowing to lead a cation process of polymerization, formation of a complex with the growing polyconjugated chains during synthesis. Practical realization of such process probably at presence of the big surplus of a strong acceptor of electrons in the reaction medium. In this case the electronic density of polyene chain falls. The probability of electron transfer from a chain on the active center decreases accordingly.

Polyacetylene films were able to synthesize on an internal surface of a reactor in an interval of temperatures from –78°C up to –198°C at addition of acetylene to arsenic pentafluoride. Strips of absorption *cis*-polyacetylene are identified in the field of 740 cm^{-1} and doped complexes in the field of 900 and 1370 cm^{-1}.

Similar in composition films were prepared by polymerization vinylacetylene in the gas phase in the presence SbF$_5$. However, to achieve a metallic state has failed. Soluble polyenes were obtained in solution AsF$_3$ in the by cationic polymerization, including soluble agents. Polymerization was carried out at the freezing temperature of acetylene. From the resulting solution were cast films with low conductivity, characteristic of weakly doped polyacetylene with 103 molecular weight. Spectral studies confirmed the presence of the polymer obtained in the *cis*-structure. Practical interest are insoluble polymers with a conductivity of 10^{-3}

($\Omega^{-1}.cm^{-1}$), obtained at $-78°C$ polymerization of acetylene in the presence AsF_5, $NaAsF$, SiF, AsF_3, BF_3, SbF_5, PF_6.

Effective co-catalyst of cationic polymerization of acetylene and its derivatives are compounds of bivalent mercury and its organic derivatives. As a result, the reactions of oxide or mercury salts with proton and aprotic acids into saturated hydrocarbon formed heterogeneous complexes. They are effectively polymerized acetylene vinylacetylene, phenylacetylene, propargyl alcohol. Polyacetylene predominantly *trans*-structure with a crystallinity of 70% was obtained in the form of films during the polymerization of acetylene on the catalyst surface. According to X-ray studies, the main reflection corresponds to the interplanar distance $d = 3.22$ Å. Homogeneous catalyst obtained in the presence of aromatic ligands. Active complex of the catalyst with acetylene is stable at low temperatures. The alkylation of solvent and its interpolymerization with acetylene is in the presence of aromatic solvents (toluene, benzene). In the infrared spectrum of the films revealed absorption strias corresponding to the aromatic cycle. Patterns of polymerization of acetylene monomers, the effect of temperature and composition of the catalyst on the structure and properties of the resulting polymers were studied. Morphology of the films showed the absence of fibrils.

Almost all of the above methods of synthesis polyacetylene with high molecular weight lead to the formation of insoluble polymers. Their insolubility due to the high intermolecular interaction and form a network structures. For the polymers obtained by Ziegler catalyst systems, the crosslink density of the NMR data of 3–5%. This is confirmed by ozonolysis. Quantum chemical calculations confirm that the isomerization process intermolecular bonds are formed.

The most accessible and promising method for synthesis of the polyacetylene of linear structure is the polymerization of acetylene in the presence of metals VIII group, in the presence with reducing agents (catalyst Luttlnger). Polyacetylene with high crystallinity was obtained by polymerization of acetylene on the catalyst system $Co(NO_3)_2$ $NabH_4$ with component ratio 1:2. Both components are injected into the substrate containing monomer, to prevent the death of a catalyst. Raising the temperature and the concentration of sodium borane leads to partial reduction of the polymer. The activity of nickel complex can be significantly improved if the polymerization leads in the presence of $NabH_2$. Crystalline polymers, obtained at low temperatures, do not contain fibrils. Crystallites have dimensions of 70 Å.

A typical reflex observed at $23.75°C$ ($d=3.74$ Å) confirms the *trans*-structure of the polymer. Palladium complexes are ineffective in obtaining high molecular weight polymers. Catalyst Luttlnger enables one to obtain linear polymers of *cis*-structure, characterized by high crystallinity. The yield of polyacetylene is 25–30 g/g catalyst.

Chlorination of the freshly prepared polymers at low temperatures allows to obtain soluble chlorpolymers about 104 molecular weight [29]. Although the authors argue that the low-temperature chlorination, in contrast to hydrogenation, there is no polymer degradation, data suggest that an appropriate choice of temperature and solvent derived chlorinated polymers have a molecular weight up to 2.5×10^5. Destruction more visible at chlorination on light and on elevated temperatures. The proof of the linear structure of polyacetylene is the fact that soluble iodinated polymers are obtained by iodination polyacetylene suspension in ethanol.

Systematic studies of methods for the synthesis of polyacetylene allowed to develop a simple and convenient method of obtaining the films on various substrates wetted by an ethereal solution of the catalyst. The disadvantages of these films, as well as films produced by Shirakava [21, 22] are difficult to clean them of residual catalyst and the dependence of properties on the film thickness. Much more manufacturable methods for obtaining films of polyacetylene spray precleaned from residues of the catalyst in a stream of polyacetylene gels inert gas or a splash of homogenized gels [24]. The properties of such films depend on the conditions of their formation. Free film thickness of 2–3 mm is filtered under pressure in an inert atmosphere containing a homogenized suspension of polyacetylene 5–10 g/L. The suspension formed in organic media at low temperatures in the presence of the catalyst $Co(NO_3)_2$ $NaBH_4$. The films obtained by spraying a stream of inert gas, homogeneous, have good adhesion to substrates made of metal, polyurethane, polyester polyethylenetereftalate, polyimide, etc.

Suspension of polyacetylene changes its properties with time significantly. Cross-linking and aggregation of fibrils observed in an inert atmosphere at temperatures above –20 °C. This leads to a decrease in the rate of oxidation and chlorination. The morphology of the films changes particularly striking during the ageing of the suspension in the presence of moisture and oxygen: increasing the diameter of the fibrils, decreasing their length, breaks and knots are formed. The suspension does not change its properties in two weeks.

Preparation of soluble polymers with a system of conjugated double bonds, and high molecular weight is practically difficult because of strong intermolecular interactions. Sufficiently high molecular weight polyacetylene were obtained in the form of fine-dispersed particles during the synthesis on a Luttinger catalyst in the presence surfactants: copolymer of styrene and polyethylene with polyethylene oxide. Acetylene was added to the catalyst and the copolymer solution in a mixture of cyclohexane – tetrahydrofuran at –60°C and heated to –30°C. Stable colloidal solutions with spherical particles in size from 40 to 2000 Å, formed. The density of selected films is 1.15 g/cm³. Colloidal solutions of polyacetylene can be obtained in the presence of other polymers that prevent aggregation of the forming molecules of polyacetylene.

Polyacetylene obtained in the presence of Group VIII metals, in combination with $NaBH_4$, has almost the same morphology, as a polymer synthesized by Shirakawa [22]. The dimensions of fibrils lay in the range 300–800 Å, and depend on the concentration of the catalyst, the synthesis temperature and medium [29].

Thermogravimetric curves for the polyacetylene, there are two exothermic peaks at 145 and 325°C [13]. The first of these corresponds to an irreversible *cis*-trans isomerization. Migration of hydrogen occurs at 325°C, open chain and cross-linking without the formation of polyacetylene volatile products. The color of the polymer becomes brown. A large number of defects appear. In the infrared spectrum there are absorption bands characteristic of the CH_2, CH_3, $-C = C-$ and $-C_2H_5$–groups [13].

Structuring polymer occurs in the temperature range 280–380°C. But 72% of initial weight of polyacetylene losses at 720°C. The main products of the decomposition of polyacetylene are benzene, hydrogen and lower hydrocarbons [12]. The crystallinity of polyacetylene reduced when heated in air to 90°C after several hours. The brown amorphous substance, similar cuprene, obtained after 70 hours.

Catalytic hydrogenation of polyacetylene leads to the formation of cross-linked product [30]. Non-cross-linked and soluble products are obtained in the case of hydrogenation of polyacetylene doped with alkali metals [27, 28]. Polyacetylenes are involved in redox reactions that occur in processing strong oxidizing and reducing agents (iodine, bromine, AsF_5, Na-naphthalene) in order to significantly increase the electrical conductivity [31].

Practical use of polyacetylene is complicated by its easy oxidation by air oxygen [32]. Oxidation is easily exposed to the polymer obtained by polymerization of acetylene [33, 34]. The *cis*- or *trans*-$(CH)_x$ in air or oxygen for about an hour exposed to the irreversible oxidative degradation [35, 36]. The limiting value of weight gain due to absorption of polyacetylene (absorption) of oxygen from air oxidation at room temperature is 35% [32]. The resulting product is characterized by the formula $[(C_2H_2)O_{0,9}]_n$. The ease of oxidation depends on the morphology of the polymer and changes in the series of crystal < amorphous component < the surface of the fibrils [34]. The absorption of oxygen begins at the surface of fibrils, and then penetrates. Polyacetylene globular morphology is more stable to the effects of O_2 than polymer fibrillar structures [37].

Polyacetylene obtained by polymerization of acetylene in Ziegler–Natta catalysts, after doping Cl_2, Br_2, I_2, AsF_5 becomes a semiconductor in the form of flexible, silvery films "organic metals" [38]. Doping with iodine increases the amorphous samples d 6×10^{-5}, and crystal – to 7×10^2 $\Omega^{-1}.cm^{-1}$ [39]. The highest electrical conductivity of the polyacetylene compared with those obtained by other methods, the authors [40] explain the presence of catalyst residues. In their view, the concentration of the structure of sp^3 – hybridized carbon atoms is relatively little effect on the conductivity as compared with the influence of catalyst residues. Doping with iodine films of polyacetylene obtained by metathesis polymerizing cyclooctatetraene leads to an increase in their electrical conductivity

10^{-8} to 50–350 $\Omega^{-1}.cm^{-1}$ [41], and have received polymerization of benzvalene with ring opening from 10^{-8}–$10^{-5}$to 10^{-4}–10^{-1} $\Omega^{-1}.cm^{-1}$ [42].

Conductivity increases when pressure is applied to polyacetylene, obtained by polymerization of acetylene [43] and interphase dehydrochlorination of PVC [44]. Anomalously large (up to ten orders of magnitude) an abrupt increase in conductivity when the load is found for iodine-doped crystalline polivinilena – conversion product of PVC [45].

Magnetic properties of polyacetylene significantly depend on the configuration of chains [46]. In the EPR spectrum of the polymerization of polyacetylene singlet line with g-factor of 2.003 [23] and a line width (ΔH) of 7 to 9.5 Oe for the *cis*-isomer [24] and from 0.28 to 5 Oe for the *trans*-isomer [25] observed. According to other reports [47], the *cis*-isomer, syn-synthesized by polymerization of acetylene at 195 K, the EPR signal with g-factor=2.0025 is not observed. This signal appears when the temperature of polymerization increases, when the *trans*-isomer in the form of short chains mainly at the ends of the molecules is 5–10 wt% [26]. The morphology of polyacetylene also has an effect on the paramagnetic properties. The concentration of PMC in the amorphous polyacetylene is ~1018 spin/g, and in the crystal – 1019 spin/g [26].

7.2   THE NATURE OF KINETIC PROPERTIES OF QUASI-ONE-DIMENSIONAL POLYMERS (THE REVIEW OF PROPERTIES AND METHODS OF THEIR STUDYING)

7.2.1   INTRODUCTION

Development of a science about kinetic properties of firm bodies and liquids always led to revealing of the big role of a stable particles in kinetic processes. In chemistry this opening of radicals and ion- radicals, development of chain processes an establishment, that catalytic properties of substances are defined by defects of structure by which energy of activation is essentially reduced. Durability, conductivity, painting and so on at crystals is defined basically by defects of structure. Properties enough greater molecules, especially kinetic properties, are connected substantially with defects and multifocal uptake. As examples polyacetylene and a number of other polymers can serve.

In the given review we shall stop only on three aspects of a problem: we shall give a general characteristic of quasi-one-dimensional systems, we shall describe the basic properties of polyacetylene and we shall formulate problems in the given area.

7.2.2   A GENERAL CHARACTERISTIC OF ONE-DIMENSIONAL SYSTEMS

A number of exact results are known for quasi-one-dimensional systems. These results are received on the basis of exact calculation of modeling systems, or are consequence of the general theorems of interaction. Results are little. The most important results are: 1) the proof of absence of phase transition at $T_c \neq 0$, that is the proof of destruction of the first order; 2) an establishment, that the account of correlations leads to collectivization of conditions; 3) the proof, that localization of conditions occurs in as much as weak casual floor; 4) an establishment of the fact, that results 1) −3) lose force already at infinitesimal interaction in three measurements.

Quasi-one-dimensional systems have high density of the raised conditions and fluctuations. It leads to a number of features, such as formation of superstructures, infringement of the concept of quasi-particles, inapplicability of a method of the self-coordinated field. We shall consider these features in more detail.

7.2.2.1   THE FORMATION OF SUPERSTRUCTURES

The formation of superstructures is universal property for one-dimensional materials: metal passes in semimetal, then in dielectric. Structural instability of an one-dimensional metal condition is proved by Peierls (1937). It leads high permittivity's $\varepsilon=10^4$ (Frelih, 1954). Peierls instability this phenomenon when at downturn of temperature in a lattice there are stationary distortions with a wave vector k, equal to the double vector of electron at the top borrowed level of metal, that is $k = 2CF$. This phenomenon is closely connected with huge anomaly−reduction of density of oscillatory conditions at a wave vector $q = 2CF$. Peierls instability is shown that frequency of the some infrared-fluctuations decreases and addresses in zero. Frequencies of fluctuations in molecules are defined by the formula:

$$\omega_q^2 = \omega_{0q}^2 \left(1 - 2\Pi(q, \omega_q) / \omega_{0q}\right)$$

ω_{0q}^2 − Not indignant frequency, $\Pi(q, \omega_q)$ − own energy of fluctuations appearing due to kernels with electrons. The equation $\left(1 - 2\Pi(q, \omega_q) / \omega_{0q}\right) = 0$ defines those frequencies and wave a vector at which before all there is an instability. It appears at $q = 2CF$. Reorganization of structure of a molecule occurs at this wave vector.

Two tendencies are distinctly shown in one-dimensional gas of electrons, kernels being a floor: superconducting pairing of electrons with opposite backs and impulses and Peierls pairing electron and holes with opposite impulses (doubling a handrail) (Brazovskiy, Dzyaloginskiy, 1974; Brazovskiy, 1981). Impurities in system suppress dielectric and superconducting transitions.

7.2.2.2   INAPPLICABILITY OF THE CONCEPT OF ONE-PARTIAL CONDITIONS

Strong collectivization of electrons occurs in one-dimensional system. It follows that speed of a sound c comparable whit speed of electrons v_F on top level (for two and three-dimensional systems they are accordingly equal $v_F / \sqrt{2}$ and $v_F / \sqrt{3}$). In model with linear spectrum $H_0 = v_F \sum p(a_{ps}^+ a_{ps} - b_{ps}^+ b_{ps})$ in one-dimensional case all degrees of freedom appear collective. While interaction is not present, it is possible to use H_0 and H_{10} – other representation H_0, written down in representation of operators of density

$$H_{10} = \frac{2\pi V_F}{L} \sum [\rho_s^{(1)}(p)\rho_s^{(1)}(-p) + \rho_s^{(2)}(p)\rho_s^{(2)}(-p)]$$

$$p_s^{(1)}(p) = \sum a_{p+k,s}^+ a_{ks} \quad p_s^{(2)}(p) = \sum b_{p+k,s}^+ b_{ks}$$

At presence of interaction $H_{int} \approx a^+ b + ab, a^+ a^+ bb, b^+ b^+ aa, a^+ a^+ aa, b^+ b^+ bb$ for one-dimensional systems the picture of the description is meaningful only H_{10}. Quasi-particle is done inapplicable at transition to one-dimensional systems. We shall consider one-partial Green's function G(q, ω). If quasi-particle the description was fair, $-\pi^{-1} JmG(0,\omega+i0) = \sum Z_i \delta(\omega-\omega_i) + b(\omega)$. This kind of function of Green allowed to enter the concept of quasi-particles because of presence $\delta(\omega-\omega_i)$. For one-dimensional systems it is received for precisely solved model of Luttinger–Tomonug (Luther, 1974)

$$-\pi^{-1} jmG(0,\omega+i0) = \frac{r}{c}(\frac{wr}{2c})^{\gamma-1}\Gamma(1-\gamma)\sin(\pi\gamma/2) + \frac{2r}{\gamma c}(\frac{\omega r}{2c})^{2r-1}[\Gamma(1-\gamma)\sin\pi\gamma]^2$$

The member of type $\delta(\omega-\omega_i)$ in this expression is not present. In this case (q, ω) has no poles. Concept of quasi-particle to enter it is impossible.

Though the quasi-particle description is impossible in case of one-dimensional systems because of strong interaction of electrons with deformations of a lattice. Thus new collective conditions arise – solitones, which are decisions of the nonlinear equations and replace one-partial, conditions at the description of properties one-dimensional and quasi-one-dimensional systems.

7.2.2.3 INAPPLICABILITY OF A METHOD OF THE SELF-COORDINATED FIELD

Used of Hartree-Fok approximation is too rough approach a one-dimensional case. It is established from comparison with exact decisions for modeling systems and some exact results received in the general approach. For example, Hartree-Fok approximation leads to temperature of phase transition distinct from zero though according to the general q^{-2}-Bogoliubov thereof it not can be.

The method of the self-coordinated field in "ladder" approach and representation one-partial excitement in which it results does not describe behavior of polyacetylene as insufficiently full considers collective phenomenes of fluctuation. Its noncritical application has led to a number of mistakes at calculation Peierls transition and one-dimensional superconductivity. Even the account of "parquet" diagrams and a method of multiplicate renormalizations lead to wrong results (1973).

From stated follows, that it is necessary to use the methods of calculation considering collective pheromones of fluctuation in a much greater degree, than in standard methods. These methods were stated by us in the beginning of a rate.

7.2.3 A GENERAL CHARACTERISTIC OF ONE-DIMENSIONAL SYSTEMS

Polyacetylene is the elementary linear interfaced polymer with stable trans–conformation and not leveled communications. Schematically its structure can be represented in the form of:

A, B – conditions of polyacetylene, C^* – division of a circuit due to instability. At big enough distance between divisions of a circuit (domain walls) conditions A and B do not differ on energies. It provides an opportunity of existence topological solitones – domain sides from a radical which extend on odd number of atoms of carbon (that the A – structure has been connected with B – structure). The domain wall can grasp electron, forming carbanion, or lose electron, forming carbocation. All these of heterogeneity are active particles and can give chemical transformations. However chemical kinetics though is of interest here it will not be considered.

Except for the structure specified above, there is still a structure which consists of two parts A and A, disconnected by a wall from even number of atoms of carbon.

Research of solitones in polyacetylene leads to the decision of equations of Bogolubov-de Jen.

$$\varepsilon_n u_n = -ic\frac{\partial}{\partial x}u_n + \Delta(x)\,\mathrm{v}_n,$$

$$\varepsilon_n\,\mathrm{v}_n = ic\frac{\partial}{\partial x}\mathrm{v}_n + \Delta(x)u_n,$$

$$\Delta(x) = -\frac{g^2}{\omega^2}\sum \mathrm{v}_n^{*}\,u_n$$

They are fair for any size Δ, Ginzburg–Landau equations follow from them at small Δ (1980).

7.2.3.1   THE BAND STRUCTURE

The Band structure without taking into account an alternation of communications has a usual appearance cosine. It is a start structure. The account of Peierls instability leads to structure of a zone, which differs a crack about an impulse equal $\pi/2a$.

The structure of a zone depends on number of electrons, falling unit, which can change at addition zones by electrons or holes. The Solitones condition lays inside of the forbidden zone, a little below its center. In polyacetylene takes place strong electron-vibrational interaction. In this case it is shown, that Peierls transition at length of circuit $N\rangle N_c = T_c^{-1}$ (Bulaevskiy, 1974).

7.2.3.2   OPTICAL PROPERTIES OF POLYACETYLENE

Anisotropy in factors of reflection is observed in many works: 80% of radiation is reflected at falling light on a plane of a molecule and 10% at falling in a perpendicular direction in the field of from 0.001eV up to 6eV. For energy of quantums of light, it is more than reflection does not occur, as in this area there is a strong absorption by plasma fluctuations. We have for dielectric polarization:

$$\varepsilon_{11}(\omega) = \varepsilon_\infty(1 - \frac{\omega_p^2}{\omega(\omega + i/\tau)}) \quad \varepsilon(\omega) = 2.2$$

where: ω_p – plasma frequency, $\omega_\infty = 2.0$, $\omega_k^2 = 2.6$, $\tau = 3\times10^{-15}$ second. The reason of the big dielectric susceptibility is the wave of charging density (VCD), cooperating with impurity (pinning), which destroys the distant order. Otherwise the susceptibility would be infinite, and due to interaction with impurity it is done final, though also big. Along the allocated axis period VCD is equal CF/2. If this period and the period of the basic lattice are commensurable, VCD can cooperate and with fluctuations of a lattice. This effect is small, if CF/2 and the period of the basic lattice are incommensurable (Larkin, 1977).

Stationary raised conditions of polyacetylene are (because of incommensurabilities) peak solitons (e=0, s=½). Solitons is conditions when wave function of electron is localized, the lattice is deformed. However, in these conditions the full density of a charge and energy is delocalize also constant on length of system (Brazovskiy, 1980). At a double commensurability (polyacetylene without impurity) can arise solitons with charges e = 0, +1, −1and spin s = 0, ½. Batching of impurity it is possible to receive even a fractional charge.

At a premise of system in a magnetic field in it there is a magnetic moment as backs are divided on subzones and on a miscellaneous they are filled.

Solitons it is possible to consider as the connected conditions of exciting and local deformation of a lattice (Davidov, 1980). The problem is reduced to the decision of the equation:

$$i\hbar \frac{\partial}{\partial t}\varphi_n + (\varepsilon_0 + \frac{\hbar^2}{2ma})\varphi_n + \frac{\hbar^2 e^{-W_n}}{2ma}(\varphi_{n-1} + \varphi_{n+1}) + G|\varphi_n|^2 \varphi_n = 0,$$

$$W_n = |\varphi_n|^4 Bf(\theta)$$

7.2.3.3 ELECTRIC PROPERTIES OF POLYACETYLENE

At kinetic processes in polyacetylene moves backs. The weight of soliton (M_s) is approximately equal to the weight of electron (m). Exact calculation shows, that M_s=6m. The charge and energy of soliton are homogeneous. They do not give the contribution in electroconductivity and heat conductivity. Kinetic display of presence of solitons is only spin diffusion. Participation of solitons leads strong nonlinearities in factors of carry.

Big time of the photoresponse is in polyacetylene (10^{-3} second). It connects with slow scattering of domain walls. Exponential growth of conductivity at border of zones is. Photoconductivity is absent in the cis-polyacetylene despite of absorption of light.

7.2.4 PROBLEMS WHICH HAVE ARISEN AT RESEARCH OF POLYACETYLENE

These problems share on two kinds: experimental and theoretical. We shall consider them more in detail.

7.2.4.1 EXPERIMENTAL PROBLEMS

1. Optical experiments are necessary for leading at different temperatures for samples of a different degree of alloying impurity that width of a crack to define, its temperature dependence and presence of conditions inside of it.
2. Character of conductivity (soliton, polaron or other) it is necessary to find out, investigate anisotropy of conductivity, photoconductivity and thermo-electro-factor.
3. The thermal capacity and its temperature dependence should be investigated in a vicinity $q = 2CF$.
4. It is necessary to investigate infrared-spectrums for studying of Konov anomaly.
5. Influence of pressure on transition metal-dielectric is necessary for investigating metal for definition of type of transition. The critical temperature of the Hardware grows with growth of pressure for Peierls change, and for Mott change falls.
6. The X-ray analysis of a monocrystal is necessary for leading to vicinities of phase transition for definition of structure.

7.2.4.2 THEORETICAL QUESTIONS

1. The basic interest represents research of dynamics in polyacetylene (quasi-one-dimensional system). To us it is not clear, whether it is possible to transfer properties of one-dimensional systems on three-dimensional with strong anisotropy.
2. Methods are necessary for developing, replacing a method of self-consistent field in approach of chaotic phases or Hartree-Foc, considering more essentially fluctuations both collective effects and leading the nonlinear equations of integrated type.
3. The theory of indignations to develop concerning the equations of integrated type. The account of fluctuations of a lattice destroys solitons and leads to non-integrated system (dynamic pinning).

These are the main questions concerning all strongly anisotropic systems. For polyacetylene it is necessary:

1. Power structure to find out in more details: presence of even lengths, what difference between *cis*- and *trans*-isomers of polyacetylene, whether is broken Peierls change at change of interaction between chains, occurrence of overlapped zones with different CF (for σ- and π-electrons), occurrence of defects.
2. Conditions to reveal when spin, also as well as charges on soliton, continuous value (fractional spin).
3. A number of formulas is necessary for receiving for kinetic factors, namely for the form of a signal of a nuclear spin induction, an electron-spin

echo, conductivity and photoconductivity. It is necessary to find out when appears paraconduction during the moment of Peierls change, connected with greater current fluctuations.

7.2.5 CONCLUSION

Almost all properties of polyacetylene are defined by an opportunity of occurrence in it solitons (domain walls, polarons, others stationary and dynamic). Mathematicians have developed methods of construction and the decision of non-linear problems for reception and researches of soliton decisions (a method of a return problem of dispersion, group methods).

The problems investigated by mathematicians, can serve as models for the description of physicochemical properties of polyacetylene and other polymers.

Properties of polyacetylene cannot be clear in terms of quasi-particles and consequently are unusual. The charge of a particle can be equal 1/3, or 1, and spin is equal 0. Properties of polyacetylene are collective and can be described by introduction of collective structures, for example, soliton-like states.

We shall not find out, what experiments should be made first of all. The main theoretical questions are the first and second of the general part and the third for polyacetylene.

7.3 THE NATURE OF KINETIC PROPERTIES OF POLYACETYLENE.

7.3.1 A STRUCTURE AND PROPERTIES OF POLYACETYLENE

The chemical formula of polyacetylene-$(CH)_x$, where C-carbon, H-hydrogen. Three external electrons of carbon form a sp^2-hybrid and give three σ-bonds. Remained electron forms π-bond. p-Clouds are focused perpendicularly to a plane in which p-bonds are located. The chain of atoms of carbon is formed as a result of polymerization. The chain has cis- or trans-form. Cis-form is formed at polymerization in usual conditions, and the trans-form turns out from it by thermalization or introduction of an impurity. We shall consider properties of trans-polyacetylene.

We shall discuss properties of polyacetylene without impurity. The length of bond C–C in a separate chain $(CH)_x$ makes approximately 1.4 Å. Distances between chains equally 3.6–4.4 Å. We can to neglect interchain interactions as a first approximation. It allows considering polyacetylene as the system consisting of separate isolated strings, bound among them. Polyacetylene is the semiconductor at a room temperature with width of a crack $2\Delta_0 = 1.4–1.8$ yM and value of a dielectric constant along chains 10–12. The crack of the order 1.6–2.0 eV turns out as a result of alternation of bonds C–C and C=C. Polyacetylene it is possible to consider as one-dimensional dielectric of Peierls. The size of displacement of atoms from balance is measured and it has appeared equal 0.03 Å.

Mobile paramagnetic centers are available in pure trans-polyacetylene with concentration of 1 spin on 3000 atoms of carbon. Their origin is not found out yet.

Data on electric and magnetic properties of alloy polyacetylene is. As impurity are used as donors (Na, K, NH_3), and (AsF_5, I, $FeCl_3$). Research of alloy polyacetylene has allowed making a number of the remarkable conclusions.

Residual conductivity is available in pure *trans*-polyacetylene. It is caused by presence of defects. Sharp increase of conductivity (in 10^{11} times) takes place with increase in concentration of impurity $y = N_1/N$ (N_1 – number of impurity, N – number of atoms of carbon). Saturation is reached at $y = 0.1$ and is equal 220 $(\Omega \cdot cm)^{-1}$.

Dependence on temperature has three areas over iodine:

a) $y < 0.05$. In this area $\sigma = \sigma_0 \exp(\frac{-B}{T})^{0.5}$, B = 10^4. One-dimensional hopping mobility takes place.

b) $0.06 < y < 0.11$ $\sigma = \sigma_0 \exp(\frac{-A}{T})^{0.25}$, A = 10^5. It is three-dimensional disordered an alloy.

c) $y > 0.15$. $\sigma = \sigma_0 T^v$, v = 0.7. It is characteristic for "dirty" metals.

The magnetic susceptibility of polyacetylene aspires to zero at y<0.05. Conductivity grows on some orders thus. Spin particles do not give the contribution to conductivity. Concentration of the paramagnetic centers falls with growth at. The magnetic susceptibility sharply increases at y<0.07. The metal behavior takes place.

Luminescence has been found out in *cis*-$(CH)_x$ at absence of a photocurrent. Luminescence is not present in *trans*-$(CH)_x$. Photoconductivity appears. Luminescence and photoconductivity is in usual semiconductors.

7.3.2   THE MODEL OF POLYACETYLENE

Dielectric properties of polyacetylene can be explained on the basis of the mechanism of Peierls. It has shown that the one-dimensional system is unstable concerning spontaneous infringement of mirror symmetry of a chain of atoms for any nonzero electron– phonon bond. Thus the crack arises in a one-electric power spectrum. The crack separates the filled and empty zones of Brillouin. One "free" π–electron on atom of carbon is available in $(CH)_x$. Doubling of the period of a lattice is preferable.

The explanation of the big number of experimental data has been received on the basis of the model using the mechanism of "internal defects" in a chain (CH) $_x$. Two approaches it is known for the description of defects.

The first approach: impurity electron or a hole will lead to infringement in alternation of communications of a chain of polyacetylene. Defect is change on 180 degrees of a phase regular polyacetylene (special case). This infringement of the order looks like a domain wall. Additional electron or a hole will borrow localized soliton-like states. The phase of lattice dimerization overturns. The elementary model describing such soliton, has Hinzburg-Landau Lagrangian.

$$L(u) = \frac{8\varepsilon_c}{(u_c\omega_\lambda)^2}[\frac{1}{2}(\frac{\partial u}{\partial t})^2 - \frac{C_0^2}{2}(\frac{\partial u}{\partial x})^2 - (\omega_\lambda u_0)^2 \, V(u)]$$

u Describes displacement of atoms of lattice,

$$V(u) = \frac{1}{8}[1-(\frac{u}{u_0})^2]^2, u_0$$

where amplitude is dimerization of Peierls. ε_c – Stabilizing energy on atom of carbon. ω_λ – Frequency of optical phonons. c_0 – the Speed describing of a lattice dispersion. $\omega_\lambda^2(q) = \omega_\lambda^2 + c_0^2 q^2$ at $|q| = \pi/2a$. a–the constant of a lattice. Model (1) has twice degenerate the basic condition and supposes the decision in the form of soliton

$$u(x-vt) = \pm u_0 \tanh\frac{x-vt}{I}, \quad I = \frac{2c_0}{\omega_\lambda}$$

where: I characterize width of soliton. Energy of soliton excitation $E_s = 4\Delta_0/3\pi$ = 0.4 electron volt, weight of soliton M_s = 6 mm – weight of electron, I = 10a. Expression (2) will enter to nuclear structure symmetrical concerning the center of soliton. The localized electronic condition appears in the center of a crack of Peierls. This condition can be borrowed by electron or a hole. Soliton gets a charge ±e. Model (1) is in the good consent with experiment, but demands care at the analysis in the field of defect x-vt→0. The model has singularity in this area.

The second approach: the topological defects dividing areas A and B, appear even during polymerization.

The area of defect remains neutral, but has not coupled backs ½ in absence of an impurity

7.3.3 THE DISCRETE MODEL

The macroscopical model including screen and electronic members looks like *The Hamiltonian:*

$$H = -\sum_{n,s} t_{n,n+1}(C_{n+1,s}^+ C_{n,s} + \text{h.C.}) + \frac{k}{2}\sum_n (y_{n+1}-y_n)^2 + \frac{M}{2}\sum_n y_n^2 \quad (4)$$

$C_{n,s}^+ (C_{n,s})$ – the operator of a birth (destruction) of electron with spin s on n (CH) – group, y_n – configuration coordinates for everyone CH– group, describing translation along a linear skeleton of a chain. M is weight of group (CH), κ is an elastic constant. The integral of overshoot electronic clouds of π – electrons $t_{n+1, n}$ can be spread out up to the first order rather nondimerization conditions $t_{n+1, n} = t_0 - \alpha(y_{n+1} - y_n)$, where t_0 – the integral of overshoot for 1D chains, α–a constant of electron –fonon bond.

We shall note that interchained bond is excluded and Coulomb interaction of π – electrons not considered in Eq. (3). Values of parameters of Hamiltonian: $t_0 = 2.5$ electron volt, $k = 21$ electron volt/A^2, $\alpha = 4.1$ electron volt/A. Length of C–C bond is equal 1.4 F. The constant of a lattice is equal $\alpha = 1.22$ Å. Reading of energy is conducted from Fermi level.

The Hamiltonian possesses symmetry to mirror reflection $y_n \rightarrow -y_n$ at $\alpha = 0$. $H(-y_n) = H(y_n)$. Infringement of symmetry of the basic condition takes place at $\alpha \neq 0$: $H(-y_n) \neq H(y_n)$.

7.3.4   *RESULTS FROM DISCRETE MODEL*

We neglect kinetic energy of phonons for definition of energy and we represent static displacement of unit n in the form of

$y_n = (-1)^n y$

The electronic spectrum looks like in a lattice without displacement (a normal phase):

$E = -2t_0 \cos qa$

$qa = 2\pi n/N$, $n = 0, \pm 1, \ldots \pm N/2$, N is number of atoms in a lattice. In a lattice with displacement

$$E_{1,2} = \pm\sqrt{(\Delta_0 \sin qa)^2 + (2 t_0 \cos qa)^2}$$

where qa = qa = $2\pi n/N$, n = 0, ±1, ... ±N/4. Brillouin zone has decreased twice and the crack arises $\Delta_0 = \alpha y$. The spectrum is symmetrical concerning Fermi energy.

Free energy of electrons in lattice is equal:

$$F(y, T) = -T \sum_{i=1,2;q} \ln[1 + \exp(-E_i(q)/T)] + 2Nky^2$$

where: T is a temperature. Free energy on one atom of carbon looks like at T = 0 and N $\rightarrow \infty$:

$$F(y, 0) = -\frac{4t_0}{\pi} E(1 - z^2) + \frac{kt_0^2 z^2}{2\alpha^2}$$

$$z = \frac{2\alpha y}{t_0}, \quad E(1 - z^2) = \int_0^{\pi/2} \sqrt{1 - (1 - z^2)\sin^2 q} \, dq$$

at $\Delta_0 \ll t_0$ $E(1 - z^2) = 1 + \frac{1}{2}(\frac{\ln 4}{|z|} - \frac{1}{2})z^2 + \ldots$

We minimize $F(y,0)$ (7) on z, we receive minima at $\pm y = \Delta_0/4\alpha$ c $\Delta_0 = 8t_0/e$ $\times\exp(-\pi kt_0/4\alpha^2)$. The local maximum is in a point $y = 0$ under the theorem of Peierls. The basic condition is doubly singular at low temperatures. Displacement $y=0.04\text{Å}$ at $\Delta_0 = 0.7$ eV and $\alpha = 4.1$ eV/A.

Presence of degeneration causes occurrence of topological solitons, dividing A- and B-phases. The numerical calculations lead within the limits of this model, have shown existence of decisions in the form of a domain wall. The parameter of the order is entered, connected with variables y_n:

$$\psi_n = (-1)^n y_n$$

Phases A and B can be defined
$\psi_{0n} = y_0$, a phase A;
$\psi_{0n} = -y_0$, a phase B.
We shall assume, that the domain wall is formed in a vicinity of unit $n = 0$. This wall divides A− and B− areas. We shall consider a birth of pair soliton−antisoliton to eliminate boundary effects. They consider far dissolved to exclude interaction between them. Model supposes the decision in the form of

$$\psi_n = y_0 \tanh(na / l)$$

Parameters of soliton have appeared the following. Width of soliton $l = 7a$, energy of a birth of soliton $E_s = 0.42$ electron volt, weight $M_s = 6m$.

In density of the conditions, caused of soliton, it is necessary to calculate changes for studying an electronic spectrum. Calculations give occurrence of one condition at $E = 0$ inside of the forbidden zone. It is connected with the advent of additional condition Φ_0 in the center of a power crack. The condition of completeness of wave functions Φ_v on each unit n means, that the integral on energy from local density $\rho_{nn}(E)$ for any unit "n" is equal to unit. We have:

$$I = \int_{-\infty}^{\infty} \rho_{nn}(E)dE = 1, \quad \rho_{nn}(E) = \sum |\phi_v(n)|^2 \delta(E - E_v)$$

The Hamiltonian is invariant concerning charging interface.

$$\rho_{nn}(E) = \rho_{nn}(-E).$$

$$I = 2\int_{-\infty}^{0} \rho_{nn}(E)dE = 2\int_{-\infty}^{0} \rho_{nn}^1(E)dE + |\phi_0(n)|^2 = 1$$

where: ρ^1 – means, that the condition of soliton is lowered.
Local deficiency of a valent zone is equal:

$$\int_{-\infty}^{0} [\rho_{nn}^{1}(E) - \rho_{nn}(E)]dE = \frac{1}{2}|\phi_0(n)|^2$$

where: ½ of condition is necessary on spin. One electron comes short in a zone. But a zone is neutral. Superfluous electron proves in condition Φ_0 with soliton ½. The charge of a condition is equal to zero. Neutral soliton appears with spin ½. If one charge will sit down on soliton from an impurity or its charge will leave in an impurity soliton begins to have soliton – zero, charge ±e.

Soliton picture allows to explain the basic properties of polyacetylene. Impurity conduction occurs due to solitons (with zero spin) if energy of a birth of soliton $E_s < \Delta_0$ solitons are Bose particles, can be condensed in superfluid state. Activation energy electric conductivity $E_a = 0/3$ electron volt it will well be coordinated with energy of bond of soliton on 0.33 electron volt. The peak 0.1 electron volt is in infrared spectrum. It will be coordinated with an oscillatory fashion of soliton. It is equal $0.07 - 0.08$ electron volt. Other properties speaking of solitons exist.

Attempts to consider electrostatic interaction are. We choose the Hamiltonian of interactions in the form of:

$$H_{ee} = \frac{V}{2}\sum \rho_{n\downarrow}\rho_{n\uparrow} + \frac{1}{2}\sum u_{nm}\rho_n\rho_m, \quad \rho_n = \sum C_{ns}^+ C_{ns}$$

Model investigated numerically. Dimerization of chain increases at $u_{nm} = 0$ with growth V up to value $V/2 = 4t_0$, and then decreases. Interaction of the first neighbors and the second also influences properties of a chain: at $u_{nm+1} < V/4$ amplifies dimerization, at $u_{nm+1} > V/4$ – it is weakened, and at u_{nm+2} – operates opposite u_{nm+1}.

7.4 CONTINUAL MODEL OF POLYACETYLENE.

7.4.1 MODEL

This model supposes the exact decision. The stationary raised conditions of such system are amplitude solitons. Quasi-classical approach we shall make on lattice variables. The Hamiltonian looks like in continual approach:

$$H = \sum_s \int dx \left\{ -iv_F (u_s^+(x)\frac{\partial}{\partial x}u_s(x) - v_s^+(x)\frac{\partial}{\partial x}v_s(x)) + \right.$$
$$\left. [\Delta^+(x) + \eta\Delta(x)]u_s^+(x)v_s(x) + [\Delta(x) + \eta\Delta^+(x)]v_s^+(x)u_s(x) \right\} +$$
$$(2\lambda\pi v_F)^{-1}\int dx\left[2\Delta^+(x)\Delta(x) + \eta\Delta^2(x) + \eta\Delta^{+2}(x)\right],$$
$$\lambda = 4\alpha^2 a / \pi v_F k, \quad v_F = 2t_0 a, \quad \hbar = c = 1, \quad \eta = 1,0$$

$$\Lambda = \frac{4\alpha^2 a}{\pi v_F k}, \quad v_F = 2t_0 a, \quad \hbar = c = 1, \quad \eta = 1.0$$

where: u(x), v(x) – characterize electronic conditions close of Fermi momentum $\pm q_F$, $\Delta^+(x)$, $\Delta(x)$ – are connected with lattice deformation (dependence from t is excluded, adiabatic approximation). The equations follow from the Hamiltonian:

$$iu_t = -iv_F u_x + \tilde{\Delta}^+ + \frac{a^2}{2}(\tilde{\Delta}^+ v_{xx} + v_x\tilde{\Delta}_x^+ + \frac{1}{2}v\tilde{\Delta}_{xx}^+),$$
$$iv_t = iv_F v_x + u\tilde{\Delta} + \frac{a^2}{2}(\tilde{\Delta} \, v_{xx} + u_x\tilde{\Delta}_x + \frac{1}{2}u\tilde{\Delta}_{xx}),$$
$$\tilde{\Delta}(x,t) = \Delta(x,t) + \eta\tilde{\Delta}(x,t),$$
$$M\tilde{\Delta}_{tt} + ka^2\tilde{\Delta}_{xx} = -4k\tilde{\Delta} - 16a\alpha^2(u^+v + \eta v^+u) - 4a^2\alpha^3\left[u^+v_{xx} + vu_{xx}^+ + \eta(uv_{xx} + v^+u_{xx})\right]$$

This problem to solve it is difficultly. We shall make a number of approaches. We shall exclude members, proportional a^2 and above, we shall lower member $M\Delta_{tt}$ (quasi-static approximation). We have:

$$iu_t = -iv_F u_x + v\tilde{\Delta}^+,$$
$$iv_t = iv_F v_x + u\tilde{\Delta},$$
$$\tilde{\Delta} = -\frac{4\alpha^2 a}{k}\sum_{k,s}(u_k^* v_k + \eta v_k^* u_k)$$

This system has structure Dirac equations in a floor. Many of such equations have analytical decisions (Bagrov V. G.). The system (4) is invariant rather Lorentz transforms (c = v_F).

$$x \rightarrow x' = \frac{x - vt}{\sqrt{1 - \frac{v^2}{v_F^2}}}, \quad t \rightarrow t' = \frac{t - \frac{xv}{v_F^2}}{\sqrt{1 - \frac{v^2}{v_F^2}}}$$

We can be limited to the decision of the stationary equations, then simply to add time, to take advantage Lorentz invariance. We search decisions in the form of:

$u(x, t) = e^{-i\omega_n t}u_n(x)$, $v(x, t) = e^{-i\omega_n t}v_n(x)$ also we shall put $\eta = 1$ ((transition in commensurable structure). We have:

$$\omega_n u_n(x) = -iv_F \frac{\partial}{\partial x} u(x) + v_n(x)\tilde{\Delta}^+(x)$$

$$\omega_n v_n(x) = -iv_F \frac{\partial}{\partial x} v_n(x) + u_n(x)\tilde{\Delta}(x)$$

$$\tilde{\Delta}(x) = -\frac{4\alpha^2 a}{k} \sum_{n,s} (u_n^*(x)v_n(x) + \eta v_n^*(x)u_n(x))$$

7.4.2 DISCUSSION OF DECISIONS OF SYSTEM OF EQUATIONS

Four classes of decisions of this system are available. Decisions are found by various methods.

The first class has the decision in the form of a flat wave, in an electronic spectrum there is a crack $\tilde{\Delta} = \Delta_0 \exp(-1/2\lambda)$. The valent zone is filled completely. Electronic wave functions in do not look like:

$$u_k(x,t) = N_k e^{i(kx - \omega t)}, \quad v_k(x,t) = -N_k e^{i(kx - \omega t)},$$

$$\omega = -\sqrt{\Delta_0^2 + k^2 v_F^2},$$

$$N_k = \frac{1}{\sqrt{8\pi}} (\frac{\omega - \Delta_0}{\omega})^{1/2} (\frac{\omega - \Delta_0 + kv_F}{kv_F})$$

$$N_k' = \frac{1}{\sqrt{8\pi}} (\frac{\omega - \Delta_0}{\omega})^{1/2} (\frac{\omega - \Delta_0 + kv_F}{kv_F})$$

Kinks are other class of decisions of system. These decisions look like:

$$u_0(x,t) = N_0 \sec h \frac{x - v_s t + x_0}{\xi_s},$$

$$v_0(x,t) = -i \frac{N_0}{v} \sec h \frac{x - v_s t + x_0}{\xi_s},$$

Where $N_0 = \sqrt{\dfrac{v\Delta_0}{4v_F}}$ width of soliton.

$$\xi_s = \frac{v_F}{\Delta_0}\sqrt{1-\beta^2}, \quad \beta = \frac{v_s}{v_F}, \quad v = \sqrt{(1+\beta)\Big/(1-\beta)}$$

In a valent zone

$$u_k(x,t) = N_k\, e^{i\theta}\left[\tanh\frac{\xi}{\xi_s} + i\frac{\omega - kv_F}{v\Delta_0}\right],$$

$$v_k(x,t) = -\frac{i\,N_k}{v}\, e^{i\theta}\left[\tanh\frac{\xi}{\xi_s} - iv\frac{\omega + kv_F}{v\Delta_0}\right],$$

$$\xi = x - v_s t + x_0,$$

$$\theta = kx + \omega t + \theta_0, \quad \omega = 2t_0\delta,$$

The additional level appears at presence of soliton in the middle of a power crack of an electronic spectrum. The level can be borrowed by one, two electrons or is not borrowed absolutely. Energy of a birth of soliton is equal 0.44 electron volt, will well be coordinated with result of discrete model. It is exact result in continual model.

The third type of excitations is polaron. Analytical expression has a complex appearance. We shall not write out dynamic decisions. We receive them by application Lorentz transform. In a static case:

$$\Delta(x) = k_0 v_F(\tanh(k_0(x - x_0)) - \tanh(k_0(x - x_0)))$$

Two symmetrically located levels appear in a power spectrum at presence of such deformation $E = +\omega_0$

$$u_0(x) = N_0[(1 - i)\sec h(k_0(x + x_0)) + (1 + i)\sec h(k_0(x + x_0))]$$

$$v_0(x) = N_0[(1 + i)\sec h(k_0(x + x_0)) + (1 - i)\sec h(k_0(x - x_0))]$$

$$N_0 = \frac{\sqrt{k_0}}{4}, \quad k_0 v_F = \sqrt{\Delta_0^2 - \omega_0^2},$$

tanh
$$k_0 x_0 = \Delta_0 - \frac{\omega_0}{k_0 v_F}$$

Wave functions are equal $u_0 = iv_0$, $v_0 = -iu_0$. In a valent zone:

$$u_-(k,x) = N_k e^{ikx}\left[(\omega+\Delta_0 - kv_F) - \gamma(1-i)t_- + \delta(1-i)t_-\right],$$

$$v_-(k,x) = -N_k e^{ikx}\left[(\omega+\Delta_0 + kv_F) - \gamma(1-i)t_+ + \delta(1+i)t_-\right],$$

$$t_\pm = \tanh[k_0(x \pm x_0)], \quad \omega = \sqrt{k^2 v_F^2 + \Delta_0^2},$$

$$\gamma = \frac{k_0 v_F}{2}\left(1 - \frac{ikv_F}{\omega - \Delta_0}\right), \quad \delta = \frac{k_0 v_F}{2}\left(1 + \frac{ikv_F}{\omega - \Delta_0}\right),$$

$$N_k = \frac{1}{2\sqrt{2\pi}}\left[\frac{\omega - \Delta_0}{\omega(k^2 v_F^2 + k_0^2 v_F^2)}\right]^{1/2}$$

The decision of the fourth type has been received in the form of bion, when:
$\Delta(x, t) = \Delta_0[1 + \delta(x, t)]$,
Energy of superincumbent ion is equal:

$$E_R = \Delta_0 \varepsilon \frac{2\sqrt{3}}{\pi}[1 - \frac{5\varepsilon^2}{9} + 0(\varepsilon^4)]$$

The contribution of ion in physical processes can be essential. Studying of ion in polyacetylene is the important problem. This studying was not spent yet.

7.4.3 SOLITON EXCITATION AND PHYSICAL PROPERTIES OF POLYACETYLENE

7.4.3.1 CHARACTERISTIC WIDTH OF POLARON

$$2x_0 = \frac{2}{k_0}\arctan h(\frac{k_0 v_F}{\Delta_0 + \omega_0}) = 10.8\,A, \quad \omega_0 = \frac{\Delta_0}{\sqrt{2}}$$

Energy of excitation of polaron is equal

$$E_p = \frac{2\sqrt{2}}{\pi}\Delta_0 = 0.9\Delta_0,$$

Polaron the decision passes in widely dissolved pair soliton – antisoliton at $x_0 \to \infty$ position of additional levels $\omega_0 \to 0$. Dependence of parameter of a crack for kink and polaron is on Figs. 5 and 6.

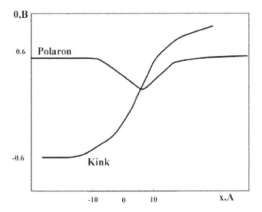

FIGURE 5   Change of a crack for kink and polaron depending on a condition.

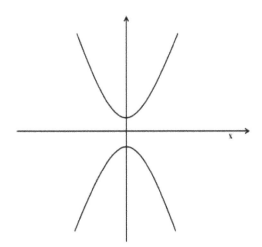

FIGURE 6   Change of a power spectrum at presence kink and polaron.

Polarons should play a greater role in the physicist of polyacetylene. Spin of system with even number of electrons should be the whole, and with odd- semi-integral value. The general number of electrons of system is kept at a birth of soliton S. Not coupled spin arises. It contradicts to Kramer theorem. Antisoliton $(-S)$

is born simultaneously and at a great distance. It is reflected on Fig. 7. The simultaneous birth of pair S(–S) demands a condition of topological stability of line. Polaron state is more favorable and charged bearers borrow polaron level. Spin of bearer is equal ½, the charge is equal to unit. But bidirectional decision any more is not stable and breaks up on S(–S) pairs. The number of charged solitons grows with increase in number of doped electrons. Charged bearers without spin appear.

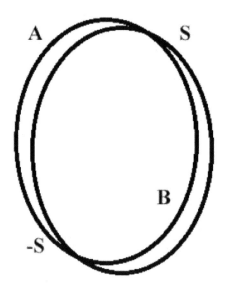

FIGURE 7   Simultaneous birth of S(–S) pair.

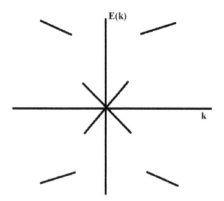

FIGURE 8   Change of an electronic spectrum depending on a degree of filling of a zone.

Formation charged or neutral soliton paires is possible depending on filling the localized levels $\pm\omega_0$. The contribution of polarons and solitons in various physical processes is well studied. Convincing acknowledgments are available in favor of soliton pictures of properties of polyacetylene.

The Soliton model continues to be specified concerning:

a) Account of Coulomb interaction,

b) Three-dimensional influence,

c) The exact decision of model is lead with any number of electrons on an elementary cell. It is important for studying of alloy polyacetylene where the deviation from a case strictly semi-occupied zone grows with growth of number of doped electrons.

   The optical crack in electronic spectrum appears continuous on parameter of filling of a zone (it is equal 1 for semi-occupied zones) and is final. Change of electronic spectrum is reduced to expansion of soliton level and to formation of two symmetrical forbidden zones (Fig. 8 instead of Fig. 6).

d) Interthreaded interaction connects solitons on the next chains and soliton can get a charge not changing backs.

e) Occurrence of a nonintegral electric charge 2/3, 1/3 (it quarks) is revealed. Nobel Prize is received in 1998.

Nobel Prize on the physicist in 1999 has been given to researchers who the first observed and has explained fractional quantum Hall-effect. It is effect in which the electric current inside of two-dimensional conductive material appears, as though created of the carriers bearing a fractional charge of electron.

Standard Hall-effect is a lateral deviation of moving carriers in a magnetic field. It was revealed in 1879 *Edwin Hall*, today a basis gives for definition of a charge and density of carriers of charges in the semiconductor (electrons and holes deviate in various directions).

Klaus von Klitzing finds in 1980, that when carriers of a charge are limited inside very thin spending film (that is move in two measurements), size of a Hall current (or conductivity of a material) does not change any more smoothly from size of magnetic field at very low temperatures.

Conductivity changes from force of a field in the form of sharp steps. Conductivity changes jump on fundamental quantum size of conductivity e^2/h (where e is a charge of electron, h is Planck's constant).

Fractional quantum Hall-effect represents a deep problem as shows change in character of fundamental particles. We can tell the same about superconductivity in which electrons draw each other, and super fluidity at which atoms of super liquid any more generate viscosity.

FQHE has been noticed *Tsui and Stormer* for carry in two-dimensional electronic gas in semi-conducting heterostructure, made *Art Gossard*. The sample cooled up to $10°K$, applied magnetic fields up to 30 tesla, observed transitions in spending condition with value e²/3h. Carriers of a charge had a fractional charge e/3. The further studying have shown that charges are 2e/5, 3e/7 and other shares (with an odd denominator).

Laughlin has offered, that the magnetic lines of a stream getting through the sample encourage carriers of a charge to be condensed in quasi-particles. It has shown, that such quasi-particles operate as if they have fractional charges with the values noticed in experiments.

Heiblum speaks: "All from us have got used to work in the semi-conductor physics with a one-electronic picture". But studying of strongly correlated electrons in the physicist of a firm body now becomes the important field of research.

This model has the direct attitude to a number of models of relativistic quantum theory fields, to Gross–Neview model, with Fermi particles. Studying of bidimentional models is more likely formal mathematical research. Results of such research are very remarkable and have the appendix in the physicist of the condensed environments.

7.5   TWO-DIMENSIONAL MODELS OF RELATIVISTIC QUANTUM THEORY OF FIELD AND OF POLYACETYLENE

7.5.1   GROSS–NEVIEW MODEL

Gross–Neview model:

$$L = \sum_{p=1}^{N} [\bar{\psi}^p (i\partial)\psi^p] + \frac{1}{2}g^2[\sum \bar{\psi}^p \psi^p]^2$$

Where ψ is a two-componential spinor, ġ is a constant.

$$\partial = \gamma_\gamma \partial^\mu, \quad \partial^0 = \frac{\partial}{\partial t}, \quad \partial^1 = \frac{\partial}{\partial x}, \quad \mu = 0.1$$

$$\gamma_0 = \sigma_3, \quad \gamma_1 = i\sigma_1,$$

This model describes N types of massless Fermi particles, cooperating in the nonlinear image. Dynamic infringement of chiral symmetry is realized in this model. Fermi particles get weight. It is similar to formation of a superconducting crack. This model is actively used in the theory of strong interactions.

The model is done renormalizable and receives properties of asymptotic freedom. This model is completely integrated. Making functional it is possible to construct:

$$Z(\eta,\bar{\eta}) = C\int d\psi d\bar{\psi} \exp(i(i\bar{\psi}\partial\psi + \frac{1}{2}g^2(\bar{\psi}\psi)^2 + \bar{\eta}\psi + \bar{\psi}\eta)) =$$

$$C'\int d\psi d\bar{\psi} d\sigma \exp(i(i\bar{\psi}\partial\psi - \frac{1}{2}\sigma^2 - g\bar{\psi}\psi\sigma + \bar{\eta}\psi + \bar{\psi}\eta))$$

And to be convinced, that Eq. (1) it is formally equivalent:

$$L = \sum_{p=1}^{N}(i\bar{\psi}^p\partial\psi^p - g\sigma\bar{\psi}^p\psi^p) - \frac{1}{2}\sigma^2,$$

Where scalar field σ has arisen which is formal object for a two-dimensional case. The equations of movement for Eq. (2) look like:

$$(i\partial - g\sigma(x))\psi^p(x) = 0,$$

$$\sigma(x) = -g\sum_{p=1}^{N}\bar{\psi}^p(x)\psi^p(x)$$

This is a condition of the self-coordination. The equation of movement becomes for stationary conditions:

$$\omega_n\psi_{1n}^p(x) = \frac{\partial}{\partial x}\psi_{2n}^p(x) + g\,\sigma(x)\psi_{1n}^p(x),$$

$$\omega_n\psi_{2n}^p(x) = \frac{\partial}{\partial x}\psi_{1n}^p(x) - g\,\sigma(x)\psi_{2n}^p(x)$$

If $\psi_1 \to \frac{(u+v)}{\sqrt{2}}$, to replace $\psi_{22} \to \frac{-i(u+v)}{\sqrt{2}}$, $q\sigma(x) \to \Delta(x)$, $x \to v_F x$, we shall receive the equations earlier considered for polyacetylene at N = 2.

$$\omega_n u_n(x) = -iv_F\frac{\partial}{\partial x}u(x) + v_n(x)\tilde{\Delta}^+(x)$$

$$\omega_n v_n(x) = iv_F\frac{\partial}{\partial x}v_n(x) + u_n(x)\tilde{\Delta}(x)$$

With a condition of the self-coordination

$$\tilde{\Delta} = -\frac{4\alpha^2 a}{k}\sum_{k,s}(u_k^* v_k + v_k^* u_k)$$

We shall stop on it in more detail. Dynamic infringement of symmetry means, that there is a decision $\sigma(x)=\sigma_0$. Well-posed interpretation a condition of the self-coordination demands performance of a parity:

$$Z(\Lambda)\sigma_0 = -q\sum \psi_n^p(x)\psi_n^p(x)$$

$Z(\Lambda)$ is ultra-violet renormalization, Λ is an impulse of trimming. Summation in above equation is carried out on all conditions with energy of less zero. The sum on p is similar to the sum on backs. $Z(\Lambda) = 1$ for model of polyacetylene, that corresponds to a choice:

$$m_\psi = q\sigma_0 = 2\Lambda\exp(-\pi/Nq^2)$$

If m_ψ is fixed also $\Lambda\rightarrow\infty$ a constant ġ aspires to zero as $1/\ln\Lambda$. The model possesses property asymptotic freedom.

The equation supposes following decisions:

1. Flat waves at $\sigma(x) = \sigma_0$ which look like

$$\psi_{1k}(x) = N_k e^{ikx}, \quad \psi_{2k}(x) = N_k e^{ikx}ik/(\omega - m_\psi)$$

$$\omega = \sqrt{k^2 + m_\psi^2}$$

For them: $\psi_k(x)\psi_k(x) = -m_\psi/2\pi\omega$, and $\sum_{p,n,\omega\langle 0} \bar{\psi}_n^p(x)\psi_n^p(x) = -\frac{N}{2\pi}\int\frac{m_\psi dk}{\sqrt{k^2 + m_\psi^2}}$

The condition of the self-coordination is executed.

2. "Kinks" at $\sigma(+\infty) = -\sigma(-\infty) = \pm\sigma_0$ when $\sigma = \sigma_0\tanh(m_\psi x + \delta_0)$, where $\tanh\delta_0 = (m_\psi - c_0)/(m_\psi + c_0)$. We come to parameters of a crack as in polyacetylene at $c_0 = m_\psi$. The Fermionic spectrum contains the connected condition with $\omega_0 = 0$. Wave functions correspond to this condition:

$$\psi_{10}(x) = \psi_{20}(x) = \sqrt{\frac{m_\psi}{4}}\sec h(m_\psi x + \delta_0)$$

and conditions with a continuous spectrum $\omega = \sqrt{k^2 + m_\psi^2}$ and wave functions:

$$\psi_1(k,x) = N_k e^{ikx}(\tanh(m_\psi x + \delta_0) - (\omega + ik)/m_\psi),$$

$$\psi_2(k,x) = N_k e^{ikx}(\tanh(m_\psi x + \delta_0) - {(\omega - ik)}/{m_\psi}),$$

$N_k = m/2\omega\sqrt{2\pi}$.

3. "Bags" is the third type of decisions. They appear under boundary conditions $\sigma(x) \to \sigma_0$, $|x| \to \infty$. Expression is the decision:

$\sigma(x) = \sigma_0 - (k_0^2/\omega_0 q)\sec h[k_0(x + x_0) + \delta_1 yc\ h[k_0(x - x_0) + \delta_1]$,

$\tanh k_0 x_0 = (m_\psi - \omega_0)/k_0$, $\tanh \delta_1 = k_0(m_\psi - \omega_0) - c_0\omega_0/ k_0(m_\psi - \omega_0) + c_0\omega_0$

$\omega_0 = \sqrt{m_\psi^2 - k_0^2}$

We come to the decision of type "polaron" at δ_0 in polyacetylene. The connected conditions arise in fermionic a spectrum with energy $\pm\omega_0$ and wave function

$$\psi_{10}(x) = \sqrt{\frac{k_0}{8}}(\sec h[k_0(x + x_0) + \delta_1] + \sec h[k_0(x - x_0) + \delta_1])$$

$$\psi_{20}(x) = \sqrt{\frac{k_0}{8}}(-\sec h[k_0(x + x_0) + \delta_1] + \sec h[k_0(x - x_0) + \delta_1])$$

Conditions with negative energy are available. They are similar to conditions u, v for polaron decisions in model of polyacetylene. Energy of excitation is:

$$E(n_0) = \frac{2}{\pi}N m_\psi \sin(\frac{\pi n_0}{2N})$$

The decision behaves as dissolved on infinity kink antikink pair in a limit $n_0 \to N$. Essential difference "sack" excitation from kink is a dependency ω_0 from number of filling n_0. Excitation is absent at $n_0 = 0$, $m_\psi = \omega_0$. For kink it is not essential, the condition $\omega_0 = 0$ is borrowed or not. $N = 2$ it is necessary to accept for comparison with model of polyacetylene. The condition with $n_0 = 1$ can be interpreted as presence of additional electron in a chain of polyacetylene. Bond with a phonon field is available; this excitation is considered as "polaron". Infinitely dissolved pair kink-antikink is available at $n_0 = 2$.

4. The Fourth type of decisions is the decision similar bion in model sine–Gordon. Periodic boundary conditions are imposed

$\sigma(x, t + T) = \sigma(x, t)$. If $g = 1$, $\sigma(x, t) = 1 + \xi f_2 + \eta f_4$,

$f_4 = f_4 \cos \Omega t$, $f_4 = (chkx + \cos \Omega t + b)^{-1}$, t and x constants are expressed in terms of $(q\sigma_0)^{-1}$. k, Ω, ξ, η, a, $–$ constants. The full analogy of the considered model and continual models of polyacetylene is available only in a static case. Dynamic properties of models are various.

The Phonon field corresponds to real fluctuations of a lattice and has dynamics in model of polyacetylene. The field σ is auxiliary that is reflected by absence of a kinetic member.

7.5.2 MODEL FIE-FOUR WITH FERMI PARTICLES

$$L = \sum_{p=1}^{N} \bar{\psi}^P (i\partial - g\phi)\psi^P + \frac{1}{2}(\partial_\mu \phi)^2 + \frac{1}{2}\mu_0^2 \phi^2 - \frac{\lambda}{4}\phi^4$$

Constants have corresponding dimension. We shall note, that the kinetic member with derivative of time is available in composed for a field.

We shall exclude effects of sea Fermi from consideration and we shall consider a discrete level in an electronic spectrum. The stationary equations for model look like:

n_0 Characterizes filling a level ω_0. The index p is lowered. Decisions of system are known: kinks, "small bags," and "double bags" – excitation of polaron type.

a. Kinks are conditions with $\omega_0 = 0$, which wave functions are given by formulas

$$\phi(x) = \frac{\mu_0}{\sqrt{\lambda}} \tanh \frac{\mu_0 x}{\sqrt{2}}, \quad \psi_{10}(x) = \psi_{20}(x) = A(ch \frac{\mu_0 x}{\sqrt{2}})^{-g\sqrt{2}/\sqrt{\lambda}}$$

Energy of kink does not depend from n_0 and in quasi-classical approach looks like:

$$E_k = \frac{2\sqrt{2}}{3} \frac{\mu_0^3}{\sqrt{\lambda}}$$

b. Decisions of polaron type

$$\phi(x) = f - \frac{k_0^2}{g\omega_0} \sec h[k_0(x+x_0)]\sec h[k_0(x-x_0)], \quad f = \frac{\mu_0}{\sqrt{\lambda}},$$

$$\psi_{10}(x,t) = \sqrt{\frac{k_0}{8}} e^{-i\omega_0 t} (\sec h[k_0(x+x_0)] + \sec h[k_0(x-x_0)]),$$

$$\psi_{20}(x,t) = \sqrt{\frac{k_0}{8}} e^{-i\omega_0 t} (-\sec h[k_0(x+x_0)] + \sec h[k_0(x-x_0)]),$$

$$k_0 = \frac{n_0 g^2}{8}, \quad \omega_0^2 = g^2 f^2 - k_0^2, \quad \tanh k_0 x_0 = (gf - \omega_0)/k_0$$

We define $\omega_0 = qf \cos\varphi$, $k_0 = qf \sin\varphi$, we come to bond $\sin 2\varphi(n_0) = \frac{n_0}{4f^2}$.

The equation has no decision at $n_0/4f^2 \rangle 1$. The equation has two decisions at $n_0/4f^2 \ll 1$ $(k_0(n_0))_+ = qf - qn_0^2/128f^3$ – "a double bag", the astable decision $(\varphi(n_0) \rangle \pi/4)$ и $(k_0(n_0))_- = qn_0/8f$ – "a fine bag", stable at $\varphi(n_0)\langle \frac{\pi}{4}$. Both decisions become equivalent at $n_0 / 4f^2 \to 1$. A necessary condition of stability of a condition is $\frac{d^2 E}{d\varphi^2} \rangle 0$. We receive $\frac{d^2 E}{d\varphi^2} = 8qf^3 \sin\varphi(n_0)\cos 2\varphi(n_0)$.

Model fie-four with Fermi particles is possible to consider as low-energy the effective theory of a field of more fundamental Gross–Neview model.

7.5.3 FINAL REMARKS

We shall discuss briefly qualitative parallels between three models considered above. All properties of solitons of polyacetylene are described by models at N=2. Crushing of the charge localized on soliton, takes place at N=1. We shall note a number of the important properties inherent in three models.

1. Dynamic infringement of symmetry of the basic condition takes place. Degeneration of the basic condition takes place as a result of spontaneous infringement of symmetry.

2. All models contain excitations in the form of topological solitons kinks, connecting vacuous vacuums. The electronic power spectrum varies at presence topological solitons a discrete level appears with zero energy. Solitons have unusual spin– charge parities.

3. All three models contain polaron the decision.

Further we shall try to investigate consequences of spontaneous infringement of symmetry in these models at final temperature of system and at presence of asymmetry of number of particles and antiparticles (electrons – holes) is as much as possible detailed. The question of crushing of the charge, localized on soliton, will be considered later.

7.6 PHASE TRANSITIONS IN BIDIMENTIONAL MODELS OF THE THEORY OF A FIELD AND POLYACETYLENE

7.6.1 INTRODUCTION

Research of systems the basic condition of which possesses the symmetry which is distinct from symmetry of Hamiltonian, is the most interesting though also a difficult problem of physics. A number of the fundamental concepts entered into the theory of many cooperating particles by Bogolyubov, underlies research of spontaneous infringement of symmetry. The model with the allocated condensate has been offered in its works for the first time. Bogolyubov's fundamental concept about "quasi-average" has huge value at studying superconductivity, superfluidity, ferromagnetics and to that similar.

The general scheme of research of phase transition consists in construction of effective potential of system at final temperature and fermionic density P(σ, T, n). The analysis of minimum P(σ, T, n) allows to receive critical values T and n_c, at which symmetry is restored. Feynman diagrams it is necessary sum up in view of replacement: zero a component of an impulse $k_0 \rightarrow i\omega_n$ and, accordingly ω_n = $(2n+1)\pi T - i\alpha(\omega_n \pm 2n\pi T)$ – Matsubara frequency. The chemical potential α is connected with presence of preservation of number of Fermi particles.

7.6.2 EFFECTIVE POTENTIAL OF GROSS–NEVIEW MODEL

We shall be limited to the account of one-loop diagrams, at construction of effective potential, and we shall use approach of an average field. We have at summation of diagrams (Fig. 9):

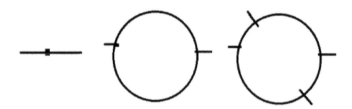

FIGURE 9   Matsubara one-loop diagrams.

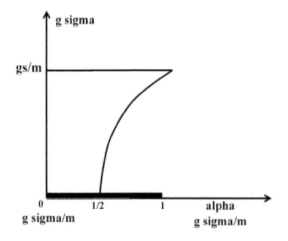

FIGURE 10   A qualitative kind of decisions.

$$P(\sigma) = \frac{1}{2}\sigma^2 - iN\sum\int\frac{dk}{(2\pi)^2}\frac{1}{n}(\frac{\lambda\sigma^2}{k^2})^n$$

N – Number of types of Fermi particles, $\lambda = Ng^2, k^2 = k_0^2 - \bar{k}_0^2$

We use decomposition and, in view of replacement, we have:

$$P(\sigma,T,\alpha) = \frac{1}{2}\sigma^2 - 2NT\int_0^\infty \frac{dk}{2\pi}\sum_{n=-\infty}^{\infty}\ln\frac{k^2+\omega_n^2+\dfrac{\lambda}{N}\sigma^2}{k^2+\omega_0^2}$$

We calculate the sum. We get

$$P(\sigma,T,\alpha) = \frac{1}{2}\sigma^2 - 2NT\int_0^\infty \frac{dk}{2\pi}\ln\frac{[1+\exp\dfrac{\alpha-\varepsilon}{T}][1+\exp\dfrac{\alpha+\varepsilon}{T}]}{[1+\exp\dfrac{\alpha-k}{T}][1+\exp\dfrac{\alpha+k}{T}]}$$

We investigate at zero value of chemical potential. In this case

$$\varepsilon = \sqrt{k^2 + \frac{\lambda}{N}\sigma^2}$$

where: Λ − impulse of trimming. Last member is final in previous expression. Effective potential renormalize at T=0. We use a following condition for this purpose (Kolumen−Vainberg):

$$\left.\frac{\partial^2 P(\sigma)}{\partial\sigma^2}\right|_{\sigma=\sigma_0} = 1$$

where: σ_c − the classical field. Meanwhile:

$$\ln\frac{\Lambda+\sqrt{\Lambda^2+(g\sigma)^2}}{g\sigma_0} = \frac{\Lambda}{\sqrt{\Lambda^2+(g\sigma)^2}}$$

We have a bond $2\Lambda = q\sigma_0 e$. A Dependence on parameter of trimming disappears as a result of renormalization of weights of a field in effective potential.

$$P(\sigma,\sigma_0) = \frac{1}{2}\sigma^2 + \frac{\lambda}{4}\pi\sigma^2[\ln(\frac{\sigma}{\sigma_0})^2 - 3]$$

The account one-loop correction has led to occurrence of an additional member $\sigma^2\ln\sigma^2$ in potential energy of system. This member gives the negative amendment at small values σ. This amendment dominates. Potential energy is positive and increases at great values σ. The theory is stable. The local maximum is in a point $\sigma = 0$. The condition of an extremum looks like

$$\frac{\partial P(\sigma,\sigma_0,T)}{\partial \sigma} = \sigma_M (1 + \frac{\lambda}{2\pi}[\ln(\frac{\sigma_M}{\sigma_0})^2 - 2] + \frac{2\lambda}{\pi}\int_0^\infty \frac{dk}{\varepsilon}(e^{\frac{\varepsilon}{T}}+1)^{-1})$$

at T = 0;

$$\sigma_M = \pm\sigma_0 \exp(1-\frac{\pi}{\lambda})$$

Dynamic infringement of symmetry occurs. Fermi particles get weight M_F. $M_F = g\sigma_M$. Crack M_F arises in a spectrum of Fermi particles. The potential has two minima.

a) We shall consider temperature effect in above relation. The integral supposes an analytical estimation at heats:

$$\int_0^\infty \frac{dk}{\sqrt{k^2+(g\sigma)^2}}[\exp(\frac{\sqrt{k^2+(g\sigma)^2}}{T})+1]^{-1} = -\frac{1}{4}\ln(\frac{g\sigma}{\pi T})^2 - \frac{\gamma_E}{2} + 0(\sigma^4)$$

γ_E = 0.577 (Euler constant). Expression (7) becomes:

$$\frac{\partial P(\sigma,\sigma_0,T)}{\partial \sigma} = \sigma[1 - \frac{\lambda}{\pi}(1+\gamma_E) + \frac{\lambda}{2\pi}\ln\frac{N}{\lambda}(\frac{\pi T}{\sigma_0})^2 + 0(\sigma^2)]$$

We receive from above expression: the minimum takes place at σ = 0, T > Tc. The maximum takes place at T < Tc.

$$T_c = \frac{1}{\pi}g\sigma_0 \exp(1-\frac{\pi}{\lambda}+\gamma_E) = \frac{\gamma M_F}{\pi}, \quad \gamma = \exp\gamma_E$$

The analogy is available with an estimation of critical temperature of the superconducting channel. The weight of Fermi particle costs instead of a crack. The last expression is connected with temperature of structural Peierls transition. Peierls transition takes place in model of polyacetylene.

b) We shall pass to the analysis at final values of chemical potential. We accept α>>T. In view of renormalization:

$$P(\sigma,\sigma_0,T) = \frac{1}{2}\sigma^2 + \frac{\lambda}{4\pi}\sigma^2[\ln(\frac{\sigma}{\sigma_0})^2 - 3] - \frac{N}{\pi}[\int_0^{\sqrt{k^2-(g\sigma)^2}} dk(\alpha-\varepsilon) - \frac{\alpha^2}{2}]$$

We shall consider two cases:
1. $\alpha \langle g\sigma$. We have

$$P(\sigma,\sigma_0) = \frac{1}{2}\sigma^2 + \frac{\lambda}{4\pi}\sigma^2[\ln(\frac{\sigma}{\sigma_0})^2 - 3] + \frac{N\alpha^2}{2\pi}$$

Condition of an extremum:

$$\frac{\partial P}{\partial \sigma} = \sigma[1 - \frac{\lambda}{\pi} + \frac{\lambda}{\pi}\ln\frac{\sigma}{\sigma_0}] = 0$$

Above expression does not contain chemical potential, at $\sigma = 0$ has a local maximum. This expression has minima at $\sigma_M = \pm\sigma_0\exp(1-\frac{\pi}{2})$.

2. The case $\alpha \rangle g\sigma$ is more interesting. We can write down;

$$P(\sigma,\sigma_0) = \frac{1}{2}\sigma^2 + \frac{\lambda}{4\pi}\sigma^2[\ln(\frac{\sigma}{\sigma_0})^2 - 3] + \frac{N}{\pi}[\frac{\alpha^2}{2} - \frac{\alpha}{2}\sqrt{\alpha^2 - (g\sigma)^2} + \frac{(g\sigma)^2}{2}\ln\frac{\alpha + \sqrt{\alpha^2 - (g\sigma)^2}}{g\sigma}]$$

Condition of an extremum:

$$\ln\frac{\alpha + \sqrt{\alpha^2 - (g\sigma)^2}}{g\sigma_M} = 0$$

The maximum takes place in $\sigma = 0$ at $\alpha\langle\frac{g\sigma}{2}$. The minimum takes place at $\alpha\rangle\frac{g\sigma}{2}$.

The decision of above expression is available in the form of:

$$\sigma^* = \sigma_M\sqrt{\frac{2\alpha}{g\sigma_M} - 1}$$

Local maximum P takes place at $\frac{g\sigma_M}{2}\langle\alpha\langle g\sigma_M$. The minimum takes place at $\alpha\langle g\sigma$. The qualitative kind of decisions is resulted on Fig. 10.

7.6.3 MODEL FIE!-FOUR WITH FERMI PARTICLES

We shall study phase transition with restoration of symmetry in this section. Potential energy of system is equal a classical case:

$$P_0(\phi) = -\frac{\mu_0^2}{2}\phi^2 + \frac{\lambda}{4}\phi^4$$

The account of quantum amendments will lead to divergence in effective potential. Divergence in case of two measurements will be logarithmic. We can eliminate these of divergence. We change weight of a field. The effective potential can be written down in case of final temperature and chemical potential in the form of:

$$P(\phi) = P_0(\phi) + \frac{1}{2}Q(T,\alpha)\phi^2$$

The factor in above expression has logarithmic divergence.

We shall define fields $\phi' = \phi - \rho$, where there is a deviation of a field from position of a minimum

$$\left.\frac{\partial P}{\partial \phi}\right|_{\phi=\rho} = 0$$

where: ρ – parameter of the order. We consider shift in initial Lagrangian. The unique one-loopback diagrams containing logarithmic divergence, are "tadpoles" and own energy of Bose particle (Fig. 11).

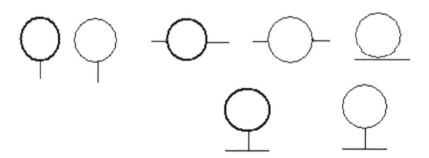

FIGURE 11   Columns, leading to divergence.

Counter-term Becomes after shift:

$$\partial L = \frac{1}{2}Q(T,\alpha)\phi'^2 + Q(T,\alpha)\rho\phi' + \frac{1}{2}Q(T,\alpha)\rho^2$$

If divergence of "tadpole" and own energy of Bose particle are reduced by counter-terms, we can define $Q(T,\alpha)$. The contribution from "tadpoles" is equal:

$$\Gamma = -i(2\pi)^2[3\lambda I_B(T,\alpha) - 2g^2 I_F(T,\alpha)]\rho,$$

where

$$I_B(T,\alpha) = \frac{1}{2\pi} \int\limits_{-\infty}^{\infty} \frac{dk}{2\omega_B} \coth\frac{\omega_B}{2T}, \quad \omega_B = \sqrt{k^2 + m_B^2(T,\alpha)},$$

$$I_F(T,\alpha) = \frac{1}{2\pi} \int\limits_{-\infty}^{\infty} \frac{dk}{2\omega_F} \frac{sh(\omega_F/T)}{ch(\omega_F/T) + ch(\alpha/T)},$$

$$\omega_F = \sqrt{k^2 + m_F^2(T,\alpha)}$$

We choose

$$Q(T,\alpha) = 3\lambda I_B(T,\alpha) - 2g^2 I_F(T,\alpha).$$

Counter-terms completely reduce divergence in own energy of Bose particle. We need to eliminate logarithmic divergence. It is possible to carry out by re-normalization weights of bosonic field, which is to subtract a member that corresponds to replacement

$$\mu_0^2 \to \mu^2 = \mu_0^2 + Q(0) = 3\lambda I_B(0) - 2g^2 I_F(0).$$

The effective potential becomes final and looks like:

$$Q(T,\alpha) = 3\lambda I_B(T,\alpha) - 2g^2 I_F(T,\alpha)$$

The condition of an extremum of effective potential becomes

$$P(\phi) = P_0(\phi, \mu_0 \to \mu) + \frac{1}{2}Q^R(T,\alpha)\phi^2,$$

$$Q^R(T,\alpha) = 3\lambda(I_B(T,\alpha) - I_B(0)) - 2g^2(I_F(T,\alpha) - I_F(0))$$

This equation has the trivial decision ρ = 0. Fermi particles are massless. The weight of bosonic field is defined from the equation

$$m_B^2(T,\alpha) = -\mu^2 + 3\lambda\rho^2 + Q^R(T,\alpha)$$

Counter-term addresses in zero at zero temperature and chemical potential. The bosonic field has negative weight. It signals about spontaneous infringement of symmetry in system. From two last expressions we can write:

$$m_B^2(T,\alpha) - 2\mu^2 + 2Q^R(T,\alpha) = 0.$$

The decision is available at zero temperature and chemical potential:

$m_B^2(T,\alpha) = 2\mu^2$,

Bose particles get physical weight, $\rho(0) = \mu/\sqrt{\lambda}$, Fermi particles become massive $m_F(0) = g\mu/\sqrt{\lambda}$.

We can believe, that our results are fair everywhere, except for area near to critical temperature. The technics of temperature functions of Green allows to receive a number of interesting results and wide application finds.

The general scheme is simple enough. The effective potential, which is simple for receiving in one-loopback approach, is thermodynamic potential of system. Small deviations from position of a minimum are considered in absence. We receive thermodynamic potential in the basic condition. Kink decision the deviation of a field from kink vacuum is considered and entered. We receive thermodynamic potential again for single-kink sectors. The thermodynamic potential for one kink Ω_K is calculated as a difference.

Realization this programme leads to energy of based kink.

$$E_K = \Omega_K - T\frac{d\Omega_K}{dT} = \frac{2\sqrt{2}\mu^3}{3\lambda} - (\frac{3}{2\pi} - \frac{1}{4\sqrt{3}})\sqrt{2}\mu + \frac{\sqrt{6}}{2}\mu N(\frac{\sqrt{6}}{2}\mu) - 3\sqrt{2}\mu f$$

At   T = 0

$$\omega_k = \sqrt{k^2 + 2\mu^2}$$

At   T >> μ

$$E_k = \frac{2\sqrt{2}\mu^3}{3\lambda} - \frac{3}{\sqrt{2\pi}} + \frac{T}{2}$$

We can generalize results in case of moving soliton, find density of number of solitones.

7.6.4 MODEL OF TRANS-POLYACETYLENE

Polyacetylene is Peierls dielectric with doubling the period. Suppression of Peierls dimerization takes place with growth of temperature. The crack disappears. Presence of an external field also leads to disappearance of a mass crack. We shall construct thermodynamic potential for system of noninteracting electrons on a deformable chain of atoms

$$\Omega(\Delta, T, \alpha) = -\frac{2T}{N}\sum_i\sum_q \ln[1+\exp(\frac{\alpha-\varepsilon_i}{T})] + \frac{\Delta^2}{4\pi\lambda t_0}$$

where $\varepsilon_{1,2} = \pm\sqrt{(\Delta\sin q)^2 + (2t_0\cos q)^2}$,

where N – Number of atoms in a chain, α – chemical potential, $\lambda = 4\alpha^2 a/\pi k v_F$. The condition of extremum looks like:

$$\frac{\partial\Omega(\Delta, T, \alpha)}{\partial\Delta} = \Delta(1 - 2\lambda\int_0^{kv_F}\frac{dq}{\varepsilon}\frac{sh(\varepsilon/T)}{ch(\varepsilon/T) + ch(\alpha/T)}) = 0$$

The crack remains to a constant depending on temperature up to 0.2 eV, and then quickly decreases up to zero at 0.4 eV. Such kind of the decision specifies presence of critical temperature.

Dependence on previous figure is received at temperature 300K (0.026 eV). Control calculations have shown that change of temperature within the limits of 250–350K does not influence qualitatively results. Numerical calculations will be coordinated with analytical.

The critical density is equal:

$n_c = \sqrt{2}\Delta/\pi v_F$

7.6.5 MODEL OF CIS-POLYACETYLENE

Model of *cis*-polyacetylene differs from *trans*-polyacetylene presence of seed crack. It is similar to introduction of seed fermion sludge. *Cis*-model gets a number of the important properties. The thermodynamic potential on one atom of carbon looks like:

$$\Omega = -\frac{2T}{N}\sum_q \ln([1+\exp(\frac{\alpha+\varepsilon}{T})][1+\exp(\frac{\alpha-\varepsilon}{T})])$$

The condition of an extremum is

$$\varepsilon = \sqrt{(\Delta\sin q)^2 + (2t_0\cos q)^2}$$

7.7   SOLITONS WITH FRACTIONAL CHARGES IN POLYACETYLENE

7.7.1   INTRODUCTION

To research of the unusual quantum numbers localized on topological solitones, many works are devoted. The physical basis of occurrence of fractional quantum numbers became clear.

Presence in system of objects with not trivial topology leads to change of structure of the basic condition of system. Quantification concerning new soliton vacuum also leads to unusual quantum numbers.

7.7.2   MODEL

We shall consider the mechanism of occurrence of fractional charges on an example of model with $N = 1$. Potential energy is equal

$$P_0(\phi) = -\frac{\mu_0^2}{2}\phi^2 + \frac{\lambda}{4}\phi^4 \tag{1}$$

and has minima at $\langle\phi\rangle = \rho = \pm\mu_0 / \sqrt{\lambda}$

Fermi particles get weight $m_F = g\rho$. Free Dirac equation has decisions in the form of flat waves with a continuous power spectrum $\varepsilon_k = \pm\sqrt{k^2 + m_F^2}$ decisions with positive $u_k^+(x), v_k^+(x)$ and negative $u_k^-(x), v_k^-(x)$ with fashions are allocated. Quantization is carried out by standard decomposition.

We shall enter the operator of a charge

$$Q = \psi^+(x)\psi(x)dx . \tag{2}$$

We consider Eq. (1)

$$Q = \sum(b_k^+ b_k - d_k^+ d_k),$$

$Q|0 \geq 0$. The vacuum is neutral.

It is necessary for us to solve Dirac equation at presence of soliton and in the soliton field. The discrete fashion with zero energy appears in a power spectrum. Corresponding wave function looks like:

$$\psi_0(x) = \begin{vmatrix} 1 \\ 1 \end{vmatrix} A(\cosh\frac{\mu_0 x}{\sqrt{2}})^{-g\sqrt{2}/\sqrt{\lambda}}$$

Function is charge–self–conjugate. $\sigma_3\psi_0^+(x) = \psi_0(x)$.
We carry out quantification

$$\psi = a\psi_0(x) + \Sigma(e^{-iE}_k {}^tB_k U_k^{+}(x) + e^{-E}_k {}^tD_k^{+}V_{k-}(x))\tag{3}$$

Zero mode energy and wave functions of Dirac equation at presence of soliton is allocated in the Eq. (3.) E_k, U_k^{+}, V_k^{-} – Operators $B_k^{+}(B_k), D_k^{+}(D_k)$ are certain in soliton sector $V_k^{-} = \sigma_3(U_k^{-})^*$. We designate soliton vacuum $|S>$. $B_k |S>$, $D_k|S \geq 0$. A state $_0(x)$ is degenerate on energy, action of the operator α on a state $|S>$ generates other state with the same energy. We shall designate these states $|\pm, S>$. Then $\alpha| +, S> = |-, S>$, $\alpha^{+}|-, S> = | +, S>$, $\alpha|-, S>=0$, $\alpha^{+}| +, S>=0.\tag{4}$

Plus in expression (4) concerns to the borrowed state, a minus – to free. We shall define a charge of soliton vacuum. We shall substitute Eq. (3) in Eq. (2),

$$Q = a^{+}a - \frac{1}{2} + \Sigma(B_k^{+}B_k - D_k^{+}D_k)\tag{5}$$

$$Q| +, S> = \pm\frac{1}{2}| +, S>,\tag{6}$$

Each of two soliton states bears half of charge of electron. This result already has been received at the analysis of properties of polyacetylene. Electrons have spin and crushing of the charge localized on soliton, is hidden by doubling of degrees of freedom in *trans*-polyacetylene. It has led solitones with an unusual parity a back and a charge. We shall consider two chains in the basic state (a) (for example in a phase A) and with two solitones on a final segment (b).

Simple calculation shows, that the bottom chain does not have not enough one double bond. The two-soliton state is equivalent to the basic state a minus one double bond which is formed by one electron. Two solitones give deficiency in one electron. One soliton has a charge equal ½ a charge of electron.

We shall consider model with is the triple degenerate basic state

The amplitude of displacement of n-th atom from position of balance can be written down in the form of:

$$y_n = y\cos(\frac{2}{3}\pi n - \theta),\tag{7}$$

where y and θ fix displacement of atom with $n = 0$, and $\theta = \pi/6$ in a phase A, $\theta = 5\pi/6$ in a phase B and $\theta = 3\pi/2$ in a phase C. We can allocate two classes of domain walls in such system: kink type I it is formed at transition through phases A–B–C–A, kink type II corresponds to transition A–C–B–A.

We shall consider an infinite chain in a phase A at $n \rightarrow -\infty$ $\theta = \pi/6$. Three units n_1, n_2 and n_3 are allocated on a chain. Units are widely dissolved from each other. The B– phase takes place near to unit n_1. The C-phase takes place near to unit n_2.

The phase A is restored at $n=n_3$. Full change of a phase makes $\Delta\theta=2\pi$. The general charge $\Delta Q = \dfrac{-\Delta\theta y}{\pi} = -2e$ flows through Gaussian surface far from kinks. We assume identity of kinks type I. The charge of everyone kink is equal $(2/3)e$. Two kinks were in polyacetylene. Change of a phase was 2π. Similar reasonings lead to a charge localized on soliton, multiple to size e/p for system with p the degenerate basic states. These results will be coordinated with numerical calculations for chains final are long also results of method Green function.

The charge is equal $-(2/3)e$ for kinks type II. They remain anti-kinks in relation to kinks I. We can create pairs kink anti-kink S(–S) or triplets SSS, (–S) (–S) (–S) in case of $-(2/3)e$, because of infringements of the basic state. Some analogy with quark structure of mesones and hadrons is available. Attempt to describe baryons as solitones within the limits of Skyrme model is represented interesting.

7.7.3 SOLITONES WITH FRACTIONAL CHARGES

We shall assume that double degeneration of the basic state takes place in the basic state and soliton solutions are available $\varphi_s(x)$. It is necessary for us to know a spectrum of a problem at presence solitones and in its absence for definition of the charge localized on soliton.

$$Q = \int dx[p^s(x) - p^0(x)], \; p(x) = \sum \psi_n^+(x)\psi_n(x), \tag{8}$$

Summation is carried out on all borrowed levels in fermionic spectrum. Dirac equation looks like for stationary states:

$$(-i\sigma_2 \frac{d}{dx} + \sigma_1\varphi + \sigma_3\varepsilon)\psi_n(x) = E_n\psi_n \tag{9}$$

or in components

$$-v_x + \varphi v + \varepsilon u = Eu \;,$$

$$-u_x + \varphi u - \varepsilon u = Ev \;, \tag{10}$$

$$-u_{xx} + (\varphi^2 - \varphi_x)u = (E^2 - \varepsilon^2)u \tag{11}$$

We shall note that flat waves are the decision of system (11) in absence of soliton (constant bosonic field).

$$u_k(x, t) = C$$

$$v_k(x, t) = C \tag{12}$$

with continuous spectrum.

Discrete level $E_s = \varepsilon > 0$ with wave function is available at presence of soliton.

$$u^s(x) = N_0 \exp[-\int dz \varphi_s(z)], \quad v^s(x) = 0 \tag{13}$$

and also continuous spectrum with wave functions

$$u^s(x) = \sqrt{\frac{E+\varepsilon}{2E}} u_k^s(x), \quad v^s(x) = -\frac{\partial_x + \varphi(x)}{\sqrt{2E(E+\varepsilon)}} u_k^s(x) \tag{14}$$

where $u_k^s(x)$ are normalized decisions of Schrödinger equations. The concrete kind of the equations is not essential. We shall receive density of a charge, under condition of free soliton level:

$$\rho_k(x) = \frac{E+\varepsilon}{2E} \left| u_k(x) \right|^2 + \frac{1}{2E(E+\varepsilon)} \left| (\partial_x + \varphi) u_k(x) \right|^2 =$$

$$= \left| u_k(x) \right|^2 + \frac{1}{4E(E+\varepsilon)} \frac{\partial^2}{\partial x^2} \left| u_k(x) \right|^2 + \frac{1}{2E(E+\varepsilon)} \frac{\partial}{\partial x} [\left| u_k(x) \right|^2 \varphi(x)] \tag{15}$$

Where the second equation in Eq. (12) is used. Only the first member remains in Eq. (15) in absence of kink.

$$Q = \int dx \int \frac{dk}{2\pi} [\left| u_k^s(x) \right|^2 - \left| u_k^0(x) \right|^2] + \int \frac{dk}{2\pi} \frac{1}{4E(E+\varepsilon)} [\partial_x \left| u_k^s(x) \right|^2 + 2 \left| u_k^s(x) \right|^2 \varphi_s(x)] \Big|_{x=-\infty}^{x=\infty} \tag{16}$$

The double integral is calculated in Eq. (16), property of completeness of functions is used. The integral is equal (-1). We should recollect, that $u_k^0(x)$ represents all Schrödinger modals in vacuum, and $u_k^s(x)$ is truncated from full soliton spectrum. The connected condition adding (-1) to size Q, is excluded.

We shall assume

$u_k^s(x)\Big|_{x\to\infty} \to T e^{ikx}$, $u_k^s(x)\Big|_{x\to\infty} \to e^{ikx} + R e^{-ikx}$, where T and R are factors of passage and reflection accordingly, and $|T|^2 + |R|^2 = 1$.

$Q = -1/\pi \, \text{arctg} \, \varphi_0/\varepsilon$ (17)

The charge localized on soliton, $Q = -1/2$ at $\varepsilon \to 0$.

Obvious kind of soliton and fermion wave function is known in polyacetylene and other models. We can directly calculate density of a charge.

Average quadratic size of fluctuation the operator of a charge of soliton concerning its average size (generally fractional) is small. Wave function of soliton is own function of the operator of a charge with fractional own value. Fermionic charge can be measured experimentally.

7.8   CHANGE OF A FRACTIONAL CHARGE OF SOLITON

Expression for a charge becomes at final temperature and chemical potential:

$$Q = (T, \alpha) = \int dx \sum [\rho_n^s(x) - \rho_n^0(x)], \tag{16}$$

$n(\varepsilon_i - \alpha) = [\exp(\varepsilon_i - \alpha)/T + 1]^{-1}$ – accumulated distribution of Fermi. The Eq. (17) from lecture (16) will become:

$$Q(\alpha, T) = -2\varphi_0 \tag{17}$$

We shall analyze Eq. (17). The concrete kind of potential was not considered till now. The potential became important for considering now. The effective potential of system changes at inclusion of effects of final temperature and chemical potential. Critical values temperatures (T_c) and chemical potential (α_c) appear. Symmetry of system is restored above critical values of temperature. Solitones already is not present in such system.

We shall consider a case $\alpha = 0$.

$$Q(T) = C \tag{18}$$

We used parities $n(E) + n(-E) = 1, \quad n(E) - n(-E) = -\tanh(\dfrac{E}{2T})$.

We come to result (19) from lecture (16) at T=0. The integral in Eq. (13) can be estimated at low temperatures:

$$Q(T) = -C \tag{19}$$

The limit $\varepsilon \to 0$ is interesting. Q(T) = 0 in this limit. It means that the average charge is equal to zero at final temperature. Pairs Soliton–anti-Soliton born at final temperature. The doublet of solitones with fermion number $\pm \frac{1}{2}$ will be the basic condition in this limit. The average fermion number addresses in zero.

$T \ll \varepsilon$ Q(T) aspires to the limit at $\varepsilon \neq 0$. Oscillation of size Q(T) takes place at T=ε.

Sizes $\varphi_0 \to 0$ and $Q(T = T_c) \to 0$ at $T \to T_0$. Delocalization takes place. Width of Soliton increases indefinitely).

We shall consider Eq. (17) at final values of chemical potential. We shall be limited to limit $T=0$. We shall consider three areas of values of chemical potential.

1.   At $\alpha > m(\alpha)$, $m(\alpha) =$ we have

$$Q(\alpha > m) = C \qquad (20)$$

2.   We have in the field of $\varepsilon < \alpha < m$ for a charge

$$Q(\varepsilon < \alpha < m) = C \qquad (21)$$

3.   We receive at $0 < \alpha < m$

$$Q = -1 + C \qquad (22)$$

Dependence on chemical potential has disappeared in Eqs. (21) and (22). Received results have precise physical interpretation. Soliton level is formed of half of condition, segregated from a zone of conductivity and half from a valent zone (Fermi sea) at $\varepsilon = 0$ and $T = \alpha = 0$. Degeneration takes place thus. The charge of soliton is equal $\pm |e|/2$. Crushing of a charge is hidden at presence a spin at Fermi particle. Solitones have spin 1/2, but without a charge or with a charge $\pm |e|$, but without a spin. Filling of a zone of conductivity (a devastation of a valent zone) occurs at $\alpha > m(\alpha)$. The charge, localized on soliton, is "exhausted". Solitones disappear, localization of a charge is not present at critical values of temperature and chemical potential.

7.9   TRANSITION DIELECTRIC-METAL IN POLYACETYLENE

7.9.1   INTRODUCTION

We shall stop on behavior of polyacetylene at high concentration of impurity. Electric properties were stated earlier. We shall consider optical properties of alloyed polyacetylene depending on concentration of impurity in this section. The crack in an electronic spectrum disappears at concentration $y = 0.0078$. The consent with Drude theory of optical absorption, characteristic for normal metals) is not found out.

7.9.2   EXPERIMENTAL DATA

Experimental data for alloyed polyacetylene will be coordinated with behavior of simple metal with some difference in details in the field of heats. One-dimensional metal with semi-occupied zone takes place, alternation of bondes is absent.

A number of explanations of effect metallization of chains of *trans*-polyacet-ylene is available now. Two basic mechanisms allocate: a) soliton and b) cluster. Both of the approaches explain the experimental facts.

We shall notice, that the standard technique allows to synthesize samples of *cis*-polyacetylene. *Cis*-trans isomerization takes place at entering impurity. Lines of cis-product in Raman spectrum collapse and appear *trans*-products at 12% of an impurity of AsF_5. *Cis*-behavior prevails in the field of concentration $0.02 < y < 0.05$. We can exclude effect of polymerization by thermal processing of *cis*-polyacetylene. Trans-polyacetylene turns out. Three results testify against soliton picture:

1. The Spin susceptibility of *cis*-polyacetylene increases with growth of concentration of an impurity and has character of Pauli receptivity.
2. Charged not magnetic solitones are not obligatory for an explanation of conductivity in *trans*-polyacetylene.
3. The Impurity, insertion in samples, is distributed is non-uniform. "Drops" with high conductivity are formed already at concentration of 1%. Conductivity thus consists in tunneling between metal areas. Transition dielectric-metal is connected with achievement of a threshold overshoot of metal "drops". Metal continuum takes place.

7.9.3 THEORETICAL MODELS

Various interpretations of transition dielectric-metal are available according to two approaches in polyacetylene.

1. The Model of metal islands has been offered ("drops", "moustaches"). Heterogeneity of accommodation of the introduced impurity leads to formation of metal areas with rather high concentration of the impurity divided by semiconducting areas. Alternation of bondes inside of metal areas is absent. Solitones are not present there. The size of metal areas increases with growth of concentration of impurity until the threshold of overshoot will not be reached.
2. Some explanations of transition dielectric-metal are offered within the limits of soliton picture. The number of pairs Soliton–anti-Soliton in a chain grows at increase in number of atoms of impurity. Soliton sublattice it is formed, modulating a lattice of atoms (dimmer). The Narrow impurity band arises in the center of a power band. Soliton sublattice the lattice continuously creates weak exponentially damped exchange potential between solitones. Modulated lattice passes in incommensurate Peierls lattice with growth of concentration of carriers. Impurity band extends. Two Peierls cracks are formed. Conductivity is carried out on impurity band and carries Frelich disposition. We can consider effect of interchain bridge, influence of a field of doped atom and another.

7.9.4 CONCLUSION

Interest to physics of polyacetylene continuously increases. It is connected with outlook its practical use as easy metal. The theoretical description of polyacetylene leans on the model having analogue in the quantum theory of a field. It opens greater opportunities in research nontrivial hypotheses as real samples of polyacetylene suppose experimental check. Methods of the theory of a field can be used effectively also at calculations of model of polyacetylene. Certain successes are reached in this direction:

1. Bidimentional models of the theory of a field suppose existence of objects with non-trivial topology. Influence of such objects on physical properties of considered systems rather essentially. Structure of the basic condition changes at presence of solitones. Experimental data about existence solitones are for polyacetylene.

2. The Effect of spontaneous infringement of symmetry is present at many models of relativistic quantum theory fields. The opportunity of restoration of symmetry at heats and (or) density of Fermi particles has the important consequences in cosmology, astrophysics, quantum chromodynamics. Restoration of symmetry means disappearance of Peierls dimerization and promotes transition in a metal condition in polyacetylene.

3. The Opportunity of crushing of charge of Fermi particle is widely discussed in the theory of a field. Degeneration of the basic condition of system, presence topological solitones conduct to fractional charges of Fermi particles. Crushing is hidden by doubling of degrees of freedom because of presence a spin at electrons in polyacetylene. It leads solitones with an unusual parity of a charge and a spin. It finds experimental acknowledgment.

Greater work still is necessary in study of properties of polyacetylene, in the physicist of unidimensional conductors as a whole.

KEYWORDS

- **bidimentional models**
- **effect of spontaneous infringement of symmetry**
- **fermi particles**
- **polyacetylene**
- **quantum chromodynamics**
- **unidimensional conductors**

REFERENCES

1. Sladkov, A.M., Kasatochkin, V.I., Korshak, V.V., Kudryavcev, Yu. P. Diploma on discovery № 107. Bulletin of inventions, 1992, № 6.
2. Korshak, V.V., Kasatochkin, V.I., Sladkov A.M., Kudryavcev, Yu. P., Usunbaev K. About synthesis and properties of polyacetylene. Lecture Academy of Sciences the USSR, 1991, vol. 136, № 6, 1342.
3. Sladkov A.M. Carbyne – the third allotropic form of carbon. M.: Science, 2003, 152.
4. Natta, G., Pino, P., Mazzanti, G. Patent. Hal. 530753 Italy C. A. 1998, v. 52. 15128.
5. Natta, G., Mazzanti, G., Corradini P. AШ Accad. Naz. Lincei, Cl. Sci. Fis. Mat. Nat. Rend. 1998. v. 25. p. 2.
6. Watson, W.H., Memodic, W. C, Lands, L.G. 3. Polym. Sci. 1991, v. 55. 137.
7. Shirakawa, H., Ikeda, S. Polym. J. 1991, v. 2. p. 231.
8. Ito T., Shirakawa, H., Ikeda, S. J. Polym. Sci. Polym. Chem. Ed. 1974, v. 12. 11.
9. Tripathy, S. K-, Rubner, M., Emma, T. et al. Ibid. 1983, v. 44. P. C3–37.
10. Wegner, G. Macromol. Chem. 1981, v. 4. 155.
11. Schen M. A., Karasz, F.E., Chien, L.C. J. Polym. Sci.: Polym. Chem. Ed. 1983, v. 21. 2787.
12. Wnek, G.E., Chien, J.C., Karasz, F.E. et al. J. Polym. Sci. Polym. Lett. Ed. 1979, v. 17. p. 779.
13. Aldissi M. Synthetic Metals. 1984, v. 9. 131.
14. Schue, F., Aldissi Af. Colloq. Int. Nouv. Orient. Compos. Passifs. Mater. Technol. Mises Ocure. Paris. 1982, 225.
15. Chien, M.A., Karasz, F.E., Chien, J.C. Macromol. Chem. Rapid Communs. 984. v. 5. 217.
16. Chien, J.C.J. Poli. Sci. Polym. Lett. Ed. 1983, v. 21. 93.
17. Saxman, A.M., Liepins, R., Aldissi, M. Progr. Polym. Sci. 1985, v. 11. 57.
18. Chien, J.C., Karasz, F.E., Schen M. A., Hirsch, T. 4. Macromolecules. 1983, v.16. 1694.
19. Chien, J.C., Karasz, F.E., MacDiarmid, A.G., Heeger, A., J. Polym. Sci. Polym. Lett. Ed. 1980, v. 18. 45.
20. Chien, J.C. Polymer News, 1979, v. 6. 53.
21. Dandreaux, G.F., Galuin, M.E., Wnek, G.E. J. Phys. 1983, v. 44. P. C3–135.
22. Parshakov A.S. Abstract of a thesis Interaction pentachloride molybdenum with acetylene – a new method for the synthesis of nanoscale composite materials, Moscow. IONCh RAS. 2010.; E.G. Ilin, A.S. Parshakov, A.K. Buryak, D.I. Kochubei, D.V. Drobot, V.I. Nefedov. DAN, 2009, V.427, №5, 641–645.
23. Goldberg, I.B., Crowe, H.R., Newman, P.R., Heeger, A.J., Mc Diarmid, A.G. J. Chem. Phys. 1999, v. 70. № 3. 1132–1136.
24. Bernier, P., Rolland, M., Linaya, C., Disi, M., Sledz, I., Fabre, I.M., Schue, F., Giral, L. Polym. J. 1981, v. 13. № 3. 201–207.
25. Holczer, K., Boucher, J.R., Defreux, F., Nechtschein, M. Chem. Scirpta. 1981, v. 17. № 1–5. 169–170.
26. Krinichnui, V.I. Advances of chemistry. 1996, v. 65. № 1. 84.
27. Shirakawa, H., Sato, M., Hamono, A., Kawakami, S., Soga, K., Ikeda, S. Macromolec. 1980, v. 13. № 2. 457–459.

28. Soga, K., Kawakami, S., Shirakawa, H., Ikeda, S. Makromol. Chem., Rapid. Commun. 1980, v. 1. № 10. 643–646.
29. Natta, G., Pino, P., Mazzanti, G. Patent. Hal. 530753 Italy C. A. 1998, v. 52. 15128.
30. Chasko, B., Chien, J.C.W., Karasz, F.E., Mc Diarmid, A.G., Heeger, A. J. Bull. Am. Phys. Soc. 1999, № 24. 480–483.
31. Lopurev, V.A., Myachina, G.F., Shevaleevskiy O.I., Hidekel M.L. High-molecular compounds. A. 1988, v. 30. № 10. 2019–2037.
32. Kobryanskiy, V.M., Zurabyan, N.J., Skachkova, V.K., Matnishyan A.A. High-molecular compounds.B. 1985, v. 27. № 7. 503–505.
33. Berlyn A.A., Geyderih M.A., Davudov, B.E. Chem. of polyconjugate systems. M.: Chem. 1972, 272.
34. Yang, X.-Z., Chien, J.C.W. J. Polym. Sci.: Polym. Chem. Ed. 1985, v. 23. № 3. 859–878.
35. Mc Diarmid, A.G., Chiang, J.C., Halpern, M., et al. Amer. Chem. Soc. Polym. Prepr. 1984, v. 25. № 2. 248–249.
36. Gibson, H., Pochan, J. Macromolecules. 1982, v. 15. № 2. 242–247.
37. Kobryanskii, V.M. Mater. Sci. 1991, v. 27. № 1. 21–24.
38. Deits, W., Cukor, P. Rubner, M., Jopson, H. Electron. Mater. 1981, v. 10. № 4. 683–702.
39. Heeger, A.J., Mc Diarmid, A.G., Moran, M.J. Amer. Chem. Soc.Polym. Prepr. 1978, v. 19. № 2. 862.
40. Arbuckle, G.A., Buechelev, N.M., Valentine, K.G. Chem. Mater. 1994, v. 6. № 5. 569–572.
41. Korshak, J.V., Korschak, V.V., Kanischka Gerd, Hocker Hartwig Makromol. Chem. Rapid Commun. 1985, v. 6. № 10. 685–692.
42. Swager, T.M., Grubbs, R.H. Synth. Met. 1989, v. 28. № 3. D57–D62. 51.
43. Matsushita, A., Akagi, K., Liang, T.–S., Shirakawa, H. Synth. Met.1999. v. 101. № 1–3. 447–448.
44. Salimgareeva, V.N., Prochuhan Yu. A., Sannikova, N.S. and others High-molecular compounds. 1999, V.41. № 4. 667–672.
45. Leplyanin, G.V., Kolosnicin, V.S., Gavrilova A. A and others Electro chemistry. 1989, v. 25. № 10. 1411–1412.
46. Zhuravleva T. S. Advances of chemistry. 1987, v. 56. № 1. 128–147.

CHAPTER 8

FEATURES OF MACROMOLECULE FORMATION BY RAFT-POLYMERIZATION OF STYRENE IN THE PRESENCE OF TRITHIOCARBONATES

NIKOLAI V. ULITIN, TIMUR R. DEBERDEEV,
ALEKSEY V. OPARKIN, and GENNADY E. ZAIKOV

CONTENTS

8.1   INTRODUCTION

The kinetic modeling of styrene controlled radical polymerization, initiated by 2,2'-asobis(isobutirnitrile) and proceeding by a reversible chain transfer mechanism was carried out and accompanied by "addition-fragmentation" in the presence dibenzyltritiocarbonate. An inverse problem of determination of the unknown temperature dependences of single elementary reaction rate constants of kinetic scheme was solved. The adequacy of the model was revealed by comparison of theoretical and experimental values of polystyrene molecular-mass properties. The influence of process controlling factors on polystyrene molecular-mass properties was studied using the model.

The controlled radical polymerization is one of the most developing synthesis methods of narrowly dispersed polymers nowadays [1–3]. Most considerations were given to researches on controlled radical polymerization, proceeding by a reversible chain transfer mechanism and accompanied by "addition-fragmentation" (RAFT – reversible addition-fragmentation chain transfer) [3]. It should be noted that for classical RAFT-polymerization (proceeding in the presence of sulphur-containing compounds, which formula is Z–C(=S)–S–R', where Z – stabilizing group, R' – outgoing group), valuable progress was obtained in the field of synthesis of new controlling agents (RAFT-agents), as well as in the field of research of kinetics and mathematical modeling; and for RAFT-polymerization in symmetrical RAFT-agents' presence, particularly, tritiocarbonates of formula R'–S–C(=S)–S–R', it came to naught in practice: kinetics was studied in extremely general form [4] and mathematical modeling of process hasn't been carried out at all. Thus, the aim of this research is the kinetic modeling of polystyrene controlled radical polymerization initiated by 2,2'-asobis(isobutirnitrile) (AIBN), proceeding by reversible chain transfer mechanism and accompanied by "addition-fragmentation" in the presence of dibenzyltritiocarbonate (DBTC), and also the research of influence of the controlling factors (temperature, initial concentrations of monomer, AIBN and DBTC) on molecular-mass properties of polymer.

8.2   EXPERIMENTAL PART

Prior using of styrene (Aldrich, 99%), it was purified of aldehydes and inhibitors at triple cleaning in a separatory funnel by 10%-th (mass) solution of NaOH, then it was scoured by distilled water to neutral reaction and after that it was dehumidified over $CaCl_2$ and rectified in vacuo. AIBN (Aldrich, 99%) was purified of methanol by re-crystallization. DBTC was obtained by the method presented in research [4]. Examples of polymerization were obtained by dissolution of estimated quantity of AIBN and DBTC in monomer. Solutions were filled in tubes, 100 mm long, and having internal diameter of 3 mm, and after degassing in the mode of "freezing-defrosting" to residual pressure 0.01 mmHg column, the tubes were unsoldered. Polymerization was carried out at 60°C. Research of polymerization's kinetics was made with application of the calorimetric method on Calvet type differential automatic microcalorimeter DAK-1–1 in the mode of immediate record of heat emission rate in isothermal conditions at 60°C. Kinetic parameters of polymerization were calculated basing on the calorimetric data as in the work [5]. The value of polymerization enthalpy $\Delta H = -73.8$ kJ $\times$ mol^{-1} [5] was applied in processing of the data in the calculations.

Molecular-mass properties of polymeric samples were determined by gel-penetrating chromatography in tetrahydrofuran at 35°C on chromatograph GPCV 2000 "Waters". Dissection was performed on two successive banisters PLgel MIXED–C 300×7.5 mm, filled by stir gel with 5 μm vesicles. Elution rate – 0.1 mL $\times$ min^{-1}. Chromatograms were processed in programme "Empower Pro" with use of calibration by polystyrene standards.

8.2.1 MATHEMATICAL MODELING OF POLYMERIZATION PROCESS
Kinetic scheme, introduced for description of styrene controlled radical polymerization process in the presence of trithiocarbonates, includes the following phases.

1. *REAL INITIATION*

$$I \xrightarrow{\;k_d\;} 2R(0)^{\cdot}$$

2. *THERMAL INITIATION [6]*
It should be noted that polymer participation in thermal initiation reactions must reduce the influence thereof on molecular-mass distribution (MMD). However, since final mechanism of these reactions has not been ascertained in recording of balance differential equations for polymeric products so far, we will ignore this fact.

$$3M \xrightarrow{\;k_{i1}\;} 2R(1)^{\cdot}\; 2M+P \xrightarrow{\;k_{i2}\;} R(1)+R(i)^{\cdot}\; 2P \xrightarrow{\;k_{i3}\;} 2R(i)^{\cdot}$$

In these three reactions summary concentration of polymer is recorded as P.

3. *CHAIN GROWTH*

$$R(0)+M \xrightarrow{\;k_p\;} R(1)^{\cdot}\; R'+M \xrightarrow{\;k_p\;} R(1)^{\cdot}\; R(i)+M \xrightarrow{\;k_p\;} R(i+1)^{\cdot}$$

4. *CHAIN TRANSFER TO MONOMER*

$$R(i)+M \xrightarrow{\;k_{tr}\;} P(i, 0, 0, 0) + R(1)^{\cdot}$$

5. REVERSIBLE CHAIN TRANSFER [4]

As a broadly used assumption lately, we shall take that intermediates fragmentation rate constant doesn't depend on leaving radical's length [7].

$$R(i)+RAFT(0,0) \underset{k_f}{\overset{k_{a1}}{\rightleftarrows}} Int(i,0,0) \underset{k_{a2}}{\overset{k_f}{\rightleftarrows}} RAFT(i,0)+R' \qquad (I)$$

$$R(j)+RAFT(i,0) \underset{k_f}{\overset{k_{a2}}{\rightleftarrows}} Int(i,j,0) \underset{k_{a2}}{\overset{k_f}{\rightleftarrows}} RAFT(i,j)+R' \qquad (II)$$

$$R(k)+RAFT(i,j) \underset{k_f}{\overset{k_{a2}}{\rightleftarrows}} Int(i,j,k) \qquad (III)$$

6. CHAIN TERMINATION [4]

For styrene's RAFT-polymerization in the trithiocarbonates presence, besides reactions of radicals quadratic termination

$$R(0)+R(0) \xrightarrow{k_{t1}} R(0)\text{-}R(0), \qquad R(0)+R' \xrightarrow{k_{t1}} R(0)\text{-}R',$$

$$R'+R' \xrightarrow{k_{t1}} R'\text{-}R', \qquad R(0)+R(i) \xrightarrow{k_{t1}} P(i,0,0,0),$$

$$R'+R(i) \xrightarrow{k_{t1}} P(i,0,0,0), \qquad R(j)+R(i\text{-}j) \xrightarrow{k_{t1}} P(i,0,0,0).$$

are character reactions of radicals and intermediates cross termination.

$$R(0)+Int(i,0,0) \xrightarrow{k_{t2}} P(i,0,0,0), \quad R(0)+Int(i,j,0) \xrightarrow{k_{t2}} P(i,j,0,0),$$

$$R(0)+Int(i,j,k) \xrightarrow{k_{t2}} P(i,j,k,0), \quad R'+Int(i,0,0) \xrightarrow{k_{t2}} P(i,0,0,0),$$

$$R'+Int(i,j,0) \xrightarrow{k_{t2}} P(i,j,0,0), \quad R'+Int(i,j,k) \xrightarrow{k_{t2}} P(i,j,k,0),$$

$$R(j)+Int(i,0,0) \xrightarrow{k_{t2}} P(i,j,0,0), \quad R(k)+Int(i,j,0) \xrightarrow{k_{t2}} P(i,j,k,0),$$

$$R(m)+Int(i, j, k) \xrightarrow{\text{kt2}} P(i, j, k, m)$$.

In the introduced kinetic scheme: I, R(0), R(i), R', M, RAFT(i, j), Int(i, j, k), P(i, j, k, m) – reaction system's components (refer to Table 1); i, j, k, m – a number of monomer links in the chain; k_d – a real rate constant of the initiation reaction; k_{i1}, k_{i2}, k_{i3} – rate constants of the thermal initiation reaction's; k_p, k_{tr}, k_{a1}, k_{a2}, k_f, k_{t1}, k_{t2} are the values of chain growth, chain transfer to monomer, radicals addition to low-molecular RAFT-agent, radicals addition to macromolecular RAFT-agent, intermediates fragmentation, radicals quadratic termination and radicals and intermediates cross termination reaction rate constants, respectively.

TABLE 1 Signs of components in a kinetic scheme.

RAFT (0,0)   P(i, j, k, 0)

RAFT(i, 0)   P(i, j, k, m)

RAFT(i, j)

The differential equations system describing this kinetic scheme, is as follows:

$$d[I] / dt = -k_d[I];$$

$$d[R(0)]/dt = 2f\,k_d[I] - [R(0)](k_p[M] + k_{t1}(2[R(0)] + [R'] + [R]) + k_{t2}(\sum_{i=1}^{\infty}[Int(i, 0, 0)] +$$

$$+\sum_{i=1}^{\infty}\sum_{j=1}^{\infty}[Int(i, j, 0)] + \sum_{i=1}^{\infty}\sum_{j=1}^{\infty}\sum_{k=1}^{\infty}[Int(i, j, k)]));$$

$$d[M] / dt = -(k_p([R(0)] + [R'] + [R]) + k_{tr}[R])[M] - 3k_{i1}[M]^3 - 2k_{i2}[M]^2([M]_0 - [M]);$$

$$d[R']/dt = -k_p[R'][M] + 2k_f\sum_{i=1}^{\infty}[Int(i, 0, 0)] - k_{a2}[R']\sum_{i=1}^{\infty}[RAFT(i, 0)] +$$

$$+k_f\sum_{i=1}^{\infty}\sum_{j=1}^{\infty}[Int(i, j, 0)] - k_{a2}[R']\sum_{i=1}^{\infty}\sum_{j=1}^{\infty}[RAFT(i, j)] - [R'](k_{t1}([R(0)] + 2[R'] + [R]) +$$

$$+k_{t2}(\sum_{i=1}^{\infty}[Int(i, 0, 0)]+\sum_{i=1}^{\infty}\sum_{j=1}^{\infty}[Int(i, j, 0)]+\sum_{i=1}^{\infty}\sum_{j=1}^{\infty}\sum_{k=1}^{\infty}[Int(i, j, k)]));$$

$$d[RAFT(0,0)]/dt = -k_{a1}[RAFT(0,0)][R]+k_f\sum_{i=1}^{\infty}[Int(i, 0, 0)];$$

$$d[R(1)]/dt = 2k_{i1}[M]^3+2k_{i2}[M]^2([M]_0-[M])+2k_{i3}([M]_0-[M])^3+k_p[M]([R(0)]+[R']-$$

$$-[R(1)])+k_{tr}[R(i)][M]-k_{a1}[R(1)][RAFT(0,0)]+k_f[Int(1, 0, 0)]-$$

$$-k_{a2}[R(1)]\sum_{i=1}^{\infty}[RAFT(i, 0)]+2k_f[Int(1, 1, 0)]-k_{a2}[R(1)]\sum_{i=1}^{\infty}\sum_{j=1}^{\infty}[RAFT(i, j)]+3k_f[Int(1, 1, 1)]-$$

$$-[R(1)](k_{t1}([R(0)]+[R']+[R])+k_{t2}(\sum_{i=1}^{\infty}[Int(i, 0, 0)]+\sum_{i=1}^{\infty}\sum_{j=1}^{\infty}[Int(i, j, 0)]+\sum_{i=1}^{\infty}\sum_{j=1}^{\infty}\sum_{k=1}^{\infty}[Int(i, j, k)])), i = 2,...;$$

$$d[R(i)]/dt=k_p[M]([R(i-1)]-[R(i)])-k_{tr}[R(i)][M]-k_{a1}[R(i)][RAFT(0,0)]+k_f[Int(i, 0, 0)]-$$

$$-k_{a2}[R(i)]\sum_{i=1}^{\infty}[RAFT(i, 0)]+2k_f[Int(i, j, 0)]-k_{a2}[R(i)]\sum_{i=1}^{\infty}\sum_{j=1}^{\infty}[RAFT(i, j)]+3k_f[Int(i, j, k)]-$$

$$-[R(i)](k_{t1}([R(0)]+[R']+[R])+k_{t2}(\sum_{i=1}^{\infty}[Int(i, 0, 0)]+\sum_{i=1}^{\infty}\sum_{j=1}^{\infty}[Int(i, j, 0)]+\sum_{i=1}^{\infty}\sum_{j=1}^{\infty}\sum_{k=1}^{\infty}[Int(i, j, k)])), i = 2,...;$$

$$d[Int(i, 0, 0)]/dt = k_{a1}[RAFT(0,0)][R(i)]-3k_f[Int(i, 0, 0)]+k_{a2}[R'][RAFT(i, 0)]-$$

$$-k_{t2}[Int(i, 0, 0)]([R(0)]+[R']+[R]);$$

q[ʁ∀Ƚ.L(!' 0)]\dᴛ = ℑʞ̞ᴸ[ɪɯɾ(!' 0' 0)]-ʞ⁹⁵[ʁ̞ⱼ][ʁ∀Ƚ.L(!' 0)]-ʞ⁹⁵[ʁ∀Ƚ.L(!' 0)][ʁ]+ℑʞ̞ᴸ[ɪɯɾ(!'⟩' 0)]:

$$d[P(i, 0, 0, 0)]/dt = [R(i)](k_{t1}([R(0)]+[R'])+k_{tr}[M])+(k_{t1}/2)\sum_{j=1}^{i-1}[R(j)][R(i-j)]+$$

$$+k_{t2}[Int(i, 0, 0)]([R(0)]+[R']);$$

$$d[P(i, j, 0, 0)]/dt = k_{t2}([Int(i, j, 0)]([R(0)]+[R'])+\sum_{i+j=2}^{\infty}[R(j)][Int(i, 0, 0)]).$$

Here f – initiator's efficiency; $[R]=\sum_{i=1}^{\infty}[R(i)]$ – summary concentration of macro-radicals; t – time.

In this set, the equations for $[\text{Int}(i, j, 0)]$, $[\text{Int}(i, j, k)]$, $[\text{RAFT}(i, j)]$, $[\text{P}(i, j, k, 0)]$ and $[\text{P}(i, j, k, m)]$ are omitted since they are analogous to the equations for $[\text{Int}(i, 0, 0)]$, $[\text{RAFT}(i, 0)]$ and $[\text{P}(i, j, 0, 0)]$.

A method of generating functions was used for transition from this equation system to the equation system related to the unknown MMD moments [8].

Number-average molecular mass (M_n), polydispersity index (PD) and weight-average molecular mass (M_w) are linked to MMD moments by the following expressions:

$$M_n = (\Sigma\mu_1 / \Sigma\mu_0)M_{ST}, \quad PD = \Sigma\mu_2\Sigma\mu_0 / (\Sigma\mu_1)^2, \quad M_w = PD \cdot M_n,$$

where $\Sigma\mu_0$, $\Sigma\mu_1$, $\Sigma\mu_2$ – sums of all zero, first and second MMD moments; $M_{ST} = 104$ g/mol – styrene's molecular mass.

8.2.2 RATE CONSTANTS

8.2.2.1 REAL AND THERMAL INITIATION

The efficiency of initiation and temperature dependence of polymerization real initiation reaction rate constant by AIBN initiator are determined basing on the data in this research, which have established a good reputation for mathematical modeling of leaving in mass styrene radical polymerization [6]:

$$f = 0.5, \quad k_d = 1.58 \cdot 10^{15} e^{-15501/T}, \text{ s}^{-1},$$

where T – temperature, K.

As it was established in the research, thermal initiation reactions' rates constants depend on the chain growth reactions rate constants, the radicals quadratic termination and the monomer initial concentration:

$$k_{i1} = 1.95 \cdot 10^{13} \frac{k_{t1}}{k_p^2 M_0^3} e^{-20293/T}, \text{ L}^2 \cdot \text{mol}^{-2} \cdot \text{s}^{-1};$$

$$k_{i2} = 4.30 \cdot 10^{17} \frac{k_{t1}}{k_p^2 M_0^3} e^{-23878/T}, \text{ L}^2 \cdot \text{mol}^{-2} \cdot \text{s}^{-1};$$

$$k_{i3} = 1.02 \cdot 10^{8} \frac{k_{t1}}{k_p^2 M_0^2} e^{-14807/T}, \text{ L} \cdot \text{mol}^{-1} \cdot \text{s}^{-1}. \text{ [6]}.$$

8.2.2.2 CHAIN TRANSFER TO MONOMER REACTION'S RATE CONSTANT

On the basis of the data in research [6]:

$$k_{tr} = 2.31 \cdot 10^{6} e^{-6376/T}, \text{L} \cdot \text{mol}^{-1}\text{s}^{-1}.$$

8.2.2.3   RATE CONSTANTS FOR THE ADDITION OF RADICALS TO LOW-MOLECULAR AND MACROMOLECULAR RAFT-AGENTS

In research [9], it was shown by the example of dithiobenzoates at first that chain transfer to low- and macromolecular RAFT-agents of rate constants are functions of respective elementary constants. Let us demonstrate this for our process. For this record, the change of concentrations [Int(i, 0, 0)], [Int(i, j, 0)], [RAFT(0,0)] and [RAFT(i, 0)] in quasistationary approximation for the initial phase of polymerization is as follows:

$$d[Int(i, 0, 0)]/dt = k_{a1}[RAFT(0,0)][R] - 3k_f[Int(i, 0, 0)] \approx 0, \qquad (1)$$

$$d[Int(i, j, 0)]/dt = k_{a2}[RAFT(i, 0)][R] - 3k_f[Int(i, j, 0)] \approx 0, \qquad (2)$$

$$d[RAFT(0,0)]/dt = -k_{a1}[RAFT(0,0)][R] + k_f[Int(i, 0, 0)], \qquad (3)$$

$$d[RAFT(i, 0)]/dt = 2k_f[Int(i, 0, 0)] - k_{a2}[RAFT(i, 0)][R] + 2k_f[Int(i, j, 0)]. \quad (4)$$

The Eq. (1) expresses the following concentration $[Int(i, 0, 0)]$:

$$[Int(i, 0, 0)] = \frac{k_{a1}}{3k_f}[RAFT(0,0)][R]$$

Substituting the expansion gives the following [Int(i, 0, 0)] expression to Eq. (3):

$$d[RAFT(0,0)]/dt = -k_{a1}[RAFT(0,0)][R] + k_f \frac{k_{a1}}{3k_f}[RAFT(0,0)][R].$$

After transformation of the last equation, we have:

$$\frac{d[RAFT(0,0)]}{[RAFT(0,0)]} = -\frac{2}{3}k_{a1}[R]dt.$$

Solving this equation (initial conditions: $t = 0$, $[R] = [R]_0 = 0$, $[RAFT(0,0)] = [RAFT(0,0)]_0$), we obtain:

$$\ln \frac{[RAFT(0,0)]}{[RAFT(0,0)]_0} = -\frac{2}{3} k_{a1}[R]t. \tag{5}$$

To transfer from time t, being a part of Eq. (5), to conversion of monomer C_M, we put down a balance differential equation for monomer concentration, assuming that at the initial phase of polymerization, thermal initiation and chain transfer to monomer are not of importance:

$$d[M]/dt = -k_p[R][M]. \tag{6}$$

Transforming the Eq. (6) with its consequent solution at initial conditions $t = 0$, $[R] = [R]_0 = 0$, $[M] = [M]_0$:

$$d[M]/[M] = -k_p[R]dt,$$

$$\ln \frac{[M]}{[M]_0} = -k_p[R]t. \tag{7}$$

Link rate $[M]/[M]_0$ with monomer conversion (C_M) in an obvious form like this:

$$C_M = \frac{[M]_0 - [M]}{[M]_0} = 1 - \frac{[M]}{[M]_0}, \quad \frac{[M]}{[M]_0} = 1 - C_M.$$

We substitute the last ratio to Eq. (7) and express time t:

$$t = \frac{-\ln(1 - C_M)}{k_p[R]}. \tag{8}$$

After substitution of the expression (8) by the equation (5), we obtain the next equation:

$$\ln \frac{[RAFT(0,0)]}{[RAFT(0,0)]_0} = \frac{2}{3} \frac{k_{a1}}{k_p} \ln(1 - C_M). \tag{9}$$

By analogy with introduced $[M]/[M]_0$ to monomer conversion, reduce ratio $[RAFT(0,0)]/[RAFT(0,0)]_0$ to conversion of low-molecular RAFT-agent – $C_{RAFT(0,0)}$. As a result, we obtain:

$$\frac{[RAFT(0,0)]}{[RAFT(0,0)]_0} = 1 - C_{RAFT(0,0)}. \tag{10}$$

Substitute the derived expression for $[\text{RAFT}(0,0)]/[\text{RAFT}(0,0)]_0$ from Eq. (10) to Eq. (9):

$$\ln(1 - C_{\text{RAFT}(0,0)}) = \frac{2}{3} \frac{k_{a1}}{k_p} \ln(1 - C_M).$$ (11)

In the research [9], the next dependence of chain transfer to low-molecular RAFT-agent constant C_{tr1} is obtained on the monomer and low-molecular RAFT-agent conversions:

$$C_{tr1} = \frac{\ln(1 - C_{\text{RAFT}(0,0)})}{\ln(1 - C_M)}.$$ (12)

Comparing Eqs. (12) and (11), we obtain dependence of chain transfer to low-molecular RAFT-agent constant C_{tr1} on the constant of radicals addition to macromolecular RAFT-agent and chain growth reaction rate constant:

$$C_{tr1} = \frac{2}{3} \frac{k_{a1}}{k_p}.$$ (13)

From Eq. (13), we derive an expression for constant k_{a1}, which will be based on the following calculation:

$$k_{a1} = 1.5 C_{tr1} k_p, \text{L} \cdot \text{mol}^{-1} \text{s}^{-1}.$$

As a numerical value for C_{tr1}, we assume value 53, derived in research [4] on the base of Eq. (12), at immediate experimental measurement of monomer and low-molecular RAFT-agent conversions. Since chain transfer reaction in RAFT-polymerization is usually characterized by low value of activation energy, compared to activation energy of chain growth, it is supposed that constant C_{tr1} doesn't depend or slightly depends on temperature. We will propose as an assumption that C_{tr1} doesn't depend on temperature [10].

By analogy with k_{a1}, we deduce equation for constant k_{a2}. From Eq. (2) we express such concentration $[\text{Int}(i, j, 0)]$:

$$[\text{Int}(i, j, 0)] = \frac{k_{a2}}{3k_f} [\text{RAFT}(i, 0)][R].$$

Substitute expressions, derived for $[\text{Int}(i, 0, 0)]$ and $[\text{Int}(i, j, 0)]$ in Eq. (4):

$$d[RAFT(i, 0)]/dt = \frac{2}{3} k_{a1}[RAFT(0,0)][R] - \frac{1}{3} k_{a2}[RAFT(i, 0)][R]. \qquad (14)$$

Since in the end it was found that constant of chain transfer to low-molecular RAFT-agent C_{tr1} is equal to divided to constant k_p coefficient before expression $[RAFT(0,0)][R]$ in the balance differential equation for $[RAFT(0,0)]$, from Eq. (14) for constant of chain transfer to macromolecular RAFT-agent, we obtain the next expression:

$$C_{tr2} = \frac{1}{3} \frac{k_{a2}}{k_p}.$$

From the last equation we obtain an expression for constant k_{a2}, which based on the following calculation:

$$k_{a2} = 3 C_{tr2} k_p, L \cdot mol^{-1} s^{-1}. \qquad (15)$$

In research [4] on the base of styrene and DBTK, macromolecular RAFT-agent was synthesized, thereafter with a view to experimentally determine constant C_{tr2}, polymerization of styrene was performed with the use of the latter. In the course of experiment, it may be supposed that constant C_{tr2} depends on monomer and macromolecular RAFT-agent conversions by analogy with equation (12). As a result directly from the experimentally measured monomer and macromolecular RAFT-agent conversions, value C_{tr2} was derived, equal to 1000. On the ground of the same considerations as for that of C_{tr1}, we assume independence of constant C_{tr2} on temperature.

8.2.2.4   RATE CONSTANTS OF INTERMEDIATES FRAGMENTATION, TERMINATION BETWEEN RADICALS AND TERMINATION BETWEEN RADICALS AND INTERMEDIATES

In research [4] it was shown, that RAFT-polymerization rate is determined by this equation:

$$(W_0 / W)^2 = 1 + \frac{k_{t2}}{k_{t1}} K[RAFT(0,0)]_0 + \frac{k_{t3}}{k_{t1}} K^2[RAFT(0,0)]_0^2,$$

where W_0 and W – polymerization rate in the absence and presence of RAFT-agent, respectively, s-¹; K – constant of equilibrium (III), L·mol⁻¹; k_{t3} – constant of termination between two intermediates reaction rate, L·mol⁻¹·s⁻¹ [11].

For initiated AIBN styrene polymerization in DBTC's presence at 80°C, it was shown that intermediates quadratic termination wouldn't be implemented and RAFT-polymerization rate was determined by equation [4]:

$$(W_0 / W)^2 = 1 + 8[RAFT(0,0)]_0.$$

Since $k_{t2}/k_{t1} \approx 1$, then at 80°C $K = 8\,L{\cdot}mol^{-1}$ [4]. In order to find dependence of constant K on temperature, we made research of polymerization kinetics at 60°C. It was found, that the results of kinetic measurements well rectify in coordinates $(W_0 / W)^2 = f([RAFT(0,0)]_0)$. At 60°C, $K = 345$ $L{\cdot}mol^{-1}$ was obtained. Finally dependence of equilibrium constant on temperature has been determined in the form of Vant–Goff's equation:

$$K = 4.85 \cdot 10^{-27}\, e^{22123/T}, L \cdot mol^{-1}. \tag{16}$$

In compliance with the equilibrium (III), the constant is equal to

$$K = \frac{k_{a2}}{3k_f}, L \cdot mol^{-1}.$$

Hence, reactions of intermediates fragmentation rate constant will be as such:

$$k_f = \frac{k_{a2}}{3K}, s^{-1}. \tag{17}$$

The reactions of intermediates fragmentation rate constant was built into the model in the form of dependence (Eq. 17) considering Eqs. (15) and (16).

As it has been noted above, ratio k_{t2}/k_{t1} equals approximately to one, therefore it will be taken, that $k_{t2} \approx k_{t1}$ [4]. For description of gel-effect, dependence as a function of monomer conversion C_M and temperature T (K) [12] was applied:

$$k_{t2} \approx k_{t1} \approx 1.255 \cdot 10^9 e^{-844/T} e^{-2(A_1 C_M + A_2 C_M^2 + A_3 C_M^3)}, L \cdot mol^{-1} s^{-1},$$

where $A_1 = 2.57 - 5.05 \cdot 10^{-3} T$; $A_2 = 9.56 - 1.76 \cdot 10^{-2} T$; $A_3 = -3.03 + 7.85 \cdot 10^{-3} T$.

8.2.2.5 RATE CONSTANT FOR CHAIN GROWTH

We made our choice on temperature dependence of the rate constant for chain growth that was derived on the ground of method of polymerization, being initiated by pulse laser radiation [13]:

$$k_p = 4.27 \cdot 10^7 e^{-3910/T}, L \cdot mol^{-1} s^{-1}. \tag{18}$$

8.2.3 MODEL'S ADEQUACY

The results of polystyrene molecular-mass properties calculations by the introduced mathematical model are presented in Fig. 1. Mathematical model of styrene RAFT-polymerization in the presence of trithiocarbonates, taking into account the radicals and intermediates cross termination, adequately describes the experimental data that prove the process mechanism, built in the model.

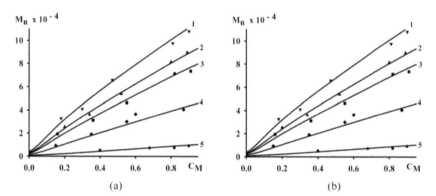

FIGURE 1 Dependence of number-average molecular mass (a) and polydispersity index (b) on monomer conversion for being initiated by AIBN ($[I]_0$=0.01 mol·L^{-1}) styrene bulk RAFT-polymerization at 60°C in the presence of DBTC (lines – estimation by model; points – experiment): $[RAFT(0,0)]_0$ = 0.005 mol·L^{-1} (1), 0.007 (2), 0.0087 (3), 0.0174 (4), 0.087 (5).

Due to adequacy of the model realization at numerical experiment it became possible to determine the influence of process controlling factors on polystyrene molecular-mass properties.

8.2.4 NUMERICAL EXPERIMENT

Research of influence of the process controlling factors on molecular-mass properties of polystyrene, synthesized by RAFT-polymerization method in the presence of AIBN and DBTC, was made in the range of initial concentrations of: initiator – 0–0.1 mol·L^{-1}, monomer – 4.35–8.7 mol·L^{-1}, DBTC – 0.001–0.1 mol·L^{-1}; and at temperatures – 60–120°C.

8.2.4.1   THE INFLUENCE OF AIBN INITIAL CONCENTRATION BY NUMERICAL EXPERIMENT

It was set forth that generally in same other conditions, with increase of AIBN initial concentration number-average, the molecular mass of polystyrene decreases (Fig. 2). At all used RAFT-agent initial concentrations, there is a linear or close to linear growth of number- average molecular mass of polystyrene with monomer conversion. This means that even the lowest RAFT-agent initial concentrations affect the process of radical polymerization. It should be noted that at high RAFT-agent initial concentrations (Fig. 3) the change of AIBN initial concentration practically doesn't have any influence on number-average molecular mass of polystyrene. But at increased temperatures, in case of high AIBN initial concentration, it is comparable to high RAFT-agent initial concentration; polystyrene molecular mass would be slightly decreased due to thermal initiation.

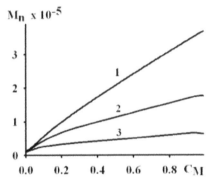

FIGURE 2   Dependence of number-average molecular mass M_n on monomer conversion C_M (60°C) $[M]_0 = 6.1$ mol·L^{-1}, $[RAFT(0, 0)]_0 = 0.001$ mol·L^{-1}, $[I]_0 = 0.001$ mol·L^{-1} (1), 0.01 (2), 0.1 (3).

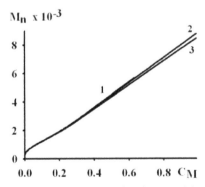

FIGURE 3   Dependence of number-average molecular mass M_n on monomer conversion C_M (60°C) $[M]_0 = 8.7$ mol·L^{-1}, $[RAFT(0, 0)]_0 = 0.1$ mol·L^{-1}, $[I]_0 = 0.001$ mol·L^{-1} (1), 0.01 (2), 0.1 (3).

Since the main product of styrene RAFT-polymerization process, proceeding in the presence of trithiocarbonates, is a narrow-dispersed high-molecular RAFT-agent (marked in kinetic scheme as RAFT(i, j)), which is formed as a result of reversible chain transfer, and widely-dispersed (minimal polydispersity – 1.5) polymer, forming by the radicals quadratic termination, so common polydispersity index of synthesizing product is their ratio. In a broad sense, with increase of AIBN initial concentration, the part of widely-dispersed polymer, which is formed as a result of the radicals quadratic termination, increase in mixture, thereafter general polydispersity index of synthesizing product increases.

However, at high temperatures this regularity can be discontinued – at low RAFT-agent initial concentrations the increase of AIBN initial concentration leads to a decrease of polydispersity index (Fig. 4, curves 3 and 4). This can be related only thereto that at high temperatures thermal initiation and elementary reactions rate constants play an important role, depending on temperature, chain growth and radicals quadratic termination reaction rate constants, monomer initial concentration in a complicated way [6]. Such complicated dependence makes it difficult to analyze the influence of thermal initiation role in process kinetics, therefore the expected width of MMD of polymer, which is expected to be synthesized at high temperatures, can be estimated in every specific case in the frame of the developed theoretical regularities.

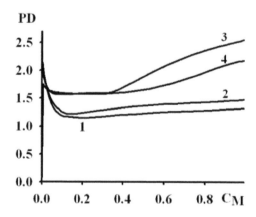

FIGURE 4   Dependence of polydispersity index PD on monomer conversion C_M (120°C) $[M]_0 = 8.7$ mol·L^{-1}, $[RAFT(0, 0)]_0 = 0.001$ mol·L^{-1}, $[I]_0 = 0$ mol·L^{-1} (1), 0.001 (2), 0.01 (3), 0.1 (4)

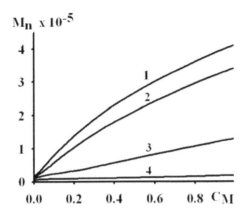

FIGURE 5   Dependence of number-average molecular mass M_n on monomer conversion C_M (120°C) $[M]_0 = 8.7$ mol·L^{-1}, $[RAFT(0, 0)]_0 = 0.001$ mol·L^{-1}, $[I]_0 = 0$ mol·L^{-1} (1), 0.001 (2), 0.01 (3), 0.1 (4)

Special attention shall be drawn to the fact that for practical objectives, realization of RAFT-polymerization process without an initiator is of great concern. In all cases at high temperatures as the result of styrene RAFT-polymerization implementation in the presence of RAFT-agent without AIBN, more high-molecular (Fig. 5) and more narrow-dispersed polymer (Fig. 4, curve 1) is built-up than in the presence of AIBN (Fig. 4, curves 2–4).

8.2.4.2   THE INFLUENCE OF MONOMER INITIAL CONCENTRATION BY NUMERICAL EXPERIMENT

In other identical conditions, the decrease of monomer initial concentration reduces the number-average molecular mass of polymer. Polydispersity index doesn't practically depend on monomer initial concentration.

8.2.4.3   INFLUENCE OF RAFT-AGENT INITIAL CONCENTRATION BY NUMERICAL EXPERIMENT

In other identical conditions, increase of RAFT-agent initial concentration reduces the number-average molecular mass and polydispersity index of polymer.

8.2.4.4   THE INFLUENCE OF TEMPERATURE BY NUMERICAL EXPERIMENT

Generally, in other identical conditions, the increase of temperature leads to a decrease of number-average molecular mass of polystyrene (Fig. 6a). Thus, polydispersity index increases (Fig. 6b). If RAFT-agent initial concentration greatly exceeds AIBN initial concentration, then the temperature practically doesn't influence the molecular-mass properties of polystyrene.

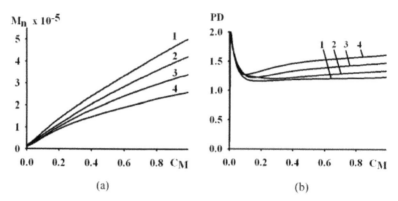

(a)   (b)

FIGURE 6   Dependence of number-average molecular mass M_n (a) and polydispersity index PD (b) on monomer conversion C_M $[I]_0 = 0.001$ mol·L^{-1}, $[M]_0 = 8.7$ mol·L^{-1}, [RAFT(0, 0)]$_0 = 0.001$ mol·L^{-1}, T = 60°C (1), 90 (2), 120 (3), 150 (4).

8.3   CONCLUSION

The kinetic model developed in this research allows an adequate description of molecular-mass properties of polystyrene, obtained by controlled radical polymerization, which proceeds by reversible chain transfer mechanism and accompanied by "addition-fragmentation." This means, that the model can be used for development of technological applications of styrene RAFT-polymerization in the presence of trithiocarbonates.

Researches were supported by Russian Foundation for Basic Research (project. no. 12-03-97050-r_povolzh'e_a).

KEYWORDS

- **controlled radical polymerization**
- **dibenzyltritiocarbonate**
- **mathematical modeling**
- **polystyrene**
- **reversible addition-fragmentation chain transfer**

REFERENCES

1. Matyjaszewski, K.: Controlled/Living Radical Polymerization: Progress in ATRP, D.C.: American Chemical Society, Washington (2009).
2. Matyjaszewski, K.: Controlled/Living Radical Polymerization: Progress in RAFT, DT, NMP and OMRP, D.C.: American Chemical Society, Washington (2009).

3. Barner–Kowollik, C.: Handbook of RAFT Polymerization, Wiley–VCH Verlag GmbH, Weinheim (2008).
4. Chernikova, E.V., Terpugova, P.S., Garina, E.S., Golubev, V.B.: Controlled radical polymerization of styrene and n-butyl acrylate mediated by tritiocarbonates. Polymer Science, Vol. 49(A), №2, 108 (2007).
5. Stephen, Z.D. Cheng: Handbook of Thermal Analysis and Calorimetry. Volume 3 – Applications to Polymers and Plastics: New York, Elsevier, (2002).
6. LI, I., Nordon, I., Irzhak, V.I., Polymer Science, 47, 1063 (2005).
7. Zetterlund, P.B., Perrier S., Macromolecules, 44, 1340 (2011).
8. Biesenberger, J.A., Sebastian D.H.: Principles of polymerization engineering, John Wiley & Sons Inc., New York (1983).
9. Chong, Y.K., Krstina, J., Le, T.P.T., Moad, G., Postma, A., Rizzardo, E., Thang, S.H., Macromolecules, 36, 2256 (2003).
10. Goto, A., Sato, K., Tsujii, Y., Fukuda, T., Moad, G., Rizzardo, E., Thang, S.H., Macromolecules, 34, 402 (2001).
11. Kwak, Y., Goto, A., Fukuda, T., Macromolecules, 37, 1219 (2004).
12. Hui, A.W., Hamielec, A.E., J. Appl. Polym. Sci, 16, 749 (1972).
13. Li, D., Hutchinson, R.A., Macromolecular Rapid Communications, 28, 1213 (2007).

CHAPTER 9

A STUDY ON PHYSICAL PROPERTIES OF COMPOSITES BASED ON EPOXY RESIN

J. ANELI, O. MUKBANIANI, E. MARKARASHVILI, G. E. ZAIKOV, and E. KLODZINSKA

CONTENTS

9.1   INTRODUCTION

Ultimate strength, softening temperature, and water absorption of the polymer composites based on epoxy resin (type ED-20) with unmodified and/or modified by tetraethoxysilane (TEOS) mineral diatomite are described. Comparison of experimental results obtained for investigated composites shows that ones containing modified filler have the better technical parameters mentioned above than composites with unmodified filler at corresponding loading. Experimentally is shown that the composites containing binary fillers diatomite and andesite at definite ratio of them possess the optimal characteristics – so called synergistic effect. Experimental *results are explained in terms of structural peculiarities of polymer composites.*

In recent time the mineral fillers attract attention as active filling agents in polymer composites [1, 2]. Thanks to these fillers many properties of the composites are improved -increases the durability and rigidity, decrease the shrinkage during hardening process and water absorption, improves thermal stability, fire proof and dielectric properties and finally the price of composites becomes cheaper [3–5]. At the same time it must be noted that the mineral fillers at high content lead to some impair of different physical properties of composites. Therefore the attention of the scientists is attracted to substances, which would be remove mentioned leaks. It is known that silicon organic substances (both low and high molecular) reveal hydrophobic properties, high elasticity and durability in wide range of filling and temperatures [6, 7].

The purpose of presented work is the investigation of effect of modify by TEOS of the mineral –diatomite as main filler and same mineral with andesite (binary filler) on some physical properties of composites based on epoxy resin.

9.2   BASIC PART

Mineral diatomite as a filler was used. The organic solvents were purified by drying and distillation. The purity of starting compounds was controlled by an LKhM-8–MD gas liquid chromatography; phase SKTF-100 (10%, the NAW chromosorb, carrier gas He, 2m column). FTIR spectra were recorded on a Jasco FTIR-4200 device.

The silanization reaction of diatomite surface with TEOS was carried out by means of three-necked flask supplied with mechanical mixer, thermometer and dropping funnel. For obtaining of modified by 3 mass % diatomite to a solution of 50 g grind finely diatomite in 80 ml anhydrous toluene the toluene solution of 1.5 g (0.0072 mole) TEOS in 5 ml toluene was added. The reaction mixture was heated at the boiling temperature of used solvent toluene. Than the solid reaction product was filtrated, the solvents (toluene and ethyl alcohol) were eliminated and the reaction product was dried up to constant mass in vacuum. Other product modified by 5% tetraethoxysilane was produced via the same method.

Following parameters were defined for obtained composites: ultimate strength (on the stretching apparatus of type "Instron"), softening temperature (Vica method), density and water absorption (at saving of the corresponding standards).

9.3 RESULTS AND DISCUSSION

High temperature condensation reaction between diatomite and TEOS from the one side and between andesite and same modifier from the other one was carried out in toluene solution (~38%). The masses of TEOS were 3 and 5% from the mass of filler. The reaction systems were heated at the solvent boiling temperature (~110°C) during 5–6 hours by stirring. The reaction proceeds according to the following scheme:

$$
\begin{array}{l}
\text{–OH} \\
\text{–OH} \;+\; \text{Si(OC}_2\text{H}_5)_4 \xrightarrow[-\text{C}_2\text{H}_5\text{OH}]{\text{T}^0\text{C}} \\
\text{–OH}
\end{array}
\qquad
\begin{array}{l}
\text{–O-Si(OC}_2\text{H}_5)_3 \\
\text{–O-Si(OC}_2\text{H}_5)_2\text{-O}\wedge\wedge \\
\text{–OH}
\end{array}
$$

The direction of reaction defined by FTIR spectra analysis shown that after reaction between mineral surface hydroxyl, $-\text{OSi(OEt)}_3$ and the $-\text{OSi(OEt)}_2\text{O}-$ groups are formed on the mineral particles surface.

In the FTIR spectra of modified diatomite one can observe absorption bands characteristic for asymmetrical valence oscillation for linear $^\circ\text{Si–O–Si}^\circ$ bonds at 1030 cm^{-1}. In the spectra one can see absorption bands characteristic for valence oscillation of $^\circ\text{Si–O–C}^\circ$ bonds at 1150 cm^{-1} and for $^\circ\text{C–H}$ bonds at 2950–3000 cm^{-1}. One can see also broadened absorption bands characteristic for unassociated hydroxyl groups.

On the basis of modified diatomite and epoxy resin (of type ED-20) the polymer composites with different content of filler were obtained after careful wet mixing of components in mixer. After the blends with hardening agent (polyethylene-polyamine) were placed to the cylindrical forms (in accordance with standards ISO) for hardening, at room temperature, during 24 h. The samples hardened later were exposed to temperature treatment at 120°C during 4 h.

The concentration of powder diatomite (average diameter up to 50 micron) was changed in the range 10–60 mass %.

The curves on the Fig. 1 show that at increasing of filler (diatomite) concentration in the composites the density of materials essentially depends on both of diatomite contain and on the degree of concentration of modify agent (TEOS). Naturally the decreasing of density of composites at increasing of filler concentration is due to increasing of micro empties because of one's localized in the filler particles (Fig. 1, curve 1). The composites with modified by TEOS diatomite contain less amount of empties as they are filled with modify agent (Fig. 1, curves 2 and 3).

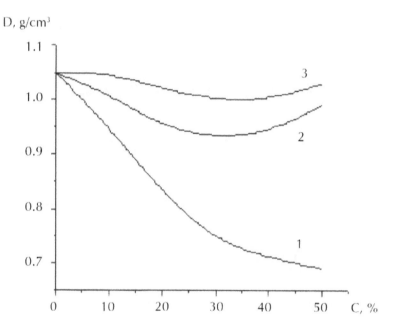

FIGURE 1 Dependence of the density of the composites based on epoxy resin on the concentration of unmodified (1), modified by 3% (2) and 5 mass % (3) tetraethoxysilane diatomite.

The dependence of ultimate strength on the content of diatomite (modified and unmodified) presented on the Fig. 2 shows that it has an extreme character. However the positions of corresponding curves maximums essentially depend on amount of modified agent TEOS. The general view of these dependences is in full conformity with well-known dependence of $\sigma - C$ [8]. The sharing of the maximum of curve for composites containing 5% of modified diatomite from the maximum for the analogous composites containing 3% modifier to some extent is due to increasing of the amount of the bonds between filler particles and macromolecules at increasing of the concentration of the filler.

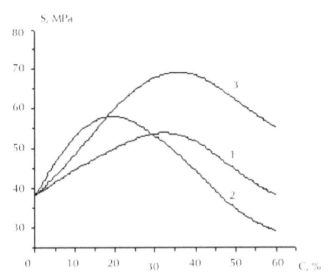

FIGURE 2   Dependence of ultimate strength of the composites based on ED-20 with unmodified (1) and modified by 3 (2), and 5 mass % (3) TEOS diatomite.

Investigation of composites softening temperature was carried out by apparatus of Vica method. Figure 3 shows the temperature dependence of the indentor deepening to the mass of the sample for composites with fixed (20 mass %) concentration of unmodified and modified by TEOS

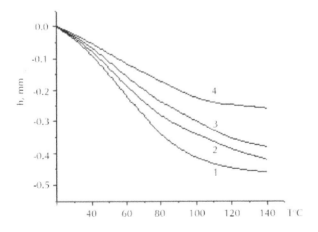

FIGURE 3   Temperature dependence of the indentor deepening in the sample for composites containing 0 (1), 20 mass % (2), 20 mass % modified by 3% TEOS (3), 20 mass % modified by 5% TEOS (4) diatomite.

Based on character of curves on the Fig. 3 it may be proposed that the composites containing diatomite modified by TEOS possesses thermo-stability higher than in case of analogous composites with unmodified filler. Probably the presence of increased interactions between macromolecules and filler particles due to modify agent leads to increasing of thermo-stability of composites with modified diatomite.

Effect of silane modifier on the investigated polymer composites reveals also in the water absorption. In accordance with Fig. 4 this parameter is increased at increasing of filler contain. However, if the composites contain the diatomite modified by TEOS this dependence becomes weak.

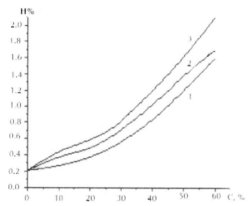

FIGURE 4 Dependence of the water-absorption on the concentration of filler in the composites based on epoxy resin containing diatomite modified by 5% (1) and 3% (2) tetraethoxysilane and unmodified (3) one.

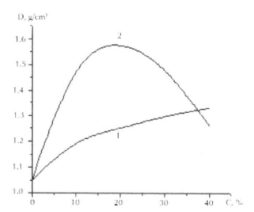

FIGURE 5 Dependence of the density on the concentration of diatomite in binary fillers with andesite. (1) unmodified and modified by 5% tetraethoxysilane (2) fillers for composites based on epoxy resin. Full concentration of binary filler in composites 50 mass %.

There were conducted the investigation of binary fillers on the properties of the composites with same polymer basis (ED-20). Two types of minerals diatomite and andesite with different ratios were used as fillers. It was interesting to establish effect both of ratio of the fillers and effect of modifier TEOS on the same properties of the polymer composites investigated above.

The curves presented on the Fig. 5 show the effect of modify agent TEOS on the dependence of the density of composites containing the binary filler diatomite and andesite on ratio of lasts when the total content of fillers is 50 mass % to which the maximal ultimate strength corresponds. The maximum of noted effect corresponds to composite, filler ratio diatomite/andesite in which is about 20/30. Probably microstructure of such composite corresponds to optimal distribution of filler particles in the polymer matrix at minimal inner energy of statistical equilibration, at which the concentration of empties is minimal because of dense disposition of the composite components. It is known that such structures consists minimal amount both of micro and macro structural defects [8].

Such approach to microstructure of composites with optimal ratio of the composite ingredients allows supposing that these composites would be possessed high mechanical properties, thermo-stability and low water-absorption. Moreover, the composites with same concentrations of the fillers modified by TEOS possess all the noted above properties better than ones for composites with unmodified by TEOS binary fillers, which may be proposed early (Figs. 6–8). Indeed the curves on the Figs. 6–8 show that the maximal ultimate strength, thermo-stability and simultaneously hydrophobicity correspond to composites with same ratio of fillers to which the maximal density corresponds.

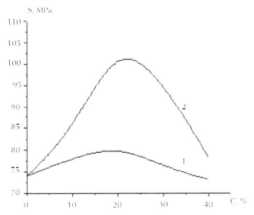

FIGURE 6 Dependence of the ultimate strength on the concentration of diatomite in binary fillers with andesite. (1) unmodified fillers and modified by 5% tetraethoxysilane (2) ones for composites based on epoxy resin Full concentration of binary filler in composites 50 mass %.

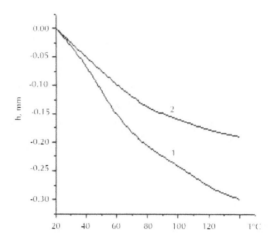

FIGURE 7 Thermo-stability of composites with binary fillers at ratio diatomite/andesite = 20/30.

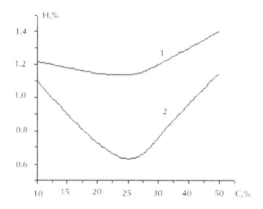

FIGURE 8 Dependence of the water-absorption of composites based on epoxy resin on the concentration of diatomite in binary fillers with andesite. (1) – unmodified and modified by 5% tetraethoxysilane (2) fillers. Total concentration of binary fillers in composites 50 mass %.

The obtained experimental results may be explained in terms of composite structure peculiarities. Silane molecules displaced on the surface of diatomite and andesite particles lead to activation of them and participate in chemical reactions between active groups of TEOS (hydroxyl) and homopolymer (epoxy group). Silane molecules create the "buffer" zones between filler and the homopolymer.

This phenomenon may be one of the reasons of increasing of strengthening of composites in comparison with composites containing unmodified fillers. The composites with modified diatomite display more high compatibility of the components than in case of same composites with unmodified filler. The modified filler has more strong contact with polymer matrix (thanks to silane modifier) than unmodified diatomite. Therefore mechanical stresses formed in composites by stretching or compressing forces absorb effectively by relatively soft silane phases, i.e. the development of micro defects in carbon chain polymer matrix of composite districts and finishes in silane part of material the rigidity of which decreases.

The structural peculiarities of composites display also in thermo-mechanical properties of the materials. It is clear that softening of composites with modified by TEOS composites begins at relatively high temperatures. This phenomenon is in good correlation with corresponding composite mechanical strength. Of course the modified filler has more strong interactions (thanks to modifier) with epoxy polymer molecules, than unmodified filler.

The amplified competition of the filler particles with macromolecules by TEOS displays well also on the characteristics of water absorption. In general loosening of micro-structure because of micro empty areas is due to the increasing of filler content. Formation of such defects in the microstructure of composite promotes the water absorption processes. Water absorption of composites with modified diatomite is lower than that for one with unmodified filler to some extent. The decreasing of water absorption of composites containing silane compound is result of hydrophobic properties of ones.

Composites with binary fillers possess so called synergistic effect- non-additive increasing of technical characteristics of composites at containing of fillers with definite ratio of them, which is due to creation of the dens distribution of ingredients in composites.

9.4 CONCLUSION

Comparison of the density, ultimate strength, softening temperature and water absorption for polymer composites based on epoxy resin and unmodified and modified by tetraethoxysilane mineral fillers diatomite and andesite leads to conclusion that modify agent stipulates the formation of heterogeneous structures with higher compatibility of ingredients and consequently to enhancing of noted above technical characteristics.

KEYWORDS

- polymer composite
- epoxy resin
- modified filler
- ultimate strength
- softening temperature
- water absorption
- synergistic effect of fillers

REFERENCES

1. Katz, H.S., Milevski, J.V. Handbook of Fillers for Plastics, RAPRA, 1987.
2. Mareri, P., Bastrole, S., Broda, N., Crespi, A. Composites Science and Technology, 1998, 58(5), pp. 747–755.
3. Tolonen, H., Sjolind, S. Mechanics of composite materials, 1996, 31(4), pp. 317–322.
4. Rothon, S.: Particulate filled polymer composites, RAPRA, NY, 2003, 205 p.
5. Lou, J., Harinath, V. Journal of Materials Processing Technology. 2004, 152(2), pp.185–193.
6. Khananashvili, L.M., Mukbaniani, O.V., Zaikov, G.E. Monograph, New Concepts in Polymer Science, "Elementorganic Monomers: Technology, Properties, Applications". Printed in Netherlands, VSP, Utrecht, (2006).
7. Aneli, J.N., Khananashvili, L.M., Zaikov, G.E. Structuring and conductivity of polymer composites. Nova Sci.Publ., New–York, 1998. 326 p.
8. Zelenev, Y.V., Bartenev, G.M. Physics of Polymers. M. Visshaya Shkola, 1978. 432 p. (in Russian).

CHAPTER 10

SYNTHESIS, STRUCTURAL PROPERTIES, DEVELOPMENT AND APPLICATIONS OF METAL-ORGANIC FRAMEWORKS IN TEXTILE

M. HASANZADEH and B. HADAVI MOGHADAM

CONTENTS

10.1 INTRODUCTION

Metal-organic frameworks (MOFs) have received increasing attention in recent years as a new class of nanoporous materials. These crystalline compounds basically consist of metal ions linked by organic bridging ligands. They have found wide range of applications including gas storage, gas separation, catalysis, luminescence, and drug delivery due to their large pore sizes, high porosity, high surface areas, and wide range of pore sizes and topologies. Textile application is one of the areas MOFs started to appear recently. Interesting chemical and physical properties of MOFs make them promising candidates for future developments in textile applications. This short review intends to introduce recent progress in application of MOFs in the field of textile engineering and some of the key advances that have been made in it.

Recently the application of nanostructured materials has garnered attention, due to their interesting chemical and physical properties. Application of nanostructured materials on the solid substrate such as fibers brings new properties to the final textile product [1]. Metal-organic frameworks (MOFs) are one of the most recognized nanoporous materials, which can be widely used for modification of fibers. These relatively crystalline materials consist of metal ions or clusters (named secondary building units, SBUs) interconnected by organic molecules called linkers, which can possess one, two or three dimensional structures [2–10]. They have received a great deal of attention, and the increase in the number of publications related to MOFs in the past decade is remarkable (Fig. 1).

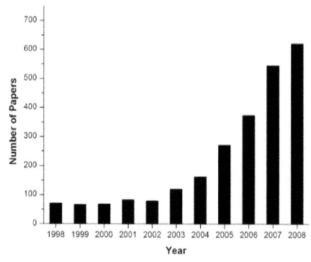

FIGURE 1   Number of publications on MOFs over the past decade, showing the increasing research interest in this topic.

These materials possess a wide array of potential applications in many scientific and industrial fields, including gas storage [11,12], molecular separation [13], catalysis [14], drug delivery [15], sensing [16], and others. This is due to the unique combination of high porosity, very large surface areas, accessible pore volume, wide range of pore sizes and topologies, chemical stability, and infinite number of possible structures [17,18].

Although other well-known solid materials such as zeolites and active carbon also show large surface area and nanoporosity, MOFs have some new and distinct advantages. The most basic difference of MOFs and their inorganic counterparts (e.g., zeolites) is the chemical composition and absence of an inaccessible volume (called dead volume) in MOFs [10]. This feature offers the highest value of surface area and porosities in MOFs materials [19]. Another difference between MOFs and other well-known nanoporous materials such as zeolites and carbon nanotubes is the ability to tune the structure and functionality of MOFs directly during synthesis [17].

The first report of MOFs dates back to 1990, when Robson introduced a design concept to the construction of 3D MOFs using appropriate molecular building blocks and metal ions. Following the seminal work, several experiments were developed in this field such as work from Yaghi and O'Keeffe [20].

In this review, synthesis and structural properties of MOFs are summarized and some of the key advances that have been made in the application of these nanoporous materials in textile fibers are highlighted.

10.2   SYNTHESIS OF MOFS

MOFs are typically synthesized under mild temperature (up to 200°C) by combination of organic linkers and metal ions (Fig. 2) in solvothermal reaction [2, 21].

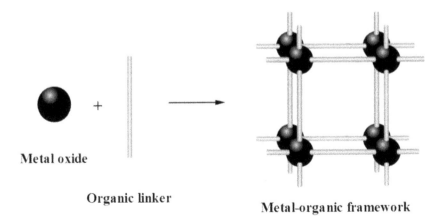

Metal oxide

Organic linker

Metal-organic framework

FIGURE 2   Formation of metal organic frameworks.

Recent studies have shown that the character of the MOF depends on many parameters including characteristics of the ligand (bond angles, ligand length, bulkiness, chirality, etc.), solubility of the reactants in the solvent, concentration of organic link and metal salt, solvent polarity, the pH of solution, ionic strength of the medium, temperature and pressure [2, 21].

In addition to this synthesis method, several different methodologies are described in the literature such as ball-milling technique, microwave irradiation, and ultrasonic approach [22].

Post-synthetic modification (PSM) of MOFs opens up further chemical reactions to decorate the frameworks with molecules or functional groups that might not be achieved by conventional synthesis. In situations that presence of a certain functional group on a ligand prevents the formation of the targeted MOF, it is necessary to first form a MOF with the desired topology, and then add the functional group to the framework [2].

10.3   STRUCTURE AND PROPERTIES OF MOFS

When considering the structure of MOFs, it is useful to recognize the secondary building units (SBUs), for understanding and predicting topologies of structures [3]. Figure 3 shows the examples of some SBUs that are commonly occurring in metal carboxylate MOFs. Figure 3(a–c) illustrates inorganic SBUs include the square paddlewheel, the octahedral basic zinc acetate cluster, and the trigonal prismatic oxo-centered trimer, respectively. These SBUs are usually reticulated into MOFs by linking the carboxylate carbons with organic units [3]. Examples of organic SBUs are also shown in Fig. 3(d–f).

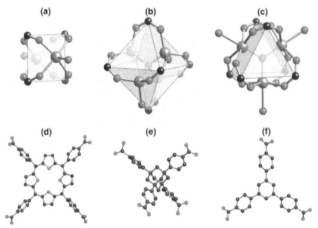

FIGURE 3   Structural representations of some SBUs, including (a–c) inorganic, and (b–f) organic SBUs. (Metals are shown as blue spheres, carbon as black spheres, oxygen as red spheres, nitrogen as green spheres).

It should be noted that the geometry of the SBU is dependent on not only the structure of the ligand and type of metal utilized, but also the metal to ligand ratio, the solvent, and the source of anions to balance the charge of the metal ion [2].

A large number of MOFs have been synthesized and reported by researchers to date. Isoreticular metal-organic frameworks (IRMOFs) denoted as IRMOF-n (n = 1 through 7, 8, 10, 12, 14, and 16) are one of the most widely studied MOFs in the literature. These compounds possess cubic framework structures in which each member shares the same cubic topology [3, 21]. Figure 4 shows the structure of IRMOF-1 (MOF-5) as simplest member of IRMOF series.

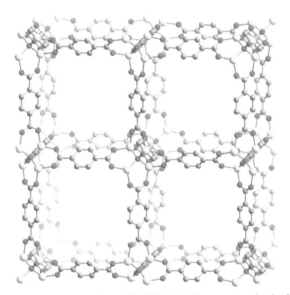

FIGURE 4   Structural representation of IRMOF-1 (Yellow, grey, and red spheres represent Zn, C, and O atoms, respectively).

10.4   APPLICATION OF MOFS IN TEXTILES

10.4.1   INTRODUCTION

There are many methods of surface modification, among which nanostructure based modifications have created a new approach for many applications in recent years. Although MOFs are one of the most promising nanostructured materials for modification of textile fibers, only a few examples have been reported to data. In this section, the first part focuses on application of MOFs in nanofibers and the second part is concerned with modifications of ordinary textile fiber with these nanoporous materials.

10.4.2 NANOFIBERS

Nanofibrous materials can be made by using the electrospinning process. Electro-spinning process involves three main components including syringe filled with a polymer solution, a high voltage supplier to provide the required electric force for stretching the liquid jet, and a grounded collection plate to hold the nanofiber mat. The charged polymer solution forms a liquid jet that is drawn towards a grounded collection plate. During the jet movement to the collector, the solvent evaporates and dry fibers deposited as randomly oriented structure on the surface of a collector [23–28]. The schematic illustration of conventional electrospinning setup is shown in Fig. 5.

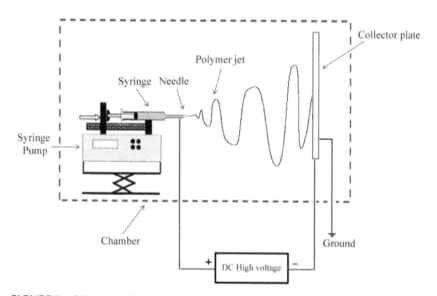

FIGURE 5   Schematic illustration of electrospinning set up.

At the present time, synthesis and fabrication of functional nanofibers represent one of the most interesting fields of nanoresearch. Combining the advanced structural features of metal-organic frameworks with the fabrication technique may generate new functionalized nanofibers for more multiple purposes.

While there has been great interest in the preparation of nanofibers, the studies on metal-organic polymers are rare. In the most recent investigation in this field, the growth of MOF (MIL-47) on electrospun polyacrylonitrile (PAN) mat was studied using in situ microwave irradiation [18]. MIL-47 consists of vanadium cations associated to six oxygen atoms, forming chains connected by terephthalate linkers (Fig. 6).

It should be mentioned that the conversion of nitrile to carboxylic acid groups is necessary for the MOF growth on the PAN nanofibers surface.

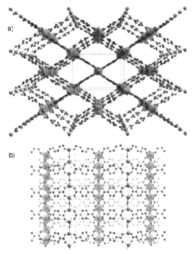

FIGURE 6   MIL-47 metal-organic framework structure: view along the b axis (a) and along the c axis (b).

The crystal morphology of MIL-47 grown on the electrospun fibers illustrated that after only 5 s, the polymer surface was partially covered with small agglomerates of MOF particles. With increasing irradiation time, the agglomerates grew as elongated anisotropic structures (Fig. 7) [18].

FIGURE 7   SEM micrograph of MIL-47 coated PAN substrate prepared from electrospun nanofibers as a function of irradiation time: (a) 5 s, (b) 30 s, (c) 3 min, and (d) 6 min.

It is known that the synthesis of desirable metal-organic polymers is one of the most important factors for the success of the fabrication of metal-organic nanofibers [29]. Among several novel microporous metal-organic polymers, only a few of them have been fabricated into metal-organic fibers.

For example, new acentric metal-organic framework was synthesized and fabricated into nanofibers using electrospinning process [29]. The two dimensional network structure of synthesized MOF is shown in Fig. 8. For this purpose, MOF was dissolved in water or DMF and saturated MOF solution was used for electrospinning. They studied the diameter and morphology of the nanofibers using an optical microscope and a scanning electron microscope (Fig. 9). This fiber display diameters range from 60 nm to 4 μm.

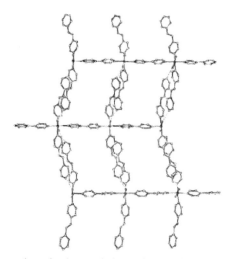

FIGURE 8   Representation of polymer chains and network structure of MOF.

FIGURE 9   SEM micrograph of electrospun nanofiber.

In 2011, Kaskel et al. [30], reported the use of electrospinning process for the immobilization of MOF particles in fibers. They used HKUST-1 and MIL-100(Fe) as MOF particles, which are stable during the electrospinning process from a suspension. Electrospun polymer fibers containing up to 80 wt% MOF particles were achieved and exhibit a total accessible inner surface area. It was found that HKUST-1/PAN gives a spider web-like network of the fibers with MOF particles like trapped flies in it, while HKUST-1/PS results in a pearl necklace-like alignment of the crystallites on the fibers with relatively low loadings.

10.4.3 ORDINARY TEXTILE FIBERS

Some examples of modification of fibers with metal-organic frameworks have verified successful. For instance, in the study on the growth of $Cu_3(BTC)_2$ (also known as HKUST-1, BTC=1,3,5-benzenetricarboxylate) MOF nanostructure on silk fiber under ultrasound irradiation, it was demonstrated that the silk fibers containing $Cu_3(BTC)_2$ MOF exhibited high antibacterial activity against the gram-negative bacterial strain *E. coli* and the gram-positive strain *S. aureus* [1]. The structure and SEM micrograph of $Cu_3(BTC)_2$ MOF is shown in Fig. 10.

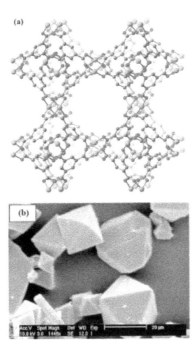

FIGURE 10 (a) The unit cell structure and (b) SEM micrograph of the Cu3(BTC)2 metal-organic framework. (Green, grey, and red spheres represent Cu, C, and O atoms, respectively).

Cu$_3$(BTC)$_2$ MOF has a large pore volume, between 62% and 72% of the total volume, and a cubic structure consists of three mutually perpendicular channels [32].

The formation mechanism of Cu$_3$(BTC)$_2$ nanoparticles upon silk fiber is illustrated in Fig. 11. It is found that formation of Cu$_3$(BTC)$_2$ MOF on silk fiber surface was increased in presence of ultrasound irradiation. In addition, increasing the concentration cause an increase in antimicrobial activity [1]. Figure 12 shows the SEM micrograph of Cu$_3$(BTC)$_2$ MOF on silk surface.

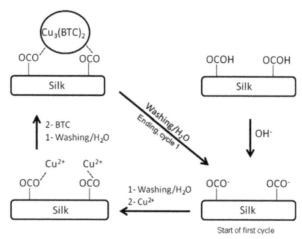

FIGURE 11 Schematic representation of the formation mechanism of Cu$_3$(BTC)$_2$ nanoparticles upon silk fiber.

FIGURE 12 SEM micrograph of Cu$_3$(BTC)$_2$ crystals on silk fibers.

The FT–IR spectra of the pure silk yarn and silk yarn containing MOF (CuBTC–Silk) are shown in Fig. 13. Owing to the reduction of the C=O bond, which is caused by the coordination of oxygen to the Cu^{2+} metal center (Fig. 11), the stretching frequency of the C=O bond was shifted to lower wavenumbers (1654 cm^{-1}) in comparison with the free silk (1664 cm^{-1}) after chelation [1].

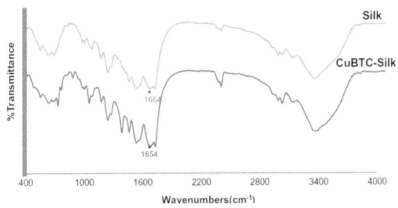

FIGURE 13   FT–IR spectra of the pure silk yarn and silk yarn containing $Cu_3(BTC)_2$.

In another study, $Cu_3(BTC)_2$ was synthesized in the presence of pulp fibers of different qualities [33]. The following pulp samples were used: a bleached and an unbleached kraft pulp, and chemithermomechanical pulp (CTMP).

All three samples differed in their residual lignin content. Indeed, owing to the different chemical composition of samples, different results regarding the degree of coverage were expected. The content of $Cu_3(BTC)_2$ in pulp samples, k-number, and single point BET surface area are shown in Table 1. k-number of pulp samples, which is indicates the lignin content indirectly, was determined by consumption of a Sulfuric permanganate solution of the selected pulp sample [33].

TABLE 1   Some characteristics of the pulp samples.

Pulp sample	MOF content^a (wt.%)	k-number^b	Surface area^c (m^2 g^{-1})
CTMP	19.95	114.5	314
Unbleached kraft pulp	10.69	27.6	165
Bleached kraft pulp	0	0.3	10

^{a}Determined by thermogravimetric analysis.
^{b}Determined according to ISO 302.
^{c}Single point BET surface area calculated at p/p0=0.3 bar.

It is found that CTMP fibers showed the highest lignin residue and largest BET surface area. As shown in the SEM micrograph (Fig. 14), the crystals are regularly distributed on the fiber surface. The unbleached kraft pulp sample provides a slightly lower content of MOF crystals and BET surface area with 165 m^2 g^{-1}. Moreover, no crystals adhered to the bleached kraft pulp, which was almost free of any lignin.

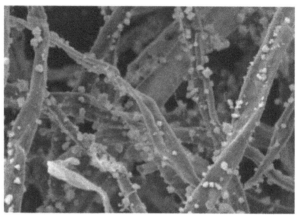

FIGURE 14   SEM micrograph of $Cu_3(BTC)_2$ crystals on the CTMP fibers.

10.5 CONCLUSION

New review on feasibility and application of several kinds of metal-organic frameworks on different substrate including nanofiber and ordinary fiber was investigated. Based on the researcher's results, the following conclusions can be drawn:

1. Metal-organic frameworks (MOF), as new class of nanoporous materials, can be used for modification of textile fibers.
2. These nanostructured materials have many exciting characteristics such as large pore sizes, high porosity, high surface areas, and wide range of pore sizes and topologies.
3. Although tremendous progress has been made in the potential applications of MOFs during past decade, only a few investigations have reported in textile engineering fields.
4. Morphological properties of the MOF/fiber composites were defined; the most advantageous, particle size distribution was shown.
5. It is concluded that the MOFs/fiber composite would be good candidates for many technological applications, such as gas separation, hydrogen storage, sensor, and others.

KEYWORDS

- development
- metal-organic frameworks
- structural properties
- synthesis
- textiles

REFERENCES

1. Abbasi, A.R., Akhbari, K., Morsali, A.: Dense coating of surface mounted CuBTC metal-organic framework nanostructures on silk fibers, prepared by layer-by-layer method under ultrasound irradiation with antibacterial activity. *Ultrasonics Sonochemistry*, 19, 846–852 (2012).

2. Kuppler, R.J., Timmons, D.J., Fang, Q.-R., Li, J.-R., Makal, T.A., Young, M.D., Yuan, D., Zhao, D., Zhuang, W., Zhou, H.-C.: Potential applications of metal-organic frameworks. *Coordination Chemistry Reviews*, 253, 3042–3066 (2009).

3. Rowsell, J.L.C., Yaghi, O.M.: Metal-organic frameworks: A new class of porous materials. *Microporous and Mesoporous Materials*, 73, 3–14 (2004).

4. An, J., Farha, O.K., Hupp, J.T., Pohl, E., Yeh, J.I., Rosi, N.L.: Metal-adeninate vertices for the construction of an exceptionally porous metal-organic framework. *Nature communications*, DOI: 10.1038/ncomms1618, (2012).

5. Morris, W., Taylor, R.E., Dybowski, C., Yaghi, O.M., Garcia–Garibay, M.A.: Framework mobility in the metal-organic framework crystal IRMOF-3: Evidence for aromatic ring and amine rotation. *Journal of Molecular Structure*, 1004, 94–101 (2011).

6. Kepert, C.J.: Metal-organic framework materials. in 'Porous Materials' (eds.: by Bruce, D.W., O'Hare, D. and Walton, R.I.) John Wiley & Sons, Chichester (2011).

7. Rowsell, J.L.C., Yaghi, O.M.: Effects of functionalization, catenation, and variation of the metal oxide and organic linking units on the low-pressure hydrogen adsorption properties of metal-organic frameworks. *Journal of the American Chemical Society*, 128, 1304–1315 (2006).

8. Rowsell, J.L.C., Yaghi, O.M.: Strategies for hydrogen storage in metal-organic frameworks. *Angewandte Chemie International Edition*, 44, 4670–4679 (2005).

9. Farha, O.K., Mulfort, K.L., Thorsness, A.M., Hupp, J.T.: Separating solids: purification of metal-organic framework materials. *Journal of the American Chemical Society*, 130, 8598–8599 (2008).

10. Khoshaman, A.H.: Application of electrospun thin films for supra-molecule based gas sensing. M.Sc. thesis, Simon Fraser University (2011).

11. Murray, L.J., Dinca, M., Long, J.R.: Hydrogen storage in metal-organic frameworks. *Chemical Society Reviews*, 38, 1294–1314 (2009).

12. Collins, D.J., Zhou, H.-C.: Hydrogen storage in metal-organic frameworks. *Journal of Materials Chemistry*, 17, 3154–3160 (2007).

13. Chen, B., Liang, C., Yang, J., Contreras, D.S., Clancy, Y.L., Lobkovsky, E.B., Yaghi, O.M., Dai, S.: A microporous metal-organic framework for gas-chromatographic separation of alkanes. *Angewandte Chemie International Edition*, 45, 1390–1393 (2006).

14. Lee, J.Y., Farha, O.K., Roberts, J., Scheidt, K.A., Nguyen, S.T., Hupp, J.T.: Metal-organic framework materials as catalysts. *Chemical Society Reviews*, 38, 1450–1459 (2009).

15. Huxford, R.C., Rocca, J.D., Lin, W.: Metal-organic frameworks as potential drug carriers. *Current Opinion in Chemical Biology*, 14, 262–268 (2010).

16. Suh, M.P., Cheon, Y.E., Lee, E.Y.: Syntheses and functions of porous metallosupramolecular networks. *Coordination Chemistry Reviews*, 252, 1007–1026 (2008).

17. Keskin, S., Kızılel, S.: Biomedical applications of metal organic frameworks. *Industrial and Engineering Chemistry Research*. 50, 1799–1812 (2011).

18. Centrone, A., Yang, Y., Speakman, S., Bromberg, L., Rutledge, G.C., Hatton, T.A.: Growth of metal-organic frameworks on polymer surfaces. *Journal of the American Chemical Society*, 132, 15687–15691 (2010).

19. Wong–Foy, A.G., Matzger, A.J., Yaghi, O.M.: Exceptional H_2 saturation uptake in microporous metal-organic frameworks. *Journal of the American Chemical Society*, 128, 3494–3495 (2006).

20. Farrusseng, D.: Metal-organic frameworks: Applications from Catalysis to Gas Storage. Wiley–VCH, Weinheim (2011).

21. Rosi, N.L., Eddaoudi, M., Kim, J., O'Keeffe, M., Yaghi, O.M.: Advances in the chemistry of metal-organic frameworks. *CrystEngComm*, 4, 401–404 (2002).

22. Zou, R., Abdel-Fattah, A.I., Xu, H., Zhao, Y., Hickmott, D.D.: Storage and separation applications of nanoporous metal-organic frameworks, *CrystEngComm*, 12, 1337–1353 (2010).

23. Reneker, D.H., Chun, I.: Nanometer diameter fibers of polymer, produced by electrospinning, *Nanotechnology*, 7, 216–223 (1996).

24. Shin, Y.M., Hohman, M.M., Brenner, M.P., Rutledge, G.C.: Experimental characterization of electrospinning: The electrically forced jet and instabilities. *Polymer*, 42, 9955–9967 (2001).

25. Reneker, D.H., Yarin, A.L., Fong, H., Koombhongse, S.: Bending instability of electrically charged liquid jets of polymer solutions in electrospinning, *Journal of Applied Physics*, 87, 4531–4547 (2000).

26. Zhang, S., Shim, W.S., Kim, J.: Design of ultra-fine nonwovens via electrospinning of Nylon 6: Spinning parameters and filtration efficiency, *Materials and Design*, 30, 3659–3666 (2009).

27. Yördem, O.S., Papila, M., Menceloğlu, Y.Z.: Effects of electrospinning parameters on polyacrylonitrile nanofiber diameter: An investigation by response surface methodology. *Materials and Design*, 29, 34–44 (2008).

28. Chronakis, I.S.: Novel nanocomposites and nanoceramics based on polymer nanofibers using electrospinning process–A review. *Journal of Materials Processing Technology*, 167, 283–293 (2005).

29. Lu, J.Y., Runnels, K.A., Norman, C.: A new metal-organic polymer with large grid acentric structure created by unbalanced inclusion species and its electrospun nanofibers. *Inorganic Chemistry*, 40, 4516–4517 (2001).

30. Rose, M., Böhringer, B., Jolly, M., Fischer, R., Kaskel, S.: MOF processing by electrospinning for functional textiles. *Advanced Engineering Materials*, 13, 356–360 (2011).
31. Basu, S., Maes, M., Cano–Odena, A., Alaerts, L., De Vos, D.E., Vankelecom, I.F.J.: Solvent resistant nanofiltration (SRNF) membranes based on metal-organic frameworks. *Journal of Membrane Science*, 344, 190–198 (2009).
32. Hopkins, J.B.: Infrared spectroscopy of H_2 trapped in metal organic frameworks. B.A. Thesis, Oberlin College Honors (2009).
33. Küsgens, P., Siegle, S., Kaskel, S.: Crystal growth of the metal-organic framework $Cu_3(BTC)_2$ on the surface of pulp fibers. *Advanced Engineering Materials*, 11, 93–95 (2009).

CHAPTER 11

THERMAL BEHAVIOR AND IONIC CONDUCTIVITY OF THE PEO/PAC AND PEO/PAC BLENDS

AMIRAH HASHIFUDIN, SIM LAI HAR, CHAN CHIN HAN,
HANS WERNER KAMMER, and SITI NOR HAFIZA MOHD YUSOFF

CONTENTS

11.1 INTRODUCTION

Solution casting technique is employed to prepare the poly(ethylene oxide) (PEO)/ polyacrylate (PAc) blends. Thermal behavior and ionic conductivity of the PEO/ PAc and PEO/PAc blends added with $LiClO_4$ were investigated using differential scanning calorimetry (DSC) and impedance spectroscopy (IS), respectively. Observations of a single composition-dependent glass transition temperature (T_g) which agrees closely with that calculated using the Fox equation, coupled with successive suppression of the melting temperature (T_m) and crystallinity of PEO with ascending PAc content, affirm the miscibility of the two constituents in the blend. The conductivity of salt-free PEO is enhanced with the addition of ≤ 25 wt% of PAc due to the reduced crystallinity of PEO in the blend. The T_g values of the blend at all compositions under study increase with the addition of $LiClO_4$. Ionic conductivity of the salt-added blend increases with increasing salt concentration. The amorphous phase of PEO forms the percolating pathway in the homogeneous $PEO/PAc/LiClO_4$ blends as blends with PEO content ≥ 25 wt% (PEO/PAc 75/25 blend) records slightly higher σ values at $LiClO_4$ concentration $Y > 0.02$. Enhancement in ionic conductivity in the blend is probably the result of increase charge carrier density and ionic dynamic of the PEO macromolecular chain.

The rapid development in advanced electrochemical and micro-ionic devices has attracted extensive research on polymer electrolytes, with the hope of applying these electrolytes in new generation high performance rechargeable batteries [1–5]. Over the last three decades, poly(ethylene oxide) (PEO) remained to be the focus in most of the researches on solid state batteries because of its strong solvating capability of wide variety of inorganic salt and its low glass transition temperature (T_g) [6–9].

It is well documented that the amorphous phase of PEO forms the percolating pathway for fast ion transport in PEO-salt system [10, 11]. Glass transition temperature results obtained in the previous studies [12–14] verified that no isotropic dispersion of Li^+ ion in different phases of a blend is demonstrated for both the immiscible blends of PEO/epoxidized natural rubber (ENR) and PEO/polyacrylate (PAc) with the addition of $LiClO_4$, instead, the Li^+ ion has a higher solubility in the amorphous phase of PEO as compared to ENR or PAc, respectively. Besides, $LiClO_4$ is found to be more soluble in the amorphous PAc than in ENR when equal amount of the salt is added to the immiscible PEO/ENR and PEO/ PAc systems of the same blend composition [12–14]. Furthermore, with the addition of salt, the T_gs of PAc in the $PEO/PAc/LiClO_4$ blend are raised to the range of 26–42°C at which conductivity of the blend is measured. Therefore, higher charge density in the PEO amorphous phase of the heterogeneous $PEO/ENR/LiClO_4$ blend accounts for the higher ionic conductivity of the blend as compared to that in the $PEO/LiClO_4$ system [12]. On the contrary, due to a reasonable amount of the salt being locked in the glassy PAc, the reduction in charge density in the PEO amorphous phase for the immiscible PEO/PAc/LiClO$_4$ blend causes the conductivity of the blend to be lower than that of the PEO/LiClO$_4$ system [15].

In the blend preparation of PEO/PAc and PEO/PAc/LiClO$_4$ electrolyte films described in the previous study [15], solution cast free standing film was dried in a vacuum oven for 48 h at 50°C. Calorimetric analysis using differential scanning calorimetry (DSC) shows that both the salt-free and the salt-added blend systems are immiscible marked by the presence of two T_gs and a relatively constant PEO crystallinity (X^*) with increasing PAc content. However, miscible PEO/PAc and PEO/PAc/LiClO$_4$ blends are obtained in the present work when the solution cast free standing film was heated at 80°C (above the melting point of PEO) for 2 h under nitrogen atmosphere before vacuum dried for another 24 h at 50°C.

A brief description of the thermal procedure used in the preparation of the homogeneous PEO/PAc blend with and without addition of the inorganic salt, LiClO$_4$ is presented. Miscibility of the two polymer components of the blend was investigated by thermal analysis using DSC. The glass transition temperature (T_g) as well as the apparent melting temperature (T_m) and the crystallinity of the as-prepared samples of the blend were studied as functions of compositions of the blend. The effect of phase behavior on the conductivity properties of selected compositions of the blend incorporated with different salt contents is discussed here.

11.2   EXPERIMENTAL

11.2.1   MATERIALS

PEO with viscosity-average molecular weight (M_v) = 3×10^5 g mol^{-1}, was purchased from Aldrich Chemical Company and used after purification. PAc, a random copolymer with weight-average molecular weight (M_w) = 1.7x10^5 g mol^{-1} estimated by gel permeation chromatography, was supplied by the Chemistry Department, Faculty of Science, University of Malaya [16, 17]. Anhydrous LiClO$_4$ with purity ≥ 99% (Acrōs Organics) was vacuum dried for 24 h at 120°C prior to application. Methanol (Fisher Scientific, Leicestershire, UK) dehydrated by molecular sieves with pore diameter of 3Å (Merck, Darmstadt, Germany) was the common solvent used for both the salt-free and the salt-added PEO/PAc blends.

11.2.2   PREPARATION OF BLENDS

Free standing films of PEO/PAc and PEO/PAc/LiClO$_4$ blends were prepared by solution casting technique. Different compositions of the 4% w/w stock solutions of PEO and PAc were mixed while for the salt-added blends, different amount in mass of LiClO$_4$ (Y_s) were added to the polymer solutions. The mixtures were stirred at 50°C for 24 h and cast into Teflon dish, left to dry overnight in a fume hood. After drying at 50°C for 24 h in an oven, the thin films were heated to 80°C under nitrogen atmosphere for 2 h. Under this thermal treatment, the components of the PEO/PAc and PEO/PAc/LiClO$_4$ blends have more time to mix, hence, enhances the interactions between the blending components in the salt-free blend and the salt in the salt-added blend resulting in the formation of miscible PEO/PAc and PEO/PAc/LiClO$_4$ blends. After thermal treatment, the blend films were vacuum dried at 50°C for 24 h. The concentration of LiClO$_4$ in the blend is defined as below:

$$Y_s = \frac{\text{mass of salt}}{\text{mass of polymer}}$$

11.2.3 DIFFERENTIAL SCANNING CALORIMETRY

The values of T_g, T_m and enthalpies of fusion (ΔH_m) of as-prepared samples of the blends were performed on TA DSC Q200, calibrated with indium standard under nitrogen atmosphere. The sample was quenched from 30°C to -90°C, annealed at this temperature for 5 min before heating up to 80°C at a rate of 10°C min^{-1}.

11.2.4 IMPEDANCE SPECTROSCOPY

Ionic conductivity (σ) at 30°C of the as-prepared samples of PEO/PAc/LiClO$_4$ blends was determined from ac-impedance measurements using a Hioki 3532–50 Hi–Tester over the frequency range between 50 Hz and 1 MHz. Thin films of the polymer electrolytes were sandwiched between two stainless steel block electrodes with a surface area of 3.142 cm². The bulk resistance (R_b) of the electrolyte were extracted from the impedance spectrum of the sample at the point of intersection between the semicircle and the real impedance axis (Z_r). Ionic conductivity (σ) was calculated from the bulk resistance (R_b) by adopting the equation $\sigma = L/(AR_b)$, where L and A represent, respectively, the thickness and the active area of the electrode.

11.3 RESULTS AND DISCUSSION

11.3.1 THERMAL ANALYSIS

The T_g for neat PAc and PEO extracted from the heating cycle of DSC traces are 16 and −54°C, respectively. A single, composition dependent T_g is observed with increasing PAc added to PEO in the salt-free PEO/PAc blends as shown in Fig. 1, indicating that the salt-free blend is miscible. The monotone dependence of T_g on composition of the binary PEO/PAc blend as depicted in Fig. 1 concurred with values calculated from the Fox equation as given in Eq. (1) [18, 19]. The Fox equation is defined as

$$\frac{1}{T_g} = \frac{W_1}{T_{g1}} + \frac{W_2}{T_{g2}} \tag{1}$$

where W_1, T_{g1} and W_2, T_{g2} refer to the weight fractions and T_gs of PEO and PAc, respectively.

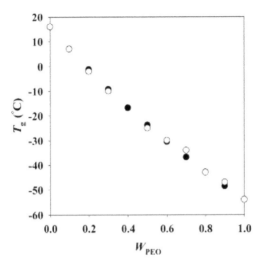

FIGURE 1 Variations of glass transition temperature of PEO/PAc blend as functions of weight fractions of PEO (W_{PEO}). Symbols (•) and (○) denote experimental value and values calculated from Fox equation, respectively.

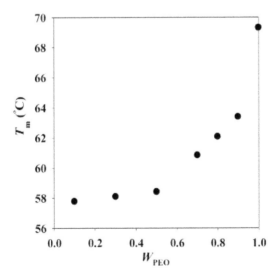

FIGURE 2 Apparent melting temperature of PEO in PEO/PAc blends as functions of weight fraction of PEO.

Figure 2 presents the variations of the apparent melting temperature (T_m) of PEO in the as-prepared samples of the PEO/PAc blend as a function of W_{PEO}. The T_m values, extracted from the first heating cycle of the DSC thermogram, descend rapidly with the addition of 10–60 wt% of PAc, then decrease gradually when more than 60 wt% of PAc is added. From the thermodynamic viewpoint, the miscibility of the PEO/PAc is affirmed by the suppression of the T_m of PEO in the presence of PAc.

PEO crystallinity (X^*) in PEO/PAc blend is calculated from the enthalpy of fusion (ΔH_m) after Eq. (2):

$$X^* = \left\{ \frac{\Delta \zeta_m}{\Delta \zeta_{ref}^o \times \left(W_{PEO} \right)} \right\} \times 100\% \tag{2}$$

where $\Delta H_{ref}^o = 188.3$ J g^{-1} is the enthalpy of fusion of 100% crystalline PEO [20]. Figure 3 demonstrates a successive decrease in the crystallinity of PEO with ascending PAc content in the blend showing the miscibility of the PEO/PAc blend. The presence of PAc in the homogeneous PEO/PAc blend greatly hinders the migration of the crystallizable material of PEO to the crystal growth front, thus, slows down the PEO crystallization rate, leading to a reduction in PEO crystallinity.

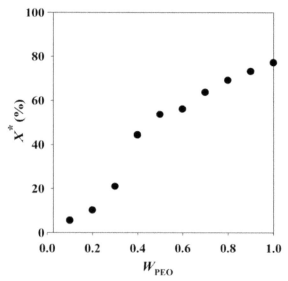

FIGURE 3   Crystallinity (X^*) of PEO in PEO/PAc blend versus W_{PEO}.

The addition of low concentrations of $LiClO_4$ ($Y_s \leq 0.02$) causes a sharp and linear increase in the T_gs of both PEO and PAc as shown in Fig. 4. However, the T_g of PAc remains relatively constant while that of PEO ascends gradually with salt concentration $0.02 < Y_s \leq 0.12$. Similar linear relationship between the T_gs of both PEO and PAc and the initial concentrations of $LiClO_4$ from $Y_s = 0$ to ~ 0.15 were also observed in the previous study by Sim et al. [15]. The difference is that the T_gs of PEO and PAc both level-off at salt concentration $Y_s \geq 0.12$ and 0.15, respectively.

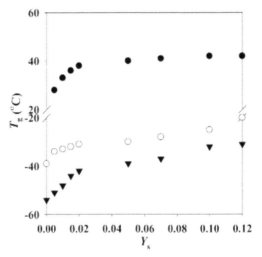

FIGURE 4   Values of T_g of (●) neat PAc, (○) PEO/PAc 75/25 blend and (▼) PEO as functions of concentrations of $LiClO_4$.

It is interesting to note that the T_gs of PEO and PAc recorded at $Y_s = 0.12$ are −31 and 42°C as compared to −37 and 25°C reported in the previous study [13]. The higher T_gs of PEO and PAc obtained at the initial salt concentration in the present work are the result of enhanced solvation of $LiClO_4$ by the polymers as the polymer films were subjected to prolong interaction at temperature above the melting point of PEO (at 80°C for 2 h under N_2 atmosphere) as described in the experimental section. The plateau observed for PAc at $Y_s \geq 0.05$ implies that no further Li^+–PAc coordination is formed whereas coordination of Li^+ ion to the ether oxygen of PEO continues, leading to a gradual increase in its T_g. This shows that the lithium salt has a higher preference in the amorphous phase of PEO than that of PAc. Under this thermal treatment, the T_g values of the homogenous PEO/PAc 75/25 blend are observed to increase linearly with the addition of $LiClO_4$ from $Y_s = 0$ to 0.12.

11.3.2 CONDUCTIVITY ANALYSIS

Ionic conductivity (σ) of the electrolyte films of the salt-free PEO/PAc blends and the blends added with 12 wt% ($Y_s = 0.12$) LiClO$_4$ are presented in Fig. 5. In the salt-free electrolyte system, neat PEO records a conductivity of 1.9×10^{-9} S cm^{-1} as compared to PAc with a lower conductivity of 3.5×10^{-11} S cm^{-1} in close proximity to 1.1×10^{-11} S cm^{-1} reported in the previous study [14]. The conductivity of the PEO/PAc blend as shown in Fig. 5 increases by 2–4 orders of magnitude with salt concentration, $Y_s = 0.12$ especially for blends with PEO content ≥ 50 wt% suggesting that the amorphous phase of PEO in the blend forms the percolating pathway for ion transport.

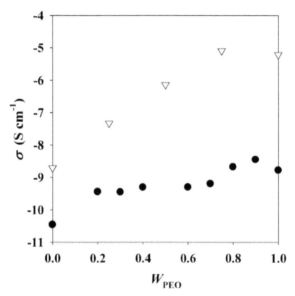

FIGURE 5 Ionic conductivity versus weight fractions of PEO in PEO/PAc blend with ($\bullet$) $Y_s = 0$ and (∇) $Y_s = 0.12$.

Both the salt-free and the salt-added PEO/PAc blends with PEO content   75 wt% attained conductivity higher than that of neat PEO. On the contrary, none of the compositions in the previous heterogeneous PEO/PAc system reported in [15] recorded higher σ values than neat PEO. The improved conductivity obtained as compared to reference [15] is attributed to the extended time in the thermal treatment during sample preparation, which allowed the polymer and the salt to interact and achieve equilibrium. This thermal treatment not only enables the formation of the miscible PEO/PAc blends but more so enhances conductivity of the neat polymers and their blends.

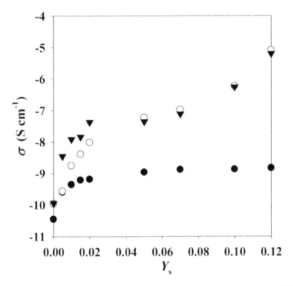

FIGURE 6 Semilogarithm plots of σ versus Y_s, for ($\bullet$) neat PAc, (o) PEO/PAc 75/25 blend and ($\blacktriangledown$) neat PEO.

Figure 6 shows the dependence of conductivity on the concentration of LiClO$_4$ incorporated into 1 gram dry weight of the neat polymers and the PEO/PAc 75/25 blend. With ascending salt content from $Y_S = 0$ to 0.02, there is a marked increase in σ followed by a level-off for PAc but a gradual increase in σ for PEO at $Y_S > 0.02$. This variation trend in σ observed for both the neat polymers correlate closely to that of T_g (c.f. Fig. 4), which in turn reflects the mode of complexation of the Li$^+$ ion to the monomer units of the macromolecular chains. Increase in the charge carrier density leads to higher ion transport through the percolating pathway of PEO whereas ion-transport through the percolation network in PAc appears to reach a maximum at $Y > 0.02$. In addition to the neat polymer, the homogeneous PEO/PAc 75/25 blend records slightly enhanced σ values at $Y_S > 0.02$ implying a synergistic effect on the charge carrier density and ionic dynamic in PEO brought about by the miscible blending of a small amount of PAc which increases the amount of amorphous phase leading to improved ion transport through the percolating pathway.

11.4 CONCLUSIONS

Films of salt-free PEO/PAc blends prepared by solution casting technique are found to be miscible after being subjected to 2 h of thermal treatment at 80°C under nitrogen atmosphere. The presence of a single, composition dependent T_g

and the suppression of the T_m of PEO in the as-prepared PEO-based blend by the addition of PAc affirm the miscibility of the PEO/PAc blend.

Addition of LiClO$_4$ from $Y_s = 0$ to 0.02 causes a marked increase in the T_g and conductivity of neat PEO and neat PAc. Enhancement in the conductivity of the two neat polymers is the result of an increase in the charge carrier density. At $Y_s >$ 0.02, both the T_g and the conductivity of PAc remains constant whereas those for PEO increase gradually. PEO/PAc blends with PEO content $\geq$ 75 wt% with and without the addition of LiClO$_4$ recorded higher σ values than neat PEO. Enhancement in the conductivity of the blend is due to improvement in the ion mobility attributed by an increase in Li$^+$ ion solvation and the addition of PAc increases the amorphous phase volume of PEO, which forms the percolating pathway for ion-transport in the blend system.

ACKNOWLEDGMENT

This study is supported by Fundamental Research Grant from Ministry of Higher Education, Malaysia, FRGS/2/2010/ST/UI/TM/02/4.

KEYWORDS

- conductivity
- electrochemical and micro-ionic devices
- lithium batteries
- PEO/PAc and PEO/PAc blends
- polymer electrolyte
- thermal behavior

REFERENCES

1. Armand, M.B.: Polymer electrolytes. in Annual Reviews Inc. 16, 245–261 (1986).
2. Eds.: Mac Callum, J.R., Vincent, C.A.: Polymer Electrolyte Reviews. Elsevier, Amsterdam (1987/1989).
3. Owen, J.R., Laskar, A.L., Chandra, S.: Superionic solids and solid state electrolytes–Recent trends. Academic Press, New York (1989).
4. Dias, F.B., Plomp, L., Veldhuis, J.B.J.: Trends in polymer electrolytes for secondary lithium batteries. Journal of Power Sources, 88, 169–191 (2000).
5. Yu, X.Y., Xiao, M., Wang, S.J., Zhao, Q.Q., Meng, Y.Z.: Fabrication and characterization of PEO/PPC polymer electrolyte for lithium-ion battery. Journal of Applied Polymer Science, 115, 2718–2722 (2010).
6. Watanabe, M., Oohashi, S.I., Sanui, K., Ogata, N., Ohtaki, Z.: Morphology and ionic conductivity of polymer complexes formed by segmented polyether poly(urethane urea) and lithium perchlorate. Macromolecules, 18, 1945–1950 (1985).

7. Rhodes, C.P., Frech, R., Local structures in crystalline and amorphous phases of diglyme–$LiCF_3SO_3$ and poly(ethylene oxide)–$LiCF_3SO_3$ systems: Implications for the mechanism of ionic transport. Macromolecules, 34, 2660–2666 (2001).
8. Caruso, T., Capoleoni, S., Cazzanelli, E., Agostino, R.G., Villano, P., Passerini, S.: Characterization of PEO–Lithium Triflate polymer electrolytes: Conductivity, DSC and Raman investigations. Ionics, 8, 36–43 (2002).
9. Gitelman, L., Israeli, M., Averbuch, A., Nathan, M., Schuss, Z., Golodnitsky, D.: Modeling and simulation of Li-ion conduction in poly(ethylene oxide). Journal of Computational Physics, 227, 1162–1175 (2007).
10. Berthier, C., Gorecki, W., Minier, M., Armand, M.B., Chabagno, J.M., Rigaud, P.: Microscopic investigation of ionic conductivity in alkali metal salts-poly(ethylene oxide) adducts. Solid State Ionics, 11, 91–95 (1983).
11. Reddy, M.J., Chu, P.P.: Optical microscopy and conductivity of poly(ethylene oxide) complexed with KI salt. Electrochimica Acta, 47, 1189–1196 (2002).
12. Chan, C.H., Kammer, H.W.: Properties of Solid Solutions of Poly(ethylene oxide)/ Epoxidized Natural Rubber Blends and $LiClO_4$. Journal of Applied Polymer Science, 110, 424–432 (2008).
13. Sim, L.H., Chan, C.H., Kammer, H.W.: Selective localization of lithium perchlorate in immiscible blends of poly(ethylene oxide) and epoxidized natural rubber. IEEE conference proceedings: International Conference on Science and Social Research (CSSR 2010). Kuala Lumpur, Malaysia, 499–503 (2011).
14. Sim, L.H., Chan, C.H., Kammer, H.W.: Melting behavior, morphology and conductivity of solid solutions of PEO/PAc blends and $LiClO_4$. Materials Research Innovations, 15, S71–S74 (2011).
15. Sim, L.H., Gan, S.N., Chan, C.H., Kammer, H.W., Yahya, R.: Compatibility and conductivity of $LiClO_4$ free and doped polyacrylate-poly(ethylene oxide) blends. Materials Research Innovations, 13, 278–281 (2009).
16. Gan, S.N.: Water-reducible Acrylic Copolymer Dipping Composition and Rubber Products Coated with Same. PI20055440, Malaysia (2005).
17. Sim, L.H., Gan, S.N., Chan, C.H., Yahya, R.: ATR–FTIR studies on ion interaction of lithium perchlorate in polyacrylate/poly(ethylene oxide) blends. Spectrochimica Acta – Part A: Molecular and Biomolecular Spectroscopy, 76, 287–292 (2010).
18. Fox, T.G.: Influence of diluent and of copolymer composition on the glass temperature of a polymer system. Bull. Am. Phys. Soc., 1, 13, 123–135 (1956).
19. Brostow, W., Chiu, R., Kalogeras, I.M., Vassilikou–Dova, A. Prediction of glass transition temperatures: Binary blends and copolymers. Materials Letters, 62, 3152–3155 (2008).
20. Cimmino, S., Pace, E.D., Martuscelli, E., Silvestre, C.: Syndiotactic polystyrene: crystallization and melting behavior. Polymer, 32, 1080–1083 (1991).

CHAPTER 12

CORRELATION BETWEEN THE STORAGE TIME OF THE NRL AND THE EFFICIENCY OF PMMA GRAFTING TO NR

YOGA SUGAMA SALIM, NUR AZIEMAH ZAINUDIN,
CHIN HAN CHAN, and KAI WENG CHAN

CONTENTS

12.1 INTRODUCTION

Natural rubber latex (NRL) is often stored for a certain period of time before the grafting of poly(methyl methacrylate) (PMMA) to natural rubber (NR). It is well known that the properties of NRL change as a function of storage time. This paper describes the influence of storage time of NRL on its mechanical stability, followed by the effect on grafting efficiency of PMMA onto the NR backbone in NRL. The mechanical stability time (MST) of NRL decreases dramatically after 30 days of storage, implying that there is an increase in volatile fatty acid content leading to lower MST. Quality control tests (i.e., total solid content (TSC), alkalinity, pH, viscosity and MST) show that the alkalinity, viscosity and pH of NRL slightly fluctuate as a function of time in both NRL that fails the MST test and the NRL after addition of potassium oleate. Both NRLs having low and high MST at fixed period of storage time were used for the grafting of PMMA onto the NR in NRL. Correlation between storage time of NRL in low and high MST and the grafting efficiencies was studied. Results suggest that there is no significant change in the grafting efficiency, which ranges from 84–88% for all samples, as well as MMA monomer conversion to PMMA for NRL with low and high MST at fixed storage time. The grafting of PMMA onto NRL backbone is confirmed using Fourier-transform infrared (FTIR) spectroscopy.

Hevea brasiliensis tree has been the object of research for the past few decades. The dry phase of latex [known as natural rubber (NR)] from the *Hevea* tree consists of 93–95 wt% *cis*-1,4-poly(isoprene) [1, 2]. NR possesses excellent physical properties such as high mechanical strength, excellent flexibility, and resistance to impact and tear [3, 4], but the unsaturated and non-polar nature of the chain makes it susceptible to flame, chemicals, solvents, ozone and weather [5, 6]. Investigations of NR modified by graft-copolymerization with a second monomer such as vinyl benzene [7, 8], methyl methacrylate (MMA) [9–11], acrylonitrile [12, 13], phosphonate [12, 14] have been reported. Among these, NR grafted with MMA [methyl-grafted (MG) rubber or NR-*g*-PMMA] has been marketed in Malaysia since 1950s under trade name "Heveaplus MG." There are three grades of MG rubber widely available: MG 30, MG 40 and MG 49 (the number denotes wt% of MMA content in the grafted NR) [9, 15]. The chemical structure of NR-*g*-PMMA is illustrated in Fig. 1.

In practical applications, NR-*g*-PMMA serves as adhesives [16], reinforcing agents and impact modifiers for thermoplastics [6, 11, 17–19]. Before the synthesis of NR-*g*-PMMA, natural rubber latex (NRL) is often stored for a certain period of time. Santipanusopon and Riyajan [20] studied the influence of ammonia content as a function of storage period of NRL on the properties of NRL including alkalinity, magnesium content and viscosity. They suggest that the magnesium content of NRL decreases while the alkalinity and viscosity of NRL increases as a function of storage time within 45 days. Furthermore, Sasidharan and coworkers [21] aged the NRL up to 300 days, and their properties were monitored at different storage intervals of 0, 20, 40, 60, 120, 180, 240, and 300 days. The MST values of all samples increase during the first 120 days of NRL storage and then begin to decrease gradually. MST is a measure of resistance against mechanical influences and it tends to increase the number of collisions between particles that are likely to coacervate the NRL. This parameter correlates to the volatile fatty acid content in NRL. Influence of storage time on properties of NRL has been numerously reported, but to our best knowledge, none have investigated the influence of storage time on grafting efficiency of PMMA onto NRL backbone. In this study, correlation between the storage time of the NRL and the efficiency of PMMA grafting to NR is discussed.

12.2   EXPERIMENTAL

12.2.1   MATERIALS

NRL with high ammonia content (alkalinity = 250 mEq) containing 60 wt% dried rubber content (DRC) was supplied by Thai Rubber Latex Ltd. (Rayong, Thailand). Reagent grade MMA monomer (purity ~ 99 wt%) was purchased from Rohm GmbH & Co. (Darmstadt, Germany) and used without further purification. The stabilizer oleic acid (Palm Oleo Sdn. Bhd., Malaysia), the activator tetraethylenepentamine (TEPA, purity ~ 99.9 wt%, Hunstman, Woodsland, USA), tetramethyl thiuramdisulphide (TMTD, conc. ~ 50 wt%, Flexy Sys, Belgium), tert-butyl hydroperoxide (TBHP, conc. ~ 70 wt%), and potassium oleate (conc. ~ 20 wt%) were used as received. Other reagents were commercially available in reagent grade and were used without further purification.

12.2.2   STORAGE PERIOD

A total of 12 kg of NRL was kept in 6 containers for 30, 60, 90, 120, 150, and 180 days at 30°C. At the specified storage time, 2 kg NRL was tested for quality control tests (TSC, alkalinity, MST, pH, viscosity), described in Section 12.2.5. Numerical subscripts in NRL refer to the numbers of storage days, for example, NRL_{30} and NRL_{90} refers to 30 days and 90 days of storage period, respectively. For NRL that passed the MST test (MST value of 800 s and above), the NRL would proceed to grafting reaction. However, data of the NRL that passed the MST test is not shown. Instead, we focus on NRL that failed the MST tests to observe the effects of storage period. For NRL that failed the MST tests, two experiments were conducted: (1) the NRL that failed the MST test was added with potassium oleate (coded as NRL_R). Approximately 20 wt% of potassium oleate was added drop-wise into every 1000 g of NRL. After 30 min of stirring at 25°C, the NRL containing potassium oleate was left overnight and the MST value was measured again on the following day. Once the MST value had been adjusted to a minimum specification (800–1000 s), it was then used for the synthesis of NR-g-PMMA. (2) the NRL that failed the MST test was used directly without any treatment for the synthesis of NR-g-PMMA (coded as NRL$_F$). The aim of the latter experiment is to study the effect of low MST on grafting efficiency of PMMA to NR.

12.2.3 SYNTHESIS OF NR-G-PMMA WITH 40 WT% OF PMMA CONTENT

Various quantities of NRL (424 g), distilled water (269 g) and ammonia solution (14.45 wt%, 43 g) were added into 1 L reactor. After 10 min of stirring at 250 rpm at 25°C, roughly 259 g MMA monomer and suitable amount of oleic acid (as a stabilizer) and TBHP were added slowly to the mixture. The mixture was maintained under stirring for 10 min before it was left 2–4 hours to obtain maximum percent grafting. After 2–4 hours, stirring was started again and suitable amount of TEPA (as an activator agent) was fed to the reactor. Addition of TEPA increased the temperature to approximately 65°C within 15–30 min. The mixture was left overnight at 25°C without stirring. Insoluble components of the reacted product (NR-g-PMMA containing ungrafted NR and free PMMA) were removed using 60 μm filter mesh. Appropriate amount of TMTD was added to the filtrated NR-g-PMMA to retard further grafting. The crude NR-g-PMMA latex was subjected to quality control tests before extraction and drying process. The ungrafted NR and free MMA in NR-g-PMMA were removed by Soxhlet extraction. Approximately 2.0 ± 0.5 g of non-purified NR-g-PMMA was extracted using light petroleum ether for 24 hours at 40°C. The residue inside the thimble was dried to a constant weight at 70°C, which was further extracted with acetone for 24 hours at 50°C and dried to a constant weight at 70°C. The flow chart for the analyses of NRL and NR-g-PMMA is shown in Fig. 2.

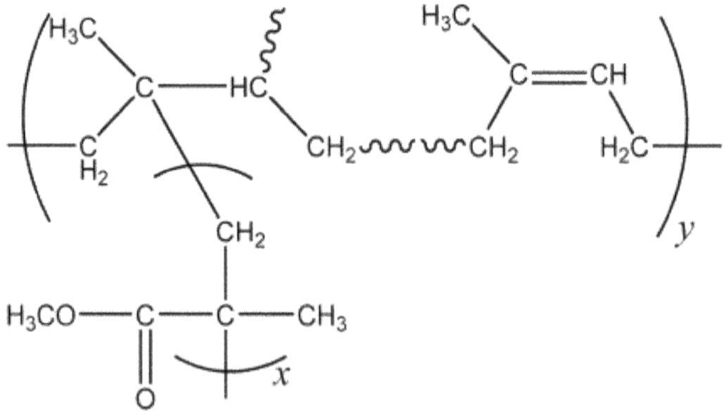

FIGURE 1 Chemical structure of NR-g-PMMA, where x and y refer to the wt% of PMMA and NR, respectively.

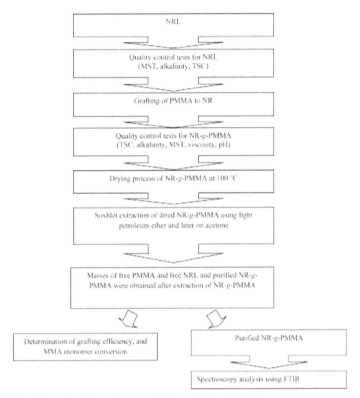

FIGURE 2   Flow chart for analyses of NRL and NR-*g*-PMMA.

12.2.4   FREE NR, FREE PMMA, GRAFTING EFFICIENCY AND PERCENTAGE OF MMA MONOMER CONVERSION

Free NRL and free PMMA can be calculated based on weight differences before and after the extraction interval, as shown in Eqs. (1) and (2). Mass difference between the initial and final mass of NR-*g*-PMMA was then used to calculate grafting efficiency (GE) according to Eq. (3) [6].

$$\text{Free NRL (wt\%)} = \frac{W_1 - W_2}{W_1} \times 100\% \tag{1}$$

$$\text{Free PMMA (wt\%)} = \frac{W_2 - W_3}{W_1} \times 100\% \tag{2}$$

Where, W_1 is the dry weight of non-purified NR-*g*-PMMA, W_2 is the weight of residue inside the thimble after Soxhlet extraction with light petroleum ether, and

W_3 is the weight of residue inside the thimble after Soxhlet extraction by acetone [6].

$$GE\ (\%) = \left[100\% - \text{free NRL (wt\%)} - \text{free PMMA (wt\%)}\right] \qquad (3)$$

Free PMMA was used to calculate the percentage of MMA monomer conversion by using Eq. (4):

$$\text{Monomer conversion (CV\%)} = \left[100\% - \text{free PMMA (wt\%)}\right] \qquad (4)$$

12.2.5 QUALITY CONTROL TESTS

The total solid content (TSC) was determined according to ISO 124:2011. In this test, NRL was placed on a weighing dish and dried at 100°C for 2 hrs. The dried sample was kept in desiccators. It is worth noting that the TSC of neat NRL was diluted from 60% to 55% with ammonium solution for the MST measurement. The dilution of NRL may increase or decrease the stability of NRL, depending on the types of NRL [22]. The TSC values were calculated according to formula given in Eq. (5). The alkalinity was determined according to ISO 125:2003, in which 5 g of NRL in 250 mL conical flask was added to 200 mL of distilled water. Subsequently, 2–3 drops of methyl red (an indicator) was added drop-wise into the mixture, and was titrated with 0.1 N of Sulfuric acid until the indicator turned pink against the white background of a slightly coagulated NRL. Viscosity tests were conducted with a Brookfield DV–I+ viscometer according to ISO 1652:2011. The stirrer of the viscometer was immersed into the NRL after it had been attached to the viscometer. The viscosity value was measured in centipoises (cPs) using a proper speed that allowed the stirrer to rotate until a stable reading was attained. The MST was determined according to ISO 35:2004. The sample was agitated at 14,000 rpm using latex testing machine (Klaxon, Secomak Ltd.) until end point was reached. This is indicated by a visual formation of aggregate and a change in the sound of stirring speed. The sound of agitation becomes loud as the sample becomes thick. All the results of the above-mentioned tests are mean values from two replicates.

$$\text{Total solid content (TSC) (\%)} = \frac{C\text{-}A}{B\text{-}A} \times 100\% \qquad (5)$$

A = Weight of weighing dish (g), B = Weight of weighing dish with NRL (g), C = Weight of weighing dish with dried NR (g).

12.2.6 FTIR CHARACTERIZATION

FTIR sample analysis was carried out using Attenuated Total Reflection (ATR) on Perkin Elmer Spectrum One spectrometer. FTIR spectra were recorded in the transmittance mode over the range of 4000–600 cm^{-1} by averaging 16 scans at maximum resolution of 2 cm^{-1} in all cases.

12.3 RESULTS AND DISCUSSION

12.3.1 MST, TSC AND ALKALINITY OF NRL AS A FUNCTION OF STORAGE TIME

Figure 3 shows the MST tests of neat NRL before and after addition of potassium oleate as a function of storage time. The MST of NRL shows great reduction from 30–180 days of storage. Conversely, Sasidharan and coworkers [21] found that the MST of NRL gradually decreases after 120 days of storage. This trend can be explained by the fact that MST correlates closely to volatile fatty acid content in NRL. Low MST in NRL means higher volatile fatty acid content in the NRL. It is suggested that the increase in volatile fatty acid content of NRL would affect the stability of NRL [22, 23]. The MST value of approximately 800–1000s is set to be the lowest standard specification for the synthesis of NR-g-PMMA; thus in this study, the amount of potassium oleate added to the system must be sufficient to increase the MST value to the expected level. Other quality control tests such as TSC seems to be constant at 61.5 ± 0.2%, while alkalinity slightly fluctuates from 354 to 417 mEq as a function of storage time (Table 1).

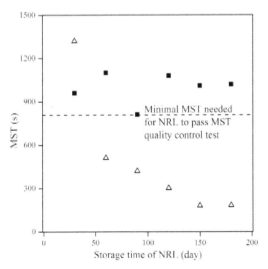

FIGURE 3   MST of NRL aged for 30 to 180 days (Δ: NRL; ■: aged NRL that failed the MST after addition of potassium oleate).

TABLE 1   TSC and alkalinity of NRL as a function of storage time.

Sample Code	TSC (%)	Alkalinity (mEq)
NRL$_{30}$	61.5	417.6
NRL$_{60}$	61.5	390.2
NRL$_{90}$	61.2	409.6
NRL$_{120}$	61.3	358.7
NRL$_{150}$	61.5	354.5
NRL$_{180}$	61.6	407.2

12.3.2   TSC, ALKALINITY, MST, AND VISCOSITY OF NR-G-PMMA LATEX

Tables 2 and 3 summarize the quality control tests and physical appearance of NR-g-PMMA. The former shows the NRL that failed the MST test was used for synthesis without addition of potassium oleate (coded as NRL$_F$), and the latter shows the NRL that passed the MST test after addition of potassium oleate was used for the synthesis of NR-g-PMMA (coded as NRL$_R$). Results show that the TSC and alkalinity values do not differ from each another in NRL$_F$ and NRL$_R$, however the viscosities of NR-g-PMMA, after addition of potassium oleate to the NRL$_R$ increase (Table 3). Higher viscosity of NR-g-PMMA may be caused by the poor colloidal stability of NRL during storage [24] or by the formation of homo-polymer PMMA through crosslinking by hydrogen bonding when potassium oleate is added [20]. Physical appearance shows a thickening of NR-g-PMMA, with a drop in meniscus of the NRL in water indicated after 1800s, in both NRL$_F$ and NRL$_R$ samples. Above 1800s, NR-g-PMMA solidifies and may harm the testing machine. The MST values of NR-g-PMMA investigated in this study are similar to that observed by Kalkornsurapranee et al. [6].

TABLE 2   Quality control tests and physical appearance of NR-g-PMMA with 40wt% of PMMA content for the aged NRL that failed MST test before the synthesis of NR-g-PMMA.

Sample code	NR-g-PMMA					
	TSC (%)	Alkalinity (mEq)	MST (s)	Viscosity (cP s)	pH	Physical appearance of NR-g-PMMA latex after MST test
NRL$_{30}$	—	—	—	—	—	
NRL$_{60/F}$	51.1	348.1	> 1800	155.0	10.6	NR-g-PMMA latex thickens up and it is stable

TABLE 2   *(Continued)*

NRL$_{90/F}$	51.1	372.3	> 1800	68.5	10.6	NR-*g*-PMMA latex thickens up and it is stable
NRL$_{120/F}$	51.0	357.8	> 1800	194.0	10.6	NR-*g*-PMMA latex thickens up and it is stable
NRL$_{150/F}$	51.1	342.9	> 1800	122.5	10.6	NR-*g*-PMMA latex thickens up and it is stable
NRL$_{180/F}$	51.4	376.3	> 1800	158.5	10. 6	NR-*g*-PMMA latex thickens up and it is stable

Subscript 'F' indicates the NRL that failed MST test and was used for the synthesis of NR-*g*-PMMA without addition of potassium oleate, and subscripted numbers indicates the storage time in days.

TABLE 3   Quality control tests and physical appearance of NR-*g*-PMMA after adding potassium oleate in the NRL, which failed MST test before the synthesis of NR-*g*-PMMA.

Sample code	NR-*g*-PMMA					
	TSC (%)	Alkalinity (mEq)	MST (s)	Viscosity (cP s)	pH	Physical appearance of NR-*g*-PMMA latex after MST test
NRL$_{30}$	—	—	—	—	—	
NRL$_{60/R}$	50.5	379.4	> 1800	220.0	10.6	NR-*g*-PMMA latex thickens up and it is stable
NRL$_{90/R}$	51.0	386.8	> 1800	259.0	10.6	NR-*g*-PMMA latex thickens up and it is stable
NRL$_{120/R}$	50.9	401.7	> 1800	157.5	10.6	NR-*g*-PMMA latex thickens up and it is stable
NRL$_{150/R}$	51.3	376.3	> 1800	209.1	10.6	NR-*g*-PMMA latex thickens up and it is stable
NRL$_{180/R}$	51.2	374.6	> 1800	83.5	10. 6	NR-*g*-PMMA latex thickens up and it is stable

Subscript 'R' indicates the NRL that passed MST test after addition of potassium oleate and was used for the synthesis of NR-*g*-PMMA, and subscripted numbers indicates the storage time in days.

TABLE 4 Effect of storage time of NRL on grafting efficiency and monomer conversion of NR-*g*-PMMA.

Sample code	Free NR (wt%)	Free PMMA (wt%)	Grafting efficiency (%)	MMA Conversion (wt%)
NRL$_{30}$	3.3	8.6	88.1	91.4
NRL$_{60/F}$	2.0	9.4	88.7	90.6
NRL$_{90/F}$	3.1	8.9	88.0	91.1
NRL$_{120/F}$	2.0	9.8	88.1	90.2
NRL$_{150/F}$	1.6	10.5	87.9	89.5
NRL$_{180/F}$	1.2	10.3	88.5	89.7
NRL$_{60/R}$	1.7	10.9	87.3	89.1
NRL$_{90/R}$	2.3	10.5	87.2	89.5
NRL$_{120/R}$	2.3	10.7	87.0	89.3
NRL$_{150/R}$	2.3	10.3	87.3	89.7
NRL$_{180/R}$	2.6	11.4	86.0	88.6

12.3.3 FREE PMMA, FREE NR, GRAFTING EFFICIENCY AND MMA MONOMER CONVERSION

The NR-*g*-PMMA products obtained from the polymerization were extracted and characterized to determine free NR, free PMMA, monomer conversion and grafting efficiency. Table 4 shows the effect of storage NRL on grafting efficiency and monomer conversion. The grafting efficiency and monomer conversion was calculated after the synthesis of NR-*g*-PMMA using Eq. (1) to Eq. (4). The grafting efficiencies of all samples range from 86 to 88 wt%, while monomer conversion ranges from 89 to 91 wt%. Free NR and free PMMA in all samples slightly fluctuate after 30 days of storage, with 1.2–3.1 wt% and 8.6–11.4 wt%, respectively. Results strongly suggest that there is no significant difference in NR-*g*-PMMA for samples obtained from different sources and different times.

12.3.4 CHARACTERIZATION OF THE NR-G-PMMA WITH 40 WT% PMMA CONTENT WITH FOURIER TRANSFORM INFRARED

Figure 4 shows representative IR spectra of NR-*g*-PMMA from the NRL that passed the quality control tests after 120 days ageing (sample code NRL$_{120/R}$). All other spectra of NR-*g*-PMMA samples from the storage study show similar adsorption band. The characteristic peaks of the saturated aliphatic sp^3 C–H bonds are observed at 2853 cm^{-1} and 2952 cm^{-1}, which corresponds to $v_{as}(CH_3)$ and $v_s(CH_2)$, respectively. A strong sharp peak located at 1727 cm^{-1} is attributed to the symmetrical stretching mode of C=O (carbonyl). The C=C stretching, CH$_3$ symmetrical deformation, and CH$_2$ twisting modes of NR are observed at 1660, 1376 and 1242 cm^{-1}, respectively. A strong peak due to O–CH$_3$ is located at 1446 cm^{-1}. The doublet peak of C–O stretching mode of PMMA can be observed at 1100–1210 cm^{-1}, with sharp and strong maximum peak at 1147 cm^{-1}. A peak at 987 cm^{-1} observed in the spectrum could be assigned to C–O–C symmetrical of PMMA [23]. FTIR results confirmed the occurrence of grafting polymerization of PMMA onto NRL.

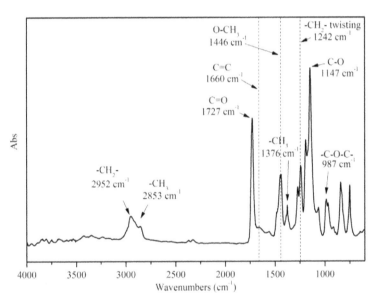

FIGURE 4   FTIR spectrum of NR-g-PMMA with 40 wt% PMMA content for NRL$_{120/R}$.

12.4   CONCLUSIONS

It is shown that the optimum storage time of NRL is approximately 30 days as the MST values drop exponentially after 30 days. Meanwhile, TSC and alkalinity of NRL are not affected by the storage time of NRL. Nevertheless, the viscosity of NR-g-PMMA latex with the aged NRL treated with potassium oleate is higher than that without treatment. The GE and monomer conversion are not affected by the storage period of NRL. Current work focuses on the effect of stabilizers on the colloidal stability of ammonia-preserved NRL.

ACKNOWLEDGMENTS

This work has been supported by Research Intensive Faculty Grant [600–RMI/DANA 5/3/RIF(636/2012)] from University Teknologi MARA (UiTM), Shah Alam, Malaysia. We are grateful for invaluable help from NR division in Synthomer (M) Sdn. Bhd., Kluang, Johor, Malaysia.

KEYWORDS

- effect of stabilizers
- efficiency of PMMA Grafting to NR
- mechanical fatigue limit
- NRL
- physical properties of natural rubber
- storage time

REFERENCES

1. Chen, H.Y. (1962) Determination of *cis*-1,4 and *trans*-1,4 contents of polyisoprenes by high resolution nuclear magnetic resonance. Anal Chem 34, 1793–1795.
2. Whelan, T. (1994) Polymer technology dictionary, Technology and engineering. Chapman & Hall, London.
3. Nasir, M., Teh, G.K. (1988) The effects of various types of crosslinks on the physical properties of natural rubber. Eur Polym J, 24, 733–736.
4. Kongparakul, S., Prasassarakich, P., Rempel, G.L. (2008) Catalytic hydrogenation of methyl methacrylate-*g*-natural rubber (MMA-*g*–NR) in the presence of homogeneous osmium catalyst $OsHCl(CO)(O_2)(PCy_3)_2$. Appl Catal, A., Gen 344, 88–97.
5. Lake, G.J., Lindley, P.B. (1965) The mechanical fatigue limit for rubber. J Appl Polym Sci 9, 1233–1251.
6. Kalkornsurapranee, E., Sahakaro, K., Kasesaman, A., Nakason, C. (2009) From a laboratory to a pilot scale production of natural rubber grafted with PMMA. J Appl Polym Sci 114, 587–597.
7. Minoura, Y., Mori, Y., Imoto, M. (1957) Vinyl polymerization XXI. Polymerization of styrene in the presence of natural rubber. Die Makromolekulare Chemie 24, 205–221.
8. Mays, J.W. (1990) Synthesis of "simple graft" poly(isoprene-*g*-styrene) by anionic polymerization. Polym Bull 23, 247–250.
9. Thiraphattaraphun, L., Kiatkamjornwong, S., Prasassarakich, P., Damronglerd, S. (2001) Natural rubber-*g*-methyl methacrylate/poly(methyl methacrylate) blends. J Appl Polym Sci 81, 428–439.
10. George, V., Britto, I.J., Sebastian, M.S. (2003) Studies on radiation grafting of methyl methacrylate onto natural rubber for improving modulus of latex film. Radiat Phys Chem 66, 367–372.

11. Kalkornsurapranee, E., Sahakaro, K., Kasesaman, A., Nakason, C. (2010) Influence of reaction volume on the propertries of natural rubber-g-methyl methacrylate. J Elastomers Plast 42, 17–34.

12. Arauj'o PHH, Sayer, C., Poco, J.G.R., Giudici, R. (2002) Techniques for reducing residual monomer content in polymers, A review. J Polym Eng Sci 42, 1442–1468.

13. Bhattacharya, A., Misra, B.N. (2004) Grafting, a versetiles means to modify polymers technique, factors and applications. Prog Polym Sci 29, 767–814.

14. Derouet, D., Intharapat, P., Quang, N.T., Gohier, F., Nakason, C. (2008) Graft copolymers of natural rubber and poly(dimethyl (acryloyloxymethyl) phosphonate) (NR-g-PDMAMP) or poly(dimethyl (methacryloyloxyethyl) phosphonate) (NR-g-PDMMEP) from photopolymerization in latex medium. Eur Polym, J., 45, 820–836.

15. Kamisan A.S., Kudin, T.I.T., Ali, A.M.M., Yahya, M.Z.A. (2011) Polymer gel electrolytes based on 49% methyl-grafted natural rubber. Sains Malaysiana 40, 49–54.

16. Rezaifard, A.H., Hodd, K.A., Tod, D.A., Barton, J.M. (1994) Toughening epoxy resins with poly(methyl methacrylate)-grafter-natural rubber, and its use in adhesive formulations. Int, J., Adhes Adhes 14, 153–159.

17. Keskkula, H., Kim, H., Paul, D.R. (2004) Impact modification of styrene-acrylonitrile copolymers by methyl methacrylate grafted rubbers. Polym Eng Sci 30, 1373–1381.

18. Keskkula, H., Paul, D.R., McCreedy, K.M., Henton, D.E. (1987) Methyl methacrylate grafted rubbers as impact modifiers for styrenic polymers. Polym 28, 2063–2069.

19. Charmondusit, K., Seeluangsawat, L. (2009) Recycling of poly(methyl methacrylate) scrap in the styrene–methyl methacrylate copolymer cast sheet process. Resour Conserv Recy 54, 97–103.

20. Santipanusopon, S., Riyajan, S.A. (2009) Effect of field natural rubber latex with different ammonia contents and storage period on physical properties of latex concentrate, stability of skim latex and dipped film. Phys Procedia 2, 127–134.

21. Sasidharan, K.K., Joseph, R., Palaty, S., Gopalakrishnan, K.S., Rajammal, G., Pillai, P.V. (2005) Effect of the vulcanization time and storage on the stability and physical properties of sulphur-prevulcanized natural rubber latex. J Appl Polym Sci 97, 1804–1811.

22. Dawson, H.G. (1949) Mechanical Stability Test for Hevea Latex. Anal Chem 21, 1066–1071.

23. Allen P.W., Merrett, F.M. (1956) Polymerization of methyl methacrylate in polyisoprene solutions. J Polym Sci 22, 193–201.

24. Blackley, D.C. (1966) High polymer Science Lattices (Vol 1 & 2). McLaren & Sons Ltd, New York.

CHAPTER 13

A STUDY ON COMPOSITE POLYMER ELECTROLYTE

TAN WINIE, N. H. A. ROSLI, M. R. AHMAD, R. H. Y. SUBBAN, and C. H. CHAN

CONTENTS

13.1 INTRODUCTION

Hexanoyl chitosan that exhibited solubility in THF was prepared by acyl modification of chitosan. Atactic polystyrene was chosen to blend with hexanoyl chitosan. $LiCF_3SO_3$ was employed as the doping salt. Untreated and HNO_3 treated TiO_2 fillers were dispersed in hexanoyl chitosan-polystyrene–$LiCF_3SO_3$ electrolyte at 4 wt.% concentration. We observed better filler dispersion in the matrix for the acid treated system. The resulting composite electrolyte films were characterized for the electrical and tensile properties. Untreated TiO_2 improved the electrolyte conductivity while HNO_3 treated TiO_2 decreased the conductivity. A model based on interaction between Lewis acid-base sites of TiO_2 with ionic species of $LiCF_3SO_3$ has been proposed to understand the conductivity mechanism brought about by the fillers. The conductivity enhancement by untreated TiO_2 is attributed to the increase in the number and mobility of Li^+ cations. HNO_3-treated TiO_2 decreased the conductivity by decreasing the anionic contribution. An enhancement in the Young's modulus and toughness was observed with the addition of TiO_2 and greater enhancement is found for the treated TiO_2. This is discussed using the percolation concept.

Polymeric electrolytes are the fastest growing and most widely investigated electrolyte system ever since the proposition and recognition of its potential application in solid-state electrochemical systems [1]. Solid polymer electrolyte (SPE) is formed by dissolving an alkali metal salt in a polymer. To date, various combinations of salts and polymers forming polymer-salt complexes have been investigated. A major drawback of these polymer-salt complexes, however, is the low ionic conductivity at ambient temperature.

In general, factors affecting the ionic conductivity of a SPE are amorphous phase, number and mobility of charge carriers. It has been shown that ionic transport takes place only in the amorphous phase [2]. The number and mobility of charge carriers are governed by the interactions between the salt and polymer matrix. Thus, two primary strategies have been adopted. The first one is to suppress crystallinity in polymer system by co-polymerization [3–6], cross-linked polymer networks [7, 8], comb formation [9, 10] and plasticizer addition [11, 12]. In these cases, the conductivity enhancement is achieved by increasing the polymer chain mobility. The second strategy to increase the number of charge carriers is by increasing salt concentration and using highly dissociable salt [13]. However, the mechanical property is often scarified by increasing salt concentration [14]. For practical applications of SPEs in various electrochemical devices, it is important for the SPE to retain good mechanical property.

One promising way to improve both the mechanical and conductivity properties of SPE is the incorporation of inorganic fillers such as TiO_2, SiO_2 and Al_2O_3 to form a kind of composite polymer electrolyte (CPE). Based on those reported in the literature, it is still not clear on the role played by the fillers as the conductivity enhancer. For example, some researchers suggested that the conductivity enhancement in CPE could not be attributed to the enhanced polymer segmental motion as no appreciable change in T_g is observed [15–17]. On the other hand, some studies have shown that T_g is affected due to the addition of fillers [18,19]. Studies by Wieczorek et al. [20] suggested that the conductivity increase is due to the increase in the number of ions. Best and co-workers [16] in their study of some CPEs have suggested that the conductivity enhancement brought about by the fillers does not come from the change in the number of ions but from the increase in the ions mobility. XRD studies on hexanoyl chitosan:polystyrene– $LiCF_3SO_{3\_}TiO_2$ have suggested that the increase in conductivity is attributable to the decrease in percentage of crystallinity [21].

It has been shown that the surface properties of fillers could affect the conductivity performance, but limited attention has been devoted to their influence on the tensile properties. In the present work, we attempt to correlate the surface properties of fillers to the electrical and tensile properties of a CPE based on hexanoyl chitosan:polystyrene- $LiCF_3SO_3$. Surface acid-base investigation of the fillers has been carried out. The variation in conductivity was discussed on the basis of number and mobility of ions. The tensile properties, specified in terms of Young's modulus and toughness were compared and discussed using the percolation concept.

13.2 EXPERIMENTAL

Hexanoyl chitosan that exhibited solubility in tetrahydrofuran (THF) were prepared by acyl modification of chitosan [22]. Polystyrene (M_w of 280,000) used in this work has an atactic chain configuration and commercially available through Sigma–Aldrich. Lithium trifluromethanesulphonate ($LiCF_3SO_3$) with purity >96% from Aldrich was dried for 24 h at 120°C prior to use. TiO_2 having particle size of 30–40 nm was acid-treated by stirring in diluted HNO_3 solution (242 mL, 0.83 v/ v% in de-ionized water) for 8 h at 80°C, rinsed by de-ionized water until the filtrates were neutralized and then dried for 12 h at 100°C. The amounts of the acidic site at the surface of TiO_2 untreated and acid-treated were determined by titration method. They are represented in terms of number of mole of H^+ and were found to be 0.06 and 0.74 mmolg^{-1} for the untreated and acid-treated, respectively. Untreated and HNO_3 treated TiO_2 were then used as the fillers.

To prepare the CPE, required amount of TiO_2 was added to hexanoyl chitosan:polystyrene (90:10) blend and $LiCF_3SO_3$ dissolved in THF. The solutions were stirred at room temperature until TiO_2 particles are dispersed homogeneously before pouring into separate glass Petri dishes. They are left to evaporate at room temperature for the films to form. For the impedance measurement, the film was sandwiched between two stainless steel electrodes. Impedance of the films was measured using HIOKI 3532–50 LCR Hi-tester impedance spectroscopy in the frequency range from 100 Hz to 1 MHz at different temperatures from 273 to 333 K. The ionic conductivity of the film was calculated using the equation

$$\sigma = \frac{t}{R_b A} \tag{1}$$

where t is the thickness of the film and A is the film-electrode contact area. The bulk resistance, R_b was obtained from the complex impedance plot. The dielectric constant, ε_r is related to the measured real, Z_r and imaginary parts, Z_i of impedance as follows:

$$\varepsilon_r = \frac{Z_i}{\omega C_o \left(Z_r^2 + Z_i^2\right)} \tag{2}$$

where $C_o = \varepsilon_o A/t$, ε_o is the permittivity of free space, symbols A and t have their usual meaning. $\omega = 2\pi f$, f being the frequency in Hz.

The Li^+ transference numbers, τ_{Li+} in the CPEs were determined by a combination of ac impedance and dc polarization methods. The film was sandwiched between two lithium metal electrodes and the measurement was performed as described in Ref. [23]. The τ_{Li+} was calculated using the following equation

$$\tau_{Li+} = \frac{I_{ss}(\Delta V - I_o R_o)}{I_o(\Delta V - I_{ss} R_{ss})} \tag{3}$$

where I_o is the initial current, I_{ss} is the steady-state current. ΔV is the applied voltage bias, R_o and R_{ss} is the initial and final interfacial resistance, respectively.

Tensile properties were studied by using Instron 3366 tensile tester. Tensile stress-strain tests were conducted at crosshead speed of 1 mm/min. The films were cut into a 0.5 mm × 7 mm × 25 mm rectangular strip samples. The gauge length was controlled to be 24 mm. The Young's modulus was determined from the slope of the initial linear part of stress-strain curve. The toughness is characterized by the area under the curve.

13.3   RESULTS

Figure 1 shows the temperature dependence of ionic conductivity for hexanoyl chitosan:polystyrene–$LiCF_3SO_3$ without and with TiO_2. Untreated TiO_2 improved the conductivity while HNO_3 treated TiO_2 decreased the electrolyte conductivity. Room temperature conductivity achieved for the untreated and HNO_3 treated TiO_2 system was 2.27 10^{-4} S cm^{-1} and 6.33 10^{-5} S cm^{-1}, respectively. Activation energies for ionic conduction, E_a were obtained from the slope of the plots in Fig. 1 and are summarized in Table 1. E_a is the energy required for an ion to begin migration from one donor site to another. This ion migration results in conduction. It can be observed that high conducting sample exhibits low value of E_a. This indicates that ions in high conducting sample require lower energy to begin migration. The experimental results of τ_{Li+} are presented in Table 1. The increase in Li$^+$ mobility is reflected in the increase value of τ_{Li+}. The τ_{Li+} for the samples differs as follows: HNO$_{3-}$treated < TiO$_2$ free < untreated TiO$_2$.

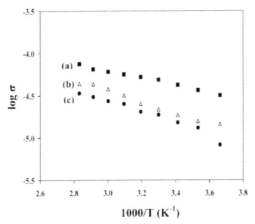

FIGURE 1 Temperature dependence of ionic conductivity for hexanoyl chitosan:polystyrene–LiCF$_3$SO$_3$ with (a) 4 wt% untreated TiO$_2$; (b) TiO$_2$ free and (c) 4 wt% HNO$_{3-}$treated TiO$_2$.

TABLE 1 Room temperature conductivity, activation energy and Li$^+$ transference number for hexanoyl chitosan: polystyrene–LiCF$_3$SO$_3$ electrolyte.

	σ_{RT} (S cm^{-1})	E_a (eV)	τ_{Li+}
TiO$_2$-free	7.21 $'$ 10^{-5}	0.11	0.41
Untreated TiO$_2$	2.27 $'$ 10^{-4}	0.05	0.46
HNO$_3$-treated TiO$_2$	6.33 $'$ 10^{-5}	0.13	0.40

Figure 2 presents the frequency dependence of dielectric constant for electrolyte system without and with TiO$_2$. The dielectric constant is found to increase in the order: TiO$_2$ free < HNO$_{3-}$treated < untreated TiO$_2$. We have deduced that increment in the number of free ions is reflected in the increment in dielectric constant [24].

In our previous work [25], the tensile properties of hexanoyl chitosan is found to increase with addition of polystyrene. The introduction of LiCF$_3$SO$_3$, on the other hand, deteriorated the tensile properties of hexanoyl chitosan:polystyrene blend. The deteriorated tensile properties are overcome with the addition of TiO$_2$. As seen in Table 2, the Young's modulus increases from 11.77 MPa to 21.53

MPa and 22.90 MPa with addition of untreated and HNO$_3$ treated TiO$_2$, respectively. Similar enhancement trend in toughness was observed. This shows that the enhancement in Young modulus and toughness brought about by the treated TiO$_2$ is greater than the untreated TiO$_2$. The HNO$_3$ treatment of TiO$_2$ improves the dispersion of filler particles in polymer matrix. In comparison with the case of untreated TiO$_2$, the particles tend to agglomerate. These observations suggest that the surface state of TiO$_2$ influences the electrical and tensile properties of the present electrolyte system, which will be discussed in the later part of this paper.

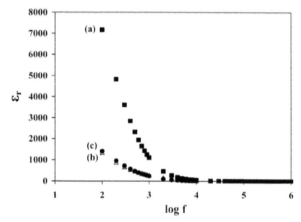

FIGURE 2 Frequency dependence of dielectric constant, ε_r for hexanoyl chitosan:polystyrene–LiCF$_3$SO$_3$ with (a) 4 wt% untreated TiO$_2$; (b) TiO$_2$ free and (c) 4 wt% HNO$_3$ treated TiO$_2$.

TABLE 2 Young's modulus and toughness for hexanoyl chitosan:polystyrene–LiCF$_3$SO$_3$ at 4 wt.% TiO$_2$.

	Young's Modulus (MPa)	Toughness (MPa)
TiO$_2$-free	11.77	0.025
Untreated TiO$_2$	21.53	0.036
HNO$_3$-treated TiO$_2$ (wt.%)	22.90	0.063

13.4 DISCUSSION

The conductivity mechanism by TiO$_2$ fillers in hexanoyl chitosan:polystyrene–LiCF$_3$SO$_3$ system is proposed in Fig. 3. Blend of hexanoyl chitosan and polystyrene is immiscible [25]. FTIR results showed that the Li$^+$ ions of LiCF$_3$SO$_3$ interacted with the donor atoms (nitrogen and oxygen atoms) of hexanoyl chitosan [26] and no interaction between LiCF$_3$SO$_3$ and polystyrene [25] as there is no complexation site for the salt in the structure of polystyrene. DSC results revealed that T_g of hexanoyl chitosan in the blend increases with increasing salt concentration whereas the T_g of polystyrene in the blend remains constant (results not shown here). We thus propose that $LiCF_3SO_3$ salt to be located in the hexanoyl chitosan phase. The hexanoyl chitosan–Li^+ interactions are electrostatic in nature. A Li^+ ion can hop from one donor site to another leaving a vacancy which will be filled by another Li^+ from a neighboring site, as illustrated in Fig. 3a. In the absence of TiO_2, this Li^+ motion is facilitated by the segmental motion of polymer chain.

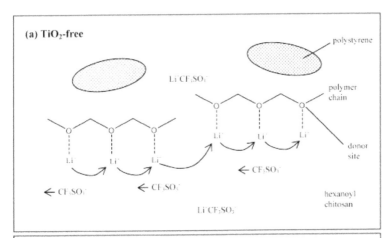

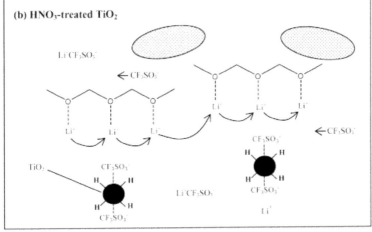

FIGURE 3   *(Continued)*

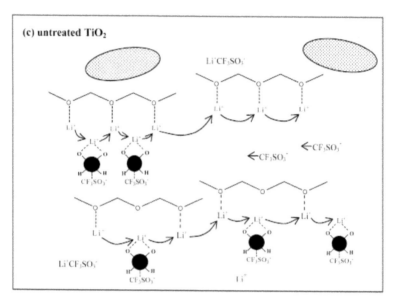

FIGURE 3 Conductivity mechanism in hexanoyl chitosan:polystyrene–LiCF$_3$SO$_3$_TiO$_2$ composite polymer electrolytes: (a) TiO$_2$_free; (b) HNO$_3$_treated TiO$_2$ and (c) untreated TiO$_2$.

No evidence is yet to show that TiO$_2$ fillers are located in the hexanoyl chitosan phase. Relevant studies on localization of TiO$_2$ in immiscible blends of hexanoyl chitosan and polystyrene are underway and are being conducted separately. Conductivity in CPE is due to the free ions from the salt. LiCF$_3$SO$_3$ is located in the hexanoyl chitosan phase. Thus, the conductivity variation as a result of interaction between LiCF$_3$SO$_3$ salt and TiO$_2$ suggests that the filler particles are most probably located in the same polymer phase with LiCF$_3$SO$_3$. Figure 3b shows the electrolyte system with HNO$_3$_treated TiO$_2$. The HNO$_3$_treated TiO$_2$ fillers have more acidic sites (0.74 mmolg^{-1}) as compared to untreated TiO$_2$ (0.06 mmolg^{-1}). The acidic site consists of OH group. For the simplicity of discussion, we denote the treated TiO$_2$ grain surface consists of solely OH groups and are marked with H in the figure. It is shown that anion of salt has larger affinity towards filler surface acid site than cation [27]. CF$_3$SO$_3$_ anions dissociated from the salt are bonded with the H of OH groups. The anions are now immobilized on TiO$_2$ grains, which would, as compared to the filler-free system where anions migration is assisted to some extent by the segmental motion of polymer chain. The cation migration along the polymer chain remains unaffected. Immobilization of anions decreases the anionic contribution to the conductivity.

Conductivity mechanism in the case of untreated TiO_2 is shown in Fig. 3c. Each untreated TiO_2 grain surface is to have equal number of acidic and basic sites. The basic site is due to the oxygen of TiO_2. Function of the acidic site will be compensated by the basic site. Li^+ cations interacted with the O atoms of surface basic groups of TiO_2. The bonds between Li^+ and TiO_2 are similar to the bonds between Li^+ with donor atoms of hexanoyl chitosan i.e. bonds are subjected to breaking and forming during Li^+ transport process. This provides additional conduction pathway for Li^+ cations, which would, otherwise, be moving only along the polymer chain. This improved Li^+ transport is reflected in a higher value of τ_{Li^+}.

In addition, TiO_2 particles may also act as transit sites for Li^+ ions to make several small jumps from one donor site in polymer chain to another. The opportunity to make several small jumps implies that the Li^+ ions need not acquire a lot of energy as is required to make one bigger jump to the next donor site. Thus more Li^+ ions can hop with ease. This explains why the activation energy for ionic conduction is reduced.

The interaction between Lewis acid-base sites of TiO_2 with ionic species of salt helps in dissociation of $LiCF_3SO_3$ and yields greater number of free ions. It is important to note that the number of free ions represented by dielectric constant is indiscriminate of speciation. The conductivity of HNO_3 treated TiO_2 system is lower than that of TiO_2 free system despite the higher number of free ions in HNO_3 treated TiO_2 system (see Fig. 2). This is attributed to the anions immobilization as discussed previously. Anions immobilization decreases the effective average mobility of carrier ions. Slight increase in the number of free ions alone is not sufficient to satisfy the requirement of conductivity enhancement.

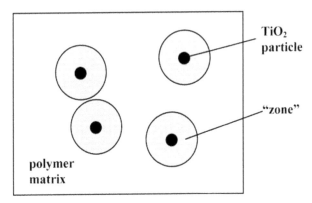

FIGURE 4 "Zone" around each TiO_2 particle.

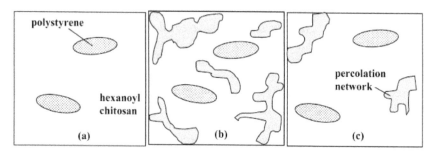

FIGURE 5   Formation of percolation networks within polymer matrix: (a) TiO_2-free; (b) HNO_3-treated TiO_2 and (c) untreated TiO_2.

The role of treated TiO_2 in modulus enhancement is discussed on the basis of the percolation concept proposed by He and Jiang [28]. Each individual filler particle is surrounded by a "zone" as presented in Fig. 4. The distances between well dispersed particles are closer. Hence, these zones tend to overlap, leading to the creation of percolation networks (see Fig. 5). Here, the continuous percolation networks develop more easily as compared to the agglomerated untreated TiO_2 particles at the same filler content. Better filler dispersion thus, leads to a greater modulus (i.e. stiffer sample film). Similar observation is reported by Svehlova and Poloucek [29].

It is generally accepted that the conductivity enhancement in CPE is attributed to the enhanced cation transport on the filler grain boundaries. However, this conductivity enhancement as a result of percolation mechanism is inconsistent with our findings. The percolation networks formed in HNO_3 treated TiO_2 system did not help in increasing the conductivity. This is because the OH functional groups on TiO_2 grains did not form the cation conduction pathway. Instead, they act as the anion trapper.

The toughness of a material is a measure of its ability to resist crack growth. The principle of the toughening process in the presence of TiO_2 is proposed as illustrated in Fig. 6. The tip of the crack is blunted by the percolation network formed by the fillers and stopped from spreading. Aggregation of untreated filler results in poorer development of continuous percolation networks in the polymer matrix. This increases the probability of crack growth (see Fig. 6c). This explains the treated TiO_2 sample film is tougher than the untreated TiO_2 film.

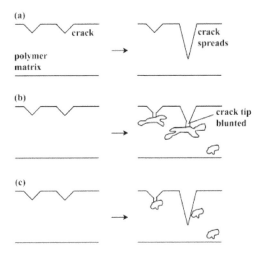

FIGURE 6   Toughening process in composite electrolyte: (a) TiO_2-free; (b) HNO_3-treated TiO_2 and (c) untreated TiO_2.

13.5   CONCLUSIONS

Untreated TiO_2 improved the electrolyte conductivity while HNO_3 treated TiO_2 decreased the conductivity. The conductivity enhancement by untreated TiO_2 comes from the increase in the number and mobility of Li^+ cations. HNO_3-treated TiO_2 decreased the conductivity by decreasing the anionic contribution. The addition of TiO_2 enhances the tensile properties of hexanoyl chitosan:polystyrene–$LiCF_3SO_3$ and greater enhancement is found for the treated TiO_2.

ACKNOWLEDGMENTS

The authors wish to thank Universiti Teknologi MARA, Malaysia for supporting this work through grant DANA 5/3 Dst (420/2011).

KEYWORDS

- characterization of polymer
- composite
- conductivity enhancement
- electrolyte
- polymer
- treated TiO_2

REFERENCES

1. Gray, F.M.: Polymer Electrolytes. The Royal Society of Chemistry, UK (1997).
2. Berthier, C., Gorecki, W., Minier, M., Armand, M.B., Chabagno, J.M., Rigand, P.: Microscopic investigation of ionic conductivity in alkali metal salts-poly (ethylene oxide) adducts. Solid State Ionics, 11(1), 91–95 (1983).
3. Fonseca, C.P., Neves, S.: Characterization of polymer electrolytes based on poly(dimethyl siloxane-co-ethylene oxide). J. Power Sources, 104, 85–89 (2002).
4. Rajendran, S., Mahendran, O., Kannan, R.: Characterisation of [(1-x)PMMA-xPVdF] polymer blend electrolyte with Li$^+$ ion. Fuel, 81, 1077–1081 (2002).
5. Park, Y.W., Lee, D.S.: The fabrication and properties of solid polymer electrolytes based on PEO/PVP blends. J. Non–Crystalline Solids, 351, 144–148 (2005).
6. Yuan, F., Chen, H.Z., Yang, H.Y., Li, H.Y., Wang, M.: PAN–PEO solid polymer electrolytes with high ionic conductivity. Mat. Chem. Phys., 89, 390–394 (2005).
7. Kobayashi, N., Kubo, N., Hirohashi, R.: Control of ionic conductivity in solid polymer electrolyte by photo irradiation. Electrochim. Acta, 37, 1515–1516 (1992).
8. Watanabe, M., Nishimoto, A.: Effects of network structures and incorporated salt spesies on electrochemical properties of polyether-based polymer electrolytes. Solid State Ionics, 79, 306–312 (1995).
9. Ding, L.M.: Synthesis, characterization and ionic conductivity of solid polymer electrolytes based on modified alternating maleic anhydride copolymer with oligo (oxyethylene) side chains. Polymer, 38(16), 4267–4273 (1997).
10. Ding, L.M., Shi, J., Yang, C.Z.: Ion-conducting polymers based on modified alternating maleic anhydride copolymer with oligo (oxyethylene) side chains. Synth. Met., 87, 157–163 (1997).
11. Rajendran, S., Mahendran, O., Kannan, R.: Lithium ion conduction in plasticized PMMA–PVdF polymer blend electrolytes. Mat. Chem. Phys., 74, 52–57 (2002).
12. Tan Winie, Ramesh, S., Arof, A.K.: Studies on the structure and transport properties of hexanoyl chitosan-based polymer electrolytes. Physica B: Condensed Matter, 404, 4308–4311 (2009).
13. Murata, K., Izuchi, S., Yoshihisa, Y.: An overview of the research and development of solid polymer electrolyte batteries. Electrochim. Acta, 45, 1501–1508 (2000).
14. Fan, L., Dang, Z., Nan, C.W., Li, M.: Thermal, electrical and mechanical properties of plasticized polymer electrolytes based on PEO/P(VDF–HFP) blends. Electrochim. Acta, 48, 205–209 (2002).
15. Krawiec, W., Scanlon, L.G., Jr., Fellner, J.P., Vaia, R.A., Vasudevan, S., Giannelis, E.P.: Polymer nanocomposites: a new strategy for synthesizing solid electrolytes for rechargeable lithium batteries. J. Power Sources, 54, 310–315 (1995).
16. Best, A.S., Adebahr, J., Jacobsson, P., MacFarlane, D.R., Forsyth, M.: Microscopic interactions in nanocomposite electrolytes. Macromolecules, 34(13), 4549–4555 (2001).
17. Chung, S.H., Wang, Y., Persi, L., Croce, F., Greenbaum, S.G., Scrosati, B., Plichita, E.: Enhancement of ion transport in polymer electrolytes by addition of nanoscale inorganic oxides. J. Power Sources, 97–98, 644–648 (2001).

18. Capiglia, C., Mustarelli, P., Quartarone, E., Tomasi, C., Magistris, A.: Effects of na-noscale SiO_2 on the thermal and transport properties of solvent-free, poly(ethylene oxide) (PEO)-based polymer electrolytes. Solid State Ionics, 118, 73–79 (1999).
19. Kim, Y.W., Lee, W., Choi, B.K.: Relation between glass transition and melting of PEO-salt complexes. Electrochim. Acta, 45, 1473–1477 (2000).
20. Wieczorek, W., Raducha, D., Zalewska, A., Stevens, J.R.: Effect of salt concentration on the conductivity of PEO-based composite polymeric electrolytes. J. Phys. Chem. B, 102(44), 8725–8731 (1998).
21. Rosli, N.H.A., Chan, C.H., Subban, R.H.Y., Tan Winie: Studies on the structural and electrical properties of hexanoyl chitosan/polystyrene-based polymer electrolytes. Physics Procedia, 25, 215–220 (2012).
22. Zong, Z., Kimura, Y., Takahashi, Yamane, M.H.: Characterization of chemical and solid state structures of acylated chitosans. Polymer, 41, 899–906 (2000).
23. Bruce, P.G., Vincent, C.A.: Steady state current flow in solid binary electrolyte cells. J. Electroanal. Chem., 225, 1–17 (1987).
24. Tan Winie, Arof, A.K.: Dielectric behavior and AC conductivity of $LiCF_3SO_3$ doped H-chitosan polymer films. Ionics, 10, 193–199 (2004).
25. Tan Winie, Rosli, N.H.A., Hanif, N.S.M., Chan, C.H., Ramesh, S.: Polymer electro-lytes based on blend of hexanoyl chitosan and polystyrene. Submitted for publication in elsewhere.
26. Tan Winie, Arof, A.K.: FT–IR studies on interactions among components in hexanoyl chitosan-based polymer electrolytes. Spectrochim. Acta A, 62, 677–684 (2006).
27. Kumar, B., Scanlon, L.G.: Polymer-ceramic composite electrolytes: conductivity and thermal history effects. Solid State Ionics, 124, 239–254 (1999).
28. He, D., Jiang, B.: The elastic modulus of filled polymer composites. J. Appl. Polym. Sci., 49, 617–621 (1993).
29. Svehlova, V., Poloucek, E.: Mechanical properties of talc-filled polypropylene. Influ-ence of filler content, filler particle size and quality of dispersion. Angew. Makromol. Chem., 214(3762), 91–99 (1994).

CHAPTER 14

A STUDY ON SOLID POLYMER ELECTROLYTES

SITI NOR HAFIZA MOHD YUSOFF, SIM LAI HAR,
CHAN CHIN HAN, AMIRAH HASHIFUDIN,
and HANS–WERNER KAMMER

CONTENTS

14.1   INTRODUCTION

Solid solutions of epoxidized natural rubber with 25 and 50 mol % epoxidation, ENR-25 and ENR-50, respectively, added with LiClO$_4$ were prepared by solution casting technique. Glass transition temperature (T_g) values obtained using differential scanning calorimetry (DSC) and the ionic conductivity evaluated from bulk resistance (R_b) determined using the impedance spectroscopy point towards higher solubility of the lithium salt in ENR-50 when the ratio of the mass of salt to the mass of polymer (Y) 0.15. This ramification correlates with spectroscopic results demonstrated in FTIR spectra. Ionic conductivity (σ) is observed to increase with ascending values of Y. When Y 0.15, ENR-50 exhibits higher ionic conductivity than ENR-25 but the σ values of ENR-25 increase sharply with increasing salt content to above that of ENR-50 when Y 0.20. Higher ion mobility, better salt molecule-chain segment correlation and higher charge carrier diffusion rate account for the higher σ value for ENR-50 at $0.00 < Y$ 0.15. However, restricted ion transport for ENR-50 and relatively flexible segmental motion for ENR-25 at Y 0.20 cause the conductivity of ENR-25 to be higher than that of ENR-50. Therefore, conductivity of the epoxidized natural rubber is primarily governed by the segmental motion of the elastomer rather than charge carrier density, since the discovery of ion-conducting polymer by Fenton et al. [1] followed by the application of polymer electrolyte in lithium batteries by Armand et al. [2], solid polymer electrolyte (SPE) has been widely studied especially on the enhancement of ionic conductivity. To date, SPE has become the focus of extensive research in pursue for a new generation of power source to cater for the latest development in electrochemical devices. Polymer electrolyte is a complex formed by dissolving an inorganic salt in a polymer matrix with polar groups acting as an immobile solvent. It is generally accepted that the charge carrier density and ion mobility are the two important parameters contributing to the ionic conductivity of a SPE [3–5]. For SPE with a semi-crystalline polymer matrix like poly(ethylene oxide) (PEO), ion mobility is attributed to the segmental motion of the amorphous phase of the macromolecular chain. Epoxidized natural rubber (ENR) is derived from natural rubber by converting different percent of the C=C bonds on the macromolecular backbone to the polar epoxy groups. ENR has good potential to be polymer host in SPE because of their distinctive characteristic such as low glass transition temperature (T_g), soft elastomeric characteristics at room temperature [5] and good electrode-electrolyte adhesion. Furthermore, the highly flexible macromolecular chain and the polar epoxy oxygen provide excellent segmental motion and coordination sites for Li$^+$ ion transport, respectively, in ENR-based polymer electrolytes [4, 6]. However, Chan and Kammer [7], in their study on the properties of PEO/ENR/lithium perchlorate (LiClO$_4$) solid solutions concluded that the solubility of the ionic salt is comparatively higher in PEO than in ENR. Therefore, the conductivity mechanism as well as the role of the oxirane ring in the dissociation of ionic salt in ENR-salt complex will be investigated in detail. Other than ENR with 25 and 50 mol percent of epoxidation, ENR-25 and ENR-50, respectively, deprotein-ized natural rubber (DPNR) is used as a polymer reference [8].

14.2 EXPERIMENTAL

14.2.1 MATERIAL

Epoxidized natural rubber (ENR-25 and ENR-50) purchased from Malaysian Rubber Board (Sungai Buloh, Malaysia) was used after purification. The chemi-cal structures of the two elastomers were shown in Fig. 1. Deproteinized natural rubber (DPNR) was supplied by Green HPSP(M) Sdn Bhd (Petaling Jaya, Malay-sia). $LiClO_4$ was purchased from Acrōs Organic Company (Geel, Belgium) and the solvent tetrahydrofuran (THF) was purchased from Merck.

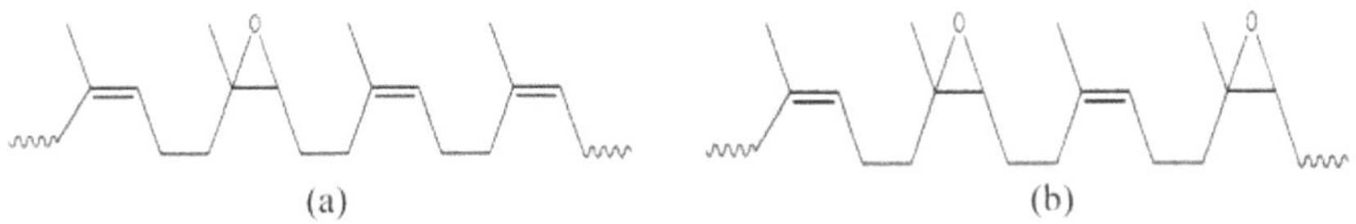

(a)	(b)

FIGURE 1   Chemical structures of (a) ENR-25, (b) ENR-50.

14.2.2 SAMPLE PREPARATION

Thin films of polymer electrolytes with ENR-25, ENR-50 and DPNR as the poly-mer host were prepared by solution casting technique. Appropriate amounts of the rubber and $LiClO_4$ were dissolved in THF by stirring with a magnetic stirrer at 50°C until a homogeneous solution was obtained. The $LiClO_4$ content (Y) added varied in a range from 0.01–0.30 of 1 gram dry weight of the polymer. It can be represented by Eq. (1)

$$Y = \frac{\text{mass of salt}}{\text{mass of polymer}} \tag{1}$$

The electrolyte solution was cast into a Teflon dish and left to dry overnight at room temperature to form a thin film. The electrolyte films were dried in an oven for 24 hours at 50°C before heated for another 24 hours in nitrogen atmosphere at 80°C. This was to ensure a good interaction between the salt and the elastomer. The free standing film was further dried in a vacuum oven at 50°C for another 24 hours before keeping in desiccators for further characterization.

14.2.3 IMPEDANCE MEASUREMENTS

The ionic conductivity measurements were performed using the HIOKI 3532-50 LCR Hi-Tester interfaced to a computer. The samples were scanned at frequencies ranging from 50 Hz to 1 MHz at room temperature. The thin film samples were sandwiched between two stainless steel block electrodes of 20 mm in diameter. The conductivity (σ) of each sample was calculated using the equation $\sigma = t/(R_b A)$, where t is the thickness of the sample, R_b is the bulk resistance and A is the cross-section area of the film. The R_b value is the intersection point between the semicircle and the x-axis in a cole–cole plot. The average of the thickness (t) was calculated from four measurements of the thickness of the thin film using Mitutoyo Digimatic Calliper (Japan).

14.2.4   DIFFERENTIAL SCANNING CALORIMETRY
Differential scanning calorimeter, TA Q200 DSC calibrated with indium standard was used to study the thermal properties of the sample. For estimation of the T_g, approximately 10–12 mg of the sample was used for each analysis. The sample was cooled down to –90°C and was heated up to 80°C at a rate of 10°C min^{-1}.

14.2.5   FOURIER TRANSFORM INFRARED SPECTROSCOPY
Infrared spectra of all samples were obtained using the Attenuated Total Reflectance (ATR) method on Perkin Elmer Spectrum 1 spectrometer at room temperature with frequency range of 4000–650 cm^{-1}. For each sample, 32 scans were taken at maximum resolution of 2 cm^{-1} using the Ge crystal plate.

14.3   RESULTS AND DISCUSSIONS

14.3.1   GLASS TRANSITION TEMPERATURE
The glass transition temperature (T_g) of a polymer, observed as an endothermic shift from the baseline of a thermogram, is determined using differential scanning calorimetry (DSC). It is governed mainly by the heating and cooling rates applied in the DSC run. The T_g values of the modified natural rubber (MNR), extracted from the second heating runs, are presented in Fig. 2. Among all the three salt-free rubber samples, DPNR records the lowest T_g value of –65°C compared to –42°C and –21°C for ENR-25 and ENR-50, respectively. It is well documented that an increase in the epoxidation level of ENR will cause a reduction in the free volume of the chain phases leading to stiffening of the molecular chain structure [9].

Figure 2 depicts that the T_g values of the ENR samples increase with ascending LiClO$_4$ from $Y = 0.00$ to 0.30 while that of the DPNR remains relatively constant in the same range of salt content. Similar result on ENR was also reported by Idris et al. [4]. Solubility of LiClO$_4$ in the ENR samples through the complexation between Li$^+$ ion and the epoxy oxygen of the oxirane group results in the stiffening of the polymer chain. ENR-50 with more epoxy oxygen in its macromolecular backbone as coordination sites for the Li$^+$ ion experience a larger increase in T_g values from –21 to 8°C as compared to ENR-25 when the salt content increases from $Y = 0.00$ to 0.30. Meanwhile, weak interaction between Li$^+$ ion and the unsaturated ethylene group (C=C) of DPNR accounts for the lower solubility of the salt in the sample leading to relatively constant T_g values obtained with ascending salt content.

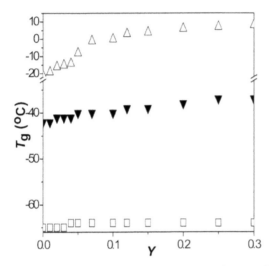

FIGURE 2   Glass transition temperature of the rubber samples as a function of Y, ($\square$) DPNR, ($\blacktriangledown$) ENR-25, and ($\triangle$) ENR-50.

14.3.2 INFRARED SPECTROSCOPY

ATR- FTIR spectroscopy is applied to investigate the effect of the ion-dipole interactions between ENR-25, ENR-50 and DPNR with the lithium salt.

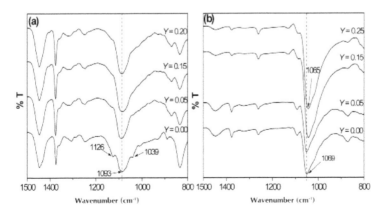

FIGURE 3   FTIR spectra in the region of 1500–800 cm⁻¹ for (a) ENR-25 and (b) ENR-50 added with different content of $LiClO_4$ (Y).

The triplet centered at 1093 cm^{-1} with two shoulders at 1039 and 1126 cm^{-1} as shown in Fig. 3a is assigned to the C–O–C stretching mode of the oxirane group in ENR-25. Addition of LiClO$_4$ from $Y = 0.00$ to 0.20 results in the disappearance of the two shoulders, indicating the coordination of Li$^+$ to the oxygen atom of the oxirane group. However, no significant shifting of the center peak at 1093 cm^{-1} is observed. On the other hand, addition of $Y = 0.01$ to 0.15 of LiClO$_4$ to ENR-50 as shown in Fig. 3b, causes an immediate downshifting of the C–O–C stretching mode from 1069 to 1065 cm^{-1}. No further shifting is observed with increasing salt content from $Y = 0.15$ to 0.30. The downshifting of the C–O–C vibration mode observed in ENR-50 and not in ENR-25 implies that more Li$^+$ ions form complexes with ENR-50 than with ENR-25. This result concurs closely with that observed in the T_g values discussed in the T_g section.

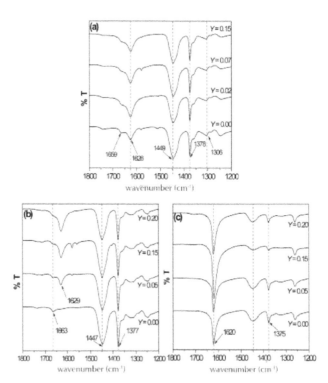

FIGURE 4   FTIR spectra for (a) DPNR, (b) ENR-25 and (c) ENR-50 doped with different salt content in the range of 1800–1200 cm^{-1}.

The unsaturated ethylene groups (C=C) which absorb moderately at 1659 cm^{-1} and strongly at 1626 cm^{-1} as shown in Fig. 4a, are the only nucleophilic sites in DPNR available for ion-dipole interactions with the Li^+ ions. No significant shifting but merging of the two absorbance peaks to form a broad peak at 1626 cm^{-1} for the C=C vibration mode is observed suggests weak ion-dipole interaction between the C=C group and Li^+ ion. This result correlates with the relatively constant T_g observed in DPNR when increasing salt content is added. Addition of epoxy oxygen to the C=C bond of natural rubber causes a shifting in the absorbance bands of the remaining C=C groups to 1663 cm^{-1} for ENR-25 and 1620 cm^{-1} for ENR-50 as shown in Figs. 4b and 4c, respectively. In Fig. 4b, one observes that the broad absorbance peak at 1663 cm^{-1} for ENR-25 continues to broaden with increasing salt content whereas the new peak at 1629 cm^{-1} which is formed on the addition of salt increases progressively in intensity. Examining the chemical structure of ENR-25, one can see that the C=C bonds experience different electron environment in the neighborhood of the epoxy groups. Therefore, coordination between Li^+ ions and the epoxy groups causes the vibration mode of the C=C bonds to downshift to different wavenumbers according to their positions relative to the epoxy group. The absorbance band at 1629 cm^{-1} is ascribed to the one next to the epoxy group while the very broad peak in the vicinity of 1663 cm^{-1} is assigned to the rest of the C=C bonds further away from the epoxy groups. It is noteworthy that apart from the increase in the Li^+-epoxy oxygen interaction, the increase in intensity with ascending salt content is also partly due to the overlapping of the C=C vibration mode at 1629 cm^{-1} and the internal vibration mode of ClO_4^- anion which absorbs at the same wavenumber [10]. With the C=C bonds alternate with the epoxy groups on the macromolecular backbone of ENR-50, the vibration mode of C=C gives a sharp and intense absorbance at 1620 cm^{-1} for the neat ENR-50 as shown in Fig. 4c. No change in the intensity and peak position of the C=C vibration mode observed in ENR-50 suggests that the Li^+ ions prefer to coordinate with epoxy oxygen than with the C=C bonds. Furthermore, it is noted in Fig. 4c that there are broadening and reduction in intensity in the bending modes of CH_2 scissoring, CH_2 wagging and CH_2 twisting at 1447, 1377 and 1309 cm^{-1}, respectively [11]. This observation points towards stronger interactions at the C–O–C of ENR-50 which influence the electron environment of the neighboring methylene (CH_2) groups, and thus closely relates to the higher increase in T_g observed with ascending $LiClO_4$ for ENR-50.

14.3.3 CONDUCTIVITY

Figure 5 demonstrates the semi-logarithm plot of σ versus Y for DPNR, ENR-25 and ENR-50 at room temperature. It is noted that the ionic conductivity of all the three rubber samples increase with ascending salt content due to an increase in the number of charge carriers. Being non- crystalline elastomers, the segmental motion of the macromolecular chains promotes ion transport of Li^+ ions within the polymer [12].

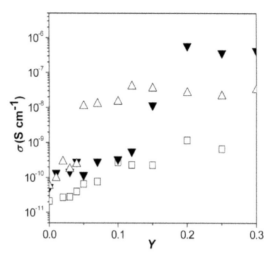

FIGURE 5   Ionic conductivity (σ) versus Y of MNR polymer electrolytes at 30°C with salt content ranging from 0.00 to 0.30; ($\square$) DPNR, ($\blacktriangledown$) ENR-25, and ($\triangle$) ENR-50, respectively.

Among the three NR-based polymer electrolytes investigated, DPNR records the lowest σ values at all salt concentrations while ENR-50 exhibits the highest σ values at low salt content from $Y = 0.00$ to 0.15. However, as the LiClO$_4$ content increases from $Y = 0.12$, ENR-25 encounters a marked increase in σ value such that when $Y \geq 0.20$, the σ value of ENR-25 is approximately 2 orders and 3 orders in magnitude higher than that of ENR-50 and DPNR, respectively. Furthermore, it is noted that both the ENR polymer electrolytes demonstrate a sharp increase in σ value (~2–3 orders in magnitude) at a certain range of LiClO$_4$, follow by a plateau with further increase in salt content. The significant jump in σ values which occurs at $Y = 0.05$–0.12 for ENR-50 and $Y = 0.12$–0.20 for ENR-25 is most probably the result of the formation of a stable percolation network in the polymer electrolytes due to polarization relaxations. Meanwhile, ENR-50 records a maximum σ value of 3.7 10^{-8} S cm^{-1} at $Y = 0.12$ whereas ENR-25 has a maximum σ value of 5.7 10^{-7} S cm^{-1} at $Y = 0.20$.

As mentioned in the discussion on T_g, flexibility and the segmental motion of the polymer chain decrease with increasing epoxidation level and salt concentration. At relatively high salt content, the macromolecular backbone of ENR-50 experiences increasing stiffness caused by higher degree of epoxidation and increasing Li$^+$-polymer complex formation [13]. In addition, ENR-25 with an optimal number of epoxy group, has an advantage over ENR-50 because it not only possesses coordination sites for Li$^+$ ions but is also able to maintain the good elastomeric characteristics of natural rubber at relatively high salt content. Therefore, ENR-50 as shown in Fig. 5 exhibits higher ionic conductivity at salt content $Y \geq$

0.15 due to higher ion mobility. However, with salt content increases to $Y \geq 0.20$, restricted segmental motion in ENR-50 and a relatively steady ion transport in the stabilize percolation network of ENR-25 account for the higher σ value for the latter electrolyte system.

At low salt concentration, the dependence of conductivity on $LiClO_4$ content (Y) can be described using the power law [14] as shown in Eq. (2).

$$\sigma_{DC} = N_A e(\alpha\mu) \frac{\rho_{MNR}}{M_{salt}} (Y)^x \qquad (2)$$

where N_A represents Avogadro's number, e is the elementary charge, μ and α denote the ion mobility and the degree of dissociation, respectively. The product of degree of dissociation and mobility ($\alpha\mu$) is referred to as the ion mobility for the following discussions. The exponent x which is determined experimentally gives the extent of correlations between the salt molecules and the MNR segments. Meanwhile, ρ_{MNR} and M_{salt} represent the density of the MNR and the molar mass of the salt molecule, respectively.

A double logarithmic plot of σ versus Y at the range of low salt content from $Y = 0.00$ to 0.15 is shown in Fig. 6. The exponent x and ion mobility ($\alpha\mu$) can be extracted from the slope and the y-intercept of the regression functions, respectively after Eq. (2) in Fig. 6. Molecular characteristics adopted for the determination of $\alpha\mu$ are M_{salt} (M_{LiClO4}: 106.5 g mol^{-1}) and ρ_{MNR} (ρ_{DPNR}: 0.920 g cm^{-3}, ρ_{ENR-25}: 0.971 g cm^{-3} and ρ_{ENR-50}: 1.027 g cm^{-3}).

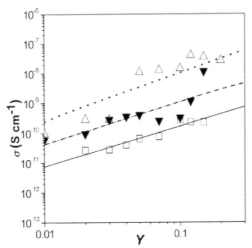

FIGURE 6 The dependence of ionic conductivity (σ) on salt concentration at 30°C, with salt content ranging from 0.00–0.15 for (□) DPNR, (▼) ENR-25, and (△) ENR-50 with solid, dashed and dotted curves, respectively, representing linear regression after Eq. (2).

The value of diffusion coefficient (D) of the charge carrier can be evaluated by adopting the Nernst's relationship as given in Eq. (3). This equation reflects the dependence of D on the absolute temperature (T) and the ion mobility ($\alpha\mu$).

$$D = \frac{k_B T \alpha\mu}{e} \tag{3}$$

where k_B is the Boltzmann constant ($k_B = 1.381 \times 10^{-23}$ m^2 kg s^{-2} K^{-1}) and T is the absolute temperature ($T = 303$ K). All the parameters determined from the regression functions after Eqs. (2) and (3) at room temperature and the salt content ranging from $Y = 0.00$ to 0.15 are listed in Table 1. The values of $\alpha\mu$ and the exponent x for ENR-50 are the highest among all the three rubber samples, and this explains its highest σ value at $Y \geq 0.15$ as shown in Figs. 5 and 6. Besides, LiClO$_4$ has the highest diffusion rate in ENR-50 at low salt content. On the contrary, DPNR displays the lowest values for ion mobility, exponent x and D which agree closely with its lowest σ value and the weak interaction between the polymer chain and the salt as described in the T_g section.

TABLE 1   Regression functions after Eq. (2), ion mobility ($\alpha\mu$); exponent x and diffusion coefficient (D) of MNR after Eq. (3) for all the MNR with LiClO$_4$ at salt content ranging from $Y = 0.00$ to 0.15.

	DPNR	ENR-25	ENR-50
Regression function	$\sigma = 3.8 \times 10^{-9}\ Y^{1.35}$	$\sigma = 2.9 \times 10^{-8}\ Y^{1.41}$	$\sigma = 4.1 \times 10^{-7}\ Y^{1.58}$
Correlation, R	0.967	0.844	0.908
Exponent x	1.35	1.41	1.58
$\alpha\mu$ (cm^2 V^{-1} s^{-1})	4.6×10^{-12}	3.3×10^{-11}	4.4×10^{-10}
D (cm^2 s^{-1})	1.2×10^{-13}	8.5×10^{-13}	1.1×10^{-11}

14.4   CONCLUSIONS

Solid polymer electrolytes of ENR-25/LiClO$_4$, ENR-50/LiClO$_4$ and DPNR/LiClO$_4$ with various concentrations of LiClO$_4$ were successfully prepared using solution casting technique. The effect of LiClO$_4$ on the conductivity, thermal properties and polymer-salt interaction of ENR-25 and ENR-50 were investigated by AC impedance spectroscopy, differential scanning calorimetry and ATR–FTIR, respectively. Ascending T_g values with increasing salt content which is higher in ENR-50 as compared to ENR-25 and DPNR correlates with increased ion-dipole interaction for ENR-50 observed in the FTIR spectra. At LiClO$_4$ content $Y \geq 0.15$, ENR-50 exhibits higher ionic conductivity than ENR-25 as a result of higher ion mobility, better correlation between salt molecules and chain segments and higher diffusion rate of charge carrier. However, at higher salt content Y 0.20, reduction in segmental motion leading to restricted ion transport for ENR-50 as compared to good ion mobility due to flexible chain movement for ENR-25 results in higher ionic conductivity in the latter electrolyte system. However, the conductivity of ENR-25 is too low for any practical application. Blending ENR with a second polymer consisting of polar groups like PEO and incorporation with inorganic fillers can be applied to enhance the conductivity of the MNR.

ACKNOWLEDGMENT

The authors would like to express our gratitude to the Research Management Institute (RMI), University Teknologi MARA for awarding the grant 600–RMI/ST/Dana 5/3/Dst (426/2011).

KEYWORDS

- **ATR–FTIR**
- **conductivity of the MNR**
- **differential scanning calorimetry**
- **electrolytes**
- **polymer-salt interaction**
- **solid polymer**
- **spectroscopy**

REFERENCES

1. D.E. Fenton, J.M. Parker, and P.V. Wright, "Complexes of alkali metal ions with poly(ethylene oxide), "*Polymer,* vol. 14, pp. 589–594, 1973.
2. M.B. Armand, M.J. Duclot, and P. Rigaud, "Polymer solid electrolytes: Stability domain, "*Solid State Ionics,* vol. 3–4, pp. 429–430, 1981.
3. F. Latif, M. Aziz, N. Katun, A.M.M. Ali, and M.Z. Yahya, "The role and impact of rubber in poly(methyl methacrylate)/lithium triflate electrolyte, "*Journal of Power Sources,* vol. 159, pp. 1401–1404, 2006.
4. R. Idris, M.D. Glasse, R.J. Latham, R.G. Linford, and W.S. Schlindwein, "Polymer electrolytes based on modified natural rubber for use in rechargeable lithium batteries, "*Journal of Power Sources,* vol. 94, pp. 206–211, 2001.
5. F. Latif, M. Aziz, A.M.M. Ali, and M.Z.A. Yahya, "The Coagulation Impact of 50% Epoxidised Natural Rubber Chain in Ethylene Carbonate–Plasticized Solid Electrolytes, "*Macromolecular Symposia,* vol. 277, pp. 62–68, 2009.

6. M.L. Hallensleben, H.R. Schmidt, and R.H. Schuster, "Epoxidation of poly(*cis*-1,4-isoprene) microgels, "*Die Angewandte Makromolekulare Chemie,* vol. 227, pp. 87–99, 1995.

7. C.H. Chan and H.W. Kammer, "Properties of solid solutions of poly(ethylene oxide)/epoxidized natural rubber blends and $LiClO_4$, "*Journal of Applied Polymer Science,* vol. 110, pp. 424–432, 2008.

8. W. Klinklai, S. Kawahara, T. Mizumo, M. Yoshizawa, Y. Isono, and H. Ohno, "Ionic conductivity of highly deproteinized natural rubber having epoxy group mixed with alkali metal salts, "*Solid State Ionics,* vol. 168, pp. 131–136, 2004.

9. I.R. Gelling, "Modification of natural rubber latex with peracetic acid, "*Rubber Chemistry and Technology,* vol. 58, pp. 86–96, 1985.

10. S. Rajendran, O. Mahendran and R. Kannan, "Ionic conductivity studies in composite solid polymer electrolytes based on PMMA, "*J. Phys. Chem. Solids*, vol. 63, pp. 303–307, 2002.

11. S.C. Ng and L.H. Gan, "Reaction of natural rubber latex with performic acid, "*European Polymer Journal,* vol. 17, pp. 1073–1077, 1981.

12. A.M.M. Ali, M.Z.A. Yahya, H. Bahron, and R.H.Y. Subban, "Electrochemical studies on polymer electrolytes based on poly(methyl methacrylate)-grafted natural rubber for lithium polymer battery, "*Ionics,* vol. 12, pp. 303–307, 2006.

13. R.H.Y. Subban, A.K. Arof, and S. Radhakrishna, "Polymer batteries with chitosan electrolyte mixed with sodium perchlorate, "*Materials Science and Engineering: B,* vol. 38, pp. 156–160, 1996.

14. N.H.A. Nasir, C.H. Chan, H.-W. Kammer, L.H. Sim, and M.Z.A. Yahya, "Ionic conductivity in solutions of poly(ethylene oxide) and lithium perchlorate, "*Macromolecular Symposia,* vol. 290, pp. 46–55, 2010.

CHAPTER 15

MODIFICATION OF PC–PHBH BLEND MONOLITH

YUANRONG XIN and HIROSHI UYAMA

CONTENTS

15.1   INTRODUCTION

A polyethylenimine (PEI)-modified blend monolith with porous structure was prepared as an effective adsorbent to remove copper ion (Cu^{2+}) in aqueous media. The polycarbonate (PC) and poly(3-hydroxybutyrate-*co*-3-hydroxyhexanoate) (PHBH) blend monolith was selected as matrix, which was fabricated *via* non-solvent induced phase separation (NIPS). PEI, a chelating agent to bind metal ions, was covalently connected onto the surface of the blend monolith by aminolysis reaction. The adsorption capacity of the PEI-modified blend monolith for Cu^{2+} was evaluated. The adsorption has been examined under various conditions such as solution pH, adsorption time, and Cu^{2+} concentration. The maximum capacity for Cu^{2+} is 55 mg/g. The present PEI-modified blend monolith has large potential for wastewater treatment.

Removal of toxic metal ions has become an urgent issue due to their serious contamination in waste effluents of many industrial processes. Copper ions are one of these toxic metal ones which could result in severe health problems. Excessive uptake of copper ions may cause damage to heart, kidney, liver, pancreas and brain [1–5]. Typical treatments of copper-containing wastewater are adsorption, chemical precipitation, ion exchange, solvent extraction, reverse osmosis and membrane separation [6–10]. Among them, adsorption technique is considered to be an attractive method because of its low cost, simple operation, potential recovery of metals, and regeneration of the adsorbent by suitable desorption process.

Monoliths, materials with open-cellular three-dimensional continuous structure, have attracted considerable attention due to their good permeability, fast mass transfer property, high stability and easy modification [11–15]. During the last decade, polymer-based monoliths have become greatly significant. The unique features such as tunable bulk and surface properties and surface functionalizations afford wide applications in various industrial fields [16–18]. These monolithic materials could provide a convenient and effective route to adsorb copper ions in an aqueous solution due to their large surface area.

Recently, we have developed a novel approach to fabricate polymer monoliths by thermally induced [19, 20] and non-solvent induced [21] phase separation techniques using the polymer itself as precursor. The fabrication process was very convenient and clean for both techniques. In the case of non-solvent induced phase separation (NIPS), a monolith was obtained by addition of a non-solvent to a homogeneous polymer solution. An appropriate selection of solvent/non-solvent and their mixed ratio enabled formation of the monolith with uniform structure. In the present work, a blend monolith composed of polycarbonate (PC) and poly(3-hydroxybutyrate-*co*-3-hydroxyhexanoate) (PHBH) fabricated *via* NIPS method was utilized as matrix for adsorption of copper ion (Cu^{2+}) [22] (Fig. 1). PC is one of the most widely used thermoplastics due to its properties such as impact resistance, heat resistance and dimensional stability [23–26]. Hence PC monolith is an ideal candidate to be used as solid substrate bearing high mechanic strength. PHBH is a kind of microbial polyesters naturally produced by microorganisms from biomass. Due to the segment of 3-hydroxyhexanoate (HH) unit in the molecular structure [27–30], PHBH could afford improved flexibility and ductility to the blend monolith.

FIGURE 1 Fabrication procedure of PC–PHBH monolith *via* NIPS method.

In order to provide the affinity towards metal ions in polymer monoliths, the attachment of functional groups on the surface of monolith is required. Branched polyethylenimine (PEI) is a water-soluble polycation consisting of primary, secondary and tertiary amine functional groups in 1/2/1 molar ratio [2–3, 31–34]. PEI is widely used for removal of toxic metal pollutants due to the excellent chelation ability towards heavy various metal ions.

In this study, the PC–PHBH blend monolith [22] was modified with PEI by aminolysis reaction for immobilization of the PEI chain *via* covalent bond on the monolith surface. The PEI-modified blend monolith was utilized as adsorbent to remove Cu^{2+} in an aqueous solution.

15.2 EXPERIMENTAL SECTION

15.2.1 MATERIALS

PC (M_n=2.3×10^4, M_w/M_n=1.4) was purchased from Sigma–Aldrich Co. PHBH with 11 mol% HH content (M_n=4.8×10^4, M_w/M_n=1.1) was supplied by Kaneka Co. PEI with molecular weight of 1x10^4 was purchased from Wako Pure Chemical Industries. Ltd. All reagents were used as received without further purification.

15.2.2 MEASUREMENTS

Scanning electron microscopic (SEM) images were recorded on a Hitachi S-3000N instrument at 15 kV. A thin gold film was sputtered on the samples. FT–IR measurement was carried out by a Perkin Elmer Spectum One System B2 spectrometer with a universal ATR sampling accessory. Metal concentrations in an aqueous solution were analyzed by induced coupled plasma-atom emission spectrometry (ICP–AES, an ICP-7510 Shimadzu sequential plasma spectrometer). pH of the aqueous solution was measured by a Horiba Compact pH meter B-211.

Elemental analysis was performed by a Yanaco CHN corder instrument (MT-5 type, Yanagimoto Mfg. Co., Ltd., Japan).

15.2.3 SYNTHESIS OF PEI-MODIFIED BLEND MONOLITH

PC–PHBH blend monolith was prepared according to the literature [22]. The PEI-modified blend monolith was synthesized through aminolysis reaction (Scheme 1). Briefly, PC–PHBH blend monolith (0.15 g, PC/PHBH=80/20 (wt)) was immersed into ethanol solution of PEI (10 mL, 0.020 g/mL), followed by shaking gently at 20°C. After 36 h, the monolith was rinsed by ethanol for 3 times and dried in vacuo.

15.2.4 ADSORPTION OF CU^{2+}

The PEI-modified blend monolith was utilized to adsorb Cu^{2+} in aqueous solutions (Scheme 1). The modified blend monolith (0.15 g) was immersed into $CuCl_2$ solution (7.5 mL) in the concentration ranging from 0.010 to 2.0 mg/mL with different pH (1.3~4.8), and the solution was shaken at 300 rpm at 20°C. pH of $CuCl_2$ solution was adjusted by 0.1 mol/L HCl or 0.1 mol/L NaOH. The adsorption procedures for other metal ions, potassium (K^+), sodium (Na^+), nickel (Ni^{2+}) and cobalt (CO_2^+) ions, were similar to those of Cu^{2+}.

The concentration of metal ions in an aqueous solution was determined by ICP–AES. The amount of metal ions adsorbed by the blend monolith was calculated using the following equation:

$$qt = (C_0 - C_t)V/m \qquad (1)$$

where qt (mg/g) is the metal adsorption capacity of the blend monolith, C_0 and C_t are the metal concentrations (mg/mL) initially and at a given time, respectively, V is the solution volume (mL), and m is the weight (g) of the PEI-modified blend monolith.

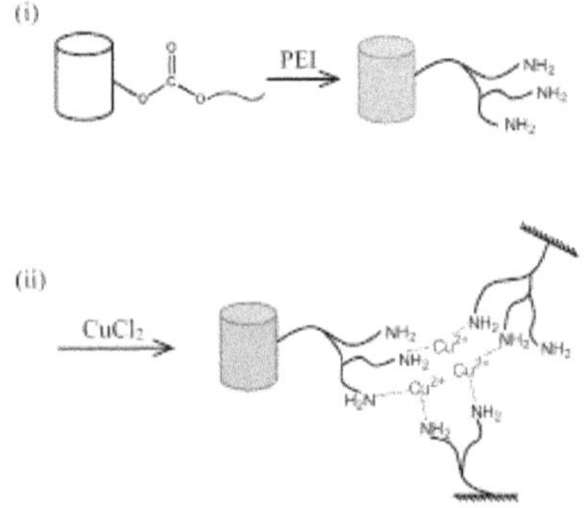

SCHEME 1   Synthesis of PEI-modified blend monolith (i) and its adsorption for Cu^{2+} (ii).

15.3 RESULTS AND DISCUSSION

15.3.1 MODIFICATION OF BLEND MONOLITH BY PEI

The PC–PHBH blend monolith was modified by aminolysis with PEI. Figure 2 shows the SEM images of the blend monolith before and after the reaction. Figure 2a shows that inside the blend monolith, PC and PHBH phases were well distributed into each other and the latter was of the round circle-shape with larger diameter, which dispersed uniformly in the continuous matrix of the former consisting of smaller pores. The sizes of the skeleton and smaller pore of the blend monolith were in the range of 0.2–0.7 and 0.9–2.3 μm, respectively. The morphology of the monolith after the aminolysis was hardly changed (Fig. 2b).

FIGURE 2   SEM images of original PC–PHBH blend monolith (a), after PEI modification (b), and after Cu^{2+} adsorption (c).

Figure 3 shows the FT–IR spectra of the unmodified (a) and modified PC (b) monoliths. For the modified monolith, the new peaks appear at 3280 and 1680 cm^{-1}, which are assigned to the stretching vibrations of N–H and amide bonds, respectively. These data strongly suggest that the aminolysis took place on the surface of the PC–PHBH blend monolith. The nitrogen content of the blend monolith determined by elemental analysis was 3.4 wt%. The PEI content in the modified monolith was estimated to be roughly 10 wt%.

15.3.2 ADSORPTION OF METAL IONS BY MONOLITH

At first, the effect of pH on the adsorption of Cu^{2+} by the blend monolith was investigated in the pH range from 1.3 to 4.8 (Fig. 4). The experiment at pH over 4.8 was not carried out because of the copper hydroxide precipitation. The adsorption capacity was strongly dependent on the pH of the solution, and the modified monolith had a maximum adsorption amount at pH of 4.8. At pH 1.3, all the amino groups of PEI on the surface of the blend monolith were protonated and repel positive metal ions, resulting in the very weak chelation ability of the monolith. At higher pH, on the other hand, the amino groups of PEI were partly or mostly deprotonated and thereby the adsorption capacity of the monolith increased. 9995632114.

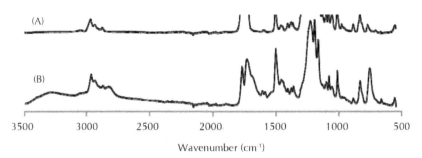

FIGURE 3   FT-IR spectra of PC-PHBH monolith before (a) and after (b) PEI modification.

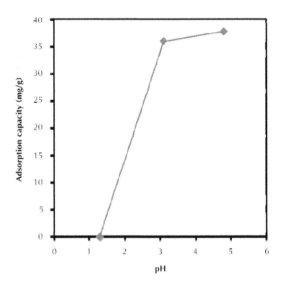

FIGURE 4   Effect of pH on the adsorption capacity of PEI-modified PC-PHBH blend monolith (initial Cu^{2+} concentration; 1.0 mg/mL; adsorption time: 14 h).

Figure 5 shows time-course in the adsorption of Cu^{2+} by the PEI-modified monolith at pH 4.8. The rapid increase of the adsorbed amount of Cu^{2+} was found from 0.5 to 12 h; afterwards, the amount gradually increased. After 14 h, the adsorption came to the equilibrium. The inside structure of the PEI-modified monolith was hardly changed through adsorption of Cu^{2+} (Fig. 2c).

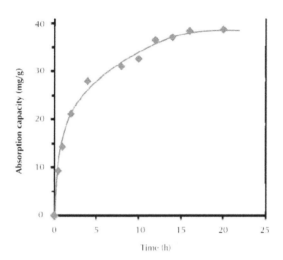

FIGURE 5 Time-course of Cu^{2+} adsorption by PEI-modified PC-PHBH blend monolith (initial Cu^{2+} concentration: 1.0 mg/mL; pH=4.8).

Figure 6 represents the relationship between the initial Cu^{2+} concentration and adsorption capacity of the PEI-modified blend monolith. The initial concentration varied from 0.010 to 2.0 mg/mL. The adsorption capacity increased from 0.24 to 55 mg/g as a function of the initial concentration. This may be because in the higher initial concentration, Cu^{2+} has higher driving force to overcome mass transfer resistance, leading to the higher adsorption capacity [6].

The adsorption of various metals by the PEI-modified monolith was examined (Fig. 7). K^+ and Na^+ ions were not adsorbed on the monolith, which is due to the very low affinity of the PEI chains towards these monovalent cations. For divalent cations of Cu^{2+}, Ni^{2+}, and CO_2^+, the adsorption capacities were 38, 16 and 11 mg/g, respectively. These data indicate that the PEI-modified blend monolith has preferable adsorption towards Cu^{2+}. The chelating order ($Cu^{2+} > Ni^{2+} > CO_2^+$) may be related to the electronegativity and size of these metals ions; metal ions of stronger electronegativity and smaller ionic radii are more easily diffused into the interior of the blend monolith and thus have larger affinity with the PEI chain of the blend monolith.

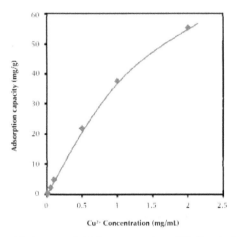

FIGURE 6   Relationship between initial concentration of Cu^{2+} and adsorption capacity of PEI-modified PC-PHBH blend monolith (adsorption time: 14 h; pH=4.8).

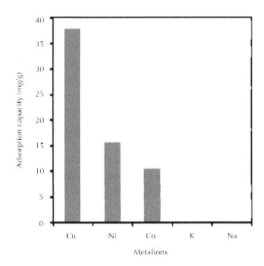

FIGURE 7   Adsorption of various metal ions by PEI-modified PC-PHBH blend monolith to (initial metal concentration: 1.0 mg/mL; adsorption time: 14 h; pH=4.8).

15.4   CONCLUSION

In this paper, the PEI-modified PC–PHBH blend monolith with continuous inter-connected structure was fabricated successfully *via* NIPS and aminolysis reaction. The resulting monolith has good chelating ability for Cu^{2+}. The maximum adsorption was found at pH 4.8. The chelating order of the monolith was $Cu^{2+} > Ni^{2+} > CO_2^+$ and no chelation was found for Na^+ and K^+.

Monoliths have characteristic properties suitable for applications of water treatment such as large surface area and high mass transfer. Furthermore, the immobilization of functional molecules on the monolith will expand their industrial applications. Therefore, the PEI-modified PC–PHBH blend monolith has large potential for various applications including removal of toxic metal ions in wastewater treatment.

ACKNOWLEDGMENT

This study is financially supported by a Grant-in-Aid for Scientific Research from the Japan Society for the Promotion of Science (No. 24655208) and the New Energy and Industrial Technology Development Organization (NEDO) of Japan. We gratefully acknowledge the gift of PHBH from Kaneka Co.

KEYWORDS

- adsorption
- aqueous media
- magnetic nanoparticles
- modification
- PC–PHBH blend
- porous structure

REFERENCES

1. Chen, J., Zeng, F., Wu, S., Su, J., Zhao, J., Tong, Z.: A facile approach for cupric ion detection in aqueous media using polyethylenimine/PMMA core-shell fluorescent nanoparticles. Nanotechnology, 20, 1–7 (2009).

2. Goon, I.Y., Zhang, C., Lim, M., Gooding, J.J., Amal, R.: Controlled fabrication of polyethylenimine-functionalized magnetic nanoparticles for the sequestration and quantification of free Cu^{2+}. Langmuir, 26, 12247–12252 (2010).

3. Wu, A., Jia, J., Luan, S.: Amphiphilic PMMA/PEI core-shell nanoparticles as polymeric adsorbents to remove heavy metal pollutants. Colloid and Surface A: Physicochem. Eng. Aspects, 384, 180–185 (2011).

4. Molinari, R., Gallo, S., Argurio, P.: Metal ions removal from wastewater or washing water from contaminated soil by ultrafiltration-complexation. Water Res., 38, 593–600 (2004).

5. Liu, M., Zhao, H., Quan, X., Chen, S., Yu, H.: Signal amplification *via* cation exchange reaction: an example in the ratiometric fluorescence probe for ultrasensitive and selective sensing of Cu(II). Chem. Commun., 46, 1144–1146 (2010).

6. Aluigi, A., Tonetti, C., Vineis, C., Tonin, C., Mazzuchetti, G.: Adsorption of copper(II) ions by keratin/PA6 blend nanofibers. Eur. Polym. J., 47, 1756–1764 (2011).

7. Sun, X.F., Liu, C., Ma, Y., Wang, S.G., Gao, B.Y., Li, X.M.: Enhanced Cu(II) and Cr(VI) biosorption capacity on poly(ethylenimine) grafted aerobic granular sludge. Colloids and Surfaces B: Biointerfaces, 82, 456–462 (2011).

8. Yin, C.Y., Aroua, M.K., Daud, W.M. A.W.: Fixed-bed adsorption of metal ions from aqueous solution on polyethylenimine-impregnated palm shell activated carbon. Chem. Eng. J., 148, 8–14 (2009).

9. Chen, Y., Pan, B., Zhang, S., Li, H., Lv, L., Zhang, W.: Immobilization of polyethylenimine nanoclusters onto a cation exchange resin through self-crosslinking for selective Cu(II) removal. J. Hazard. Mater., 190, 1037–1044 (2011).

10. Chen, Y., Pan, B., Li, H., Zhang, W., Lv, L., Wu, J.: Selective removal of Cu(II) ions by using cation-exchange resin-supported polyethylenimine (PEI) nanoclusters. Environ. Sci. Technol., 44, 3508–3513 (2010).

11. Buchmeiser, M.R.: Polymeric monolithic materials: syntheses, properties, functionalization and applications. Polymer, 48, 2187–2198 (2007).

12. Svec, F.: Porous polymer monoliths: amazingly wide variety of techniques enabling their preparation. J. Chromatogr. A, 1217, 902–924 (2010).

13. Courtois, J., Byström, E., Irgum, K.: Novel monolithic materials using poly(ethylene glycol) as porogen for protein separation. Polymer, 47, 2603–2611 (2006).

14. Yang, H., Liu, Z., Gao, H., Xie, Z.: Synthesis and catalytic performances of hierarchical SAPO-34 monolith. J. Mater. Chem., 20, 3227–3231 (2010).

15. Unger, K.K., Skudas, R., Schulte, M.M.: Particle packed columns and monolithic columns in high-performance liquid chromatography-comparison and critical appraisal. J. Chromatogr. A, 1184, 393–415 (2008).

16. Wei, S., Zhang, Y.L., Ding, H., Liu, J., Sun, J., He, Y., Li, Z., Xiao, F.S.: Solvothermal fabrication of adsorptive polymer monolith with large nanopores towards biomolecules immobilization. Colloids and Surfaces A: Physicochem. Eng. Aspects, 380, 29–34 (2011).

17. Nordborg, A., Hilder, E.F.: Recent advances in polymer monoliths for ion-change chromatography. Anal. Bioanal. Chem., 394, 71–84 (2009).

18. Svec, F., Fréchet, J.M. J.: Continuous rods of macroporous polymer as high-performance liquid chromatography separation media. Anal. Chem., 64, 820–822 (1992).

19. Okada, K., Nandi, M., Maruyama, J., Oka, T., Tsujimoto, T., Kondoh, K., Uyama, H.: Fabrication of mesoporous polymer monolith: a template-free approach. Chem. Commun., 47, 7422–7424 (2011).

20. Nandi, M., Okada, K., Uyama, H.: Functional mesoporous polymer monolith for application in ion-exchange and catalysis. Func. Mater. Lett., 4, 407–410 (2011).

21. Xin, Y., Fujimoto, T., Uyama, H.: Facile fabrication of polycarbonate monolith by non-solvent induced phase separation method. Polymer, 53, 2847–2853 (2012).

22. Xin. Y., Uyama, H.: Fabrication of polycarbonate and poly(3-hydroxybutyrate-co-3-hydroxyhexanoate) blend monolith via non-solvent induced phase separation method. Chem. Lett., in press.

23. Woo, B.G., Choi, K.Y., Song, K.H., Lee, S.H.: Melt polymerization of bisphenol–A and diphenol carbonate in a semibatch reactor. J. Appl. Polym. Sci., 80, 1253–1266 (2001).

24. Haba, O., Itakura, I., Ueda, M., Kuze, S.: Synthesis of polycarbonate from dimethyl carbonate and bisphenol–A through a non-phosgene process. J. Polym. Sci. Part A: Polym. Chem., 37, 2087–2093 (1999).

25. Fukuoka, S., Tojo, M., Hachiya, H., Aminaka, M., Hasegawa, K.: Green and sustainable chemistry in practice: development and industrialization of a novel process for polycarbonate production from CO_2 without using phosgene. Polym. J., 39, 91–114 (2007).

26. Okuyama, K., Sugiyama, J., Nagahata, R., Asai, M., Ueda, M., Takeuchi, K.: Direct synthesis of polycarbonate from carbon monoxide and bisphenol A catalyzed by Pd-carbene complex. Macromolecules, 36, 6953–6955 (2003).

27. Asrar, J., Valentin, H.E., Berger, P.A., Tran, M., Padgette, S.R., Garbow, J.R.: Biosynthesis and properties of poly(3-hydroxybutyrate-co-3-hydroxyhexanoate) polymers. Biomacromolecules, 3, 1006–1012 (2002).

28. Iwata, T.: Strong fibers and films of microbial polyesters. Macromol. Biosci., 5, 689–701 (2005).

29. Yang, Y., Ke, S., Ren, L., Wang, Y., Li, Y., Huang, H.: Dielectric spectroscopy of biodegradable poly(3-hydroxybutyrate-co-3-hydroxyhexanoate) films. Eur. Polym. J., 48, 79–85 (2012).

30. Abe, H., Ishii, N., Sato, S., Tsuge, T.: Thermal properties and crystallization behaviors of medium-chain-length poly(3-hydroxyalkanoate)s. Polymers, 53, 3026–3034 (2012).

31. Wang, X., Min, M., Liu, Z., Yang, Y., Zhou, Z., Zhu, M., Chen, Y., Hsiao, B.S.: Poly(ethylenimine) nanofibrous affinity membrane fabricated via one step wet-electrospinning from poly(vinyl alcohol)-doped poly(ethylenimine) solution system and its application. J. Membrane Sci., 379, 191–199 (2011).

32. Chen, Z., Deng, M., Chen, Y., He, G., Wu, M., Wang, J.: Preparation and performance of cellulose acetate/polyethylenimine blend microfiltration membranes and their applications. J. Membrane Sci., 235, 73–86 (2004).

33. Maketon, W., Ogden, K.L.: Synergistic effects of citric acid and polyethylenimine to remove copper from aqueous solutions. Chemosphere, 75, 206–211 (2009).

34. Bessbousse, H., Rhlalou, T., Verchère, J.F., Lebrun, L.: Removal of heavy metal ions from aqueous solutions by filtration with a novel complexing membrane containing poly(ethylenimine) in a poly(vinyl alcohol) matrix. J. Membrane Sci., 307, 249–259 (2008).

INDEX

Milton Keynes UK
Ingram Content Group UK Ltd.
UKHW022054141024
449569UK00031B/1631